1001 Problems in Classical Number Theory

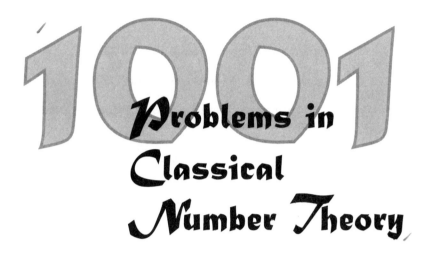

Problems in Classical Number Theory

Jean-Marie De Koninck, 1948-
Armel Mercier

Translated by Jean-Marie De Koninck

AMS
AMERICAN MATHEMATICAL SOCIETY
Providence, Rhode Island

This work was originally published in French by ELLIPSES under the title: *1001 problèmes en théorie classique des nombres*, © 2004 Édition Marketing S.A. The present translation was created under license for the American Mathematical Society and is published by permission.

Translated from the original by Jean-Marie De Koninck.

Cover image by Jean-Sébastien Bérubé.

2000 *Mathematics Subject Classification*. Primary 11A05, 11A07, 11A15, 11A25, 11A41, 11A51, 11A55, 11D04, 11N05, 11N25.

For additional information and updates on this book, visit
www.ams.org/bookpages/pint

Library of Congress Cataloging-in-Publication Data

Koninck, J. M. de, 1948–
    [1001 problèmes en théorie classique des nombres. English]
    1001 problems in classical number theory / Jean-Marie De Koninck and Armel Mercier.
       p. cm.
    ISBN 978-0-8218-4224-9 (alk. paper)
    1. Number theory—Problems, exercises, etc.   I. Mercier, Armel.   II. Title.

QA241 .K685   2007
512.7—dc22   2006052730

À ma mère qui m'a montré le chemin.

À Daphnée, un rayon de soleil dans ma vie.

Contents

Distribution of the Problems according to Their Topics

Preface

Number theory is one of the few areas of mathematics for which most problems can be understood by just about anyone, or at least by all those who are familiar with very basic notions of algebra, combinatorics and analysis. Every teacher knows the importance of practicing problem solving: indeed it turns out to be a great way to learn how to reason, no matter the area of mathematics the problems come from. Number theory is quite appropriate for this kind of exercise. For these reasons, a collection of problems in elementary or classical number theory seems in our opinion to be a complementary pedagogical tool for any learning process in mathematics. Moreover, a clever choice of problems can greatly help to raise the curiosity of those who try to solve them.

Unfortunately, very few books are entirely dedicated to problems in number theory. These include the classical work of the great master W. Sierpinski entitled *250 Problems in Elementary Number Theory* and published in Varsovie in 1970, a book which is not well known and unfortunately out of print. Hence, our manuscript does fill an important gap in this area and moreover it has the advantage of having been written to reach a large audience. One can also see it as a practical complement of an earlier book of the authors, that is *Introduction à la théorie des nombres* published by MODULO (2nd edition, 1997), or to any other introductory book in number theory.

Nevertheless, we must admit that our main motivation for writing this book has been our passion for number theory, namely this branch of mathematics which distinguishes itself by its beauty and its numerous mysteries, by its simplicity and its complexity, that is from the proof that there are infinitely many primes to the recently established proof of Fermat's Last Theorem.

This book obviously contains many problems from elementary number theory. Some of these are well known and can be found here and there in introductory books in number theory, while others are not so common. This is namely the case of several problems which we picked from the lesser known manuscript of Sierpinski mentioned above. Our book also contains some problems submitted to the readers of three well known journals: *American Mathematical Monthly*, *Mathematics Magazine* and *The College Mathematics Journal*. Finally, our book contains some 300 new problems never published before.

The choice of problems is obviously subjective; hence, it is no coincidence that the section on arithmetical functions is the longest! In any event, an effort has been made to cover, or at least brush, each of the classical themes of elementary number theory. On the other hand, since more and more students now have to use computers and software to do mathematics, our book can certainly help them in this task. Indeed, many of the problems encourage the reader to use computer software and at times, while searching for a solution, indicate how to write the program that will bring about the solution to the problem.

Although most problems presented here use basic results which can be found in just about any elementary book in number theory, we chose to include a section which provides the basic definitions and the main theorems one needs to handle the various subjects covered in the book. This "tool box" has the advantage that the reader does not have to search here and there for the basic notions needed to solve the problems. Finally, we found it convenient to include in this section a list of the main arithmetic functions with their definitions, as well as a list of the constants and symbols most frequently used in the text.

Our presentation is as follows: the first section provides the basic theory relevant for the understanding and the resolution of the stated problems; the second section gathers the statements of the problems; while the third section lists all the solutions. At the end of the book, the reader will find a bibliography, a terminology index and an index of authors.

We want to thank all those who, by their remarks and suggestions, contributed to the realization of this manuscript. In particular, our thanks go to Jean-Lou De Carufel (Québec), Nicolas Doyon (Québec), David Gill (Québec), Jacques Grah (Québec), Nicolas Guay (Québec), Aleksandar Ivić (Belgrade), Imre Kátai (Budapest), Claude Levesque (Québec), Marc-Hubert Nicole (Québec), Erik Pronovost (Québec) and Guy Robin (Limoges).

Part of this work was done in 2003 and 2004 while the first author was on sabbatical in Tucson, Arizona. This author is grateful to the Department of Mathematics of the University of Arizona, in particular to Professor William Yslas Velez, for making this possible.

Jean-Marie De Koninck

Armel Mercier

Part 1

Key Elements from the Theory

Let $\mathbb{N}$, $\mathbb{Z}$, $\mathbb{Q}$, $\mathbb{R}$ and $\mathbb{C}$ stand respectively for the set of positive integers (also called natural numbers), the set of integers, the set of rational numbers, the set of real numbers and the set of complex numbers.

Unless indicated otherwise,

- the letters $a, b, c, d, i, j, k, \ell, m, n, r$ and s stand for integers,
- the letters p and q stand for prime numbers,
- the letters $p_1, p_2, p_3, p_4, p_5, p_6, p_7, \ldots$ represent the sequence of prime numbers $2, 3, 5, 7, 11, 13, 17, \ldots$,
- by twin primes, we mean a pair of prime numbers $\{p, q\}$ such that $q = p+2$.

Given an integer $n \geq 2$, we often write

$$n = q_1^{\alpha_1} q_2^{\alpha_2} \cdots q_r^{\alpha_r}$$

for its *canonical representation* as a product of distinct prime powers: here the q_i's are the primes dividing n written in increasing order and the exponents α_i's are positive integers (see Theorem 11).

Some Classical Forms of Argument

THEOREM 1 (Induction Principle). *Let S be a set of natural numbers having the following two properties:*

(i) $1 \in S$,
(ii) *if $k \in S$, then $k + 1 \in S$.*

Then $S = \mathbb{N}$.

THEOREM 2 (Pigeonhole Principle). *If more than n objects are distributed amongst n boxes, then one of the boxes must contain at least two objects.*

THEOREM 3 (Inclusion-Exclusion Principle). *Let A be a set containing N elements and let $P_1, P_2, \ldots, P_r$ be distinct properties that each element of A must satisfy. If $n(P_{i_1}, P_{i_2}, \ldots, P_{i_k})$ stands for the number of elements of A having all the properties $P_{i_1}, P_{i_2}, \ldots, P_{i_k}$, then the number of elements of A having none of the r properties is equal to*

$$N - \Big(n(P_1) + n(P_2) + \cdots + n(P_r)\Big) + \Big(n(P_1, P_2) + n(P_1, P_3) + \cdots + n(P_{r-1}, P_r)\Big)$$

$$- \Big(n(P_1, P_2, P_3) + n(P_1, P_2, P_4) + \cdots + n(P_{r-2}, P_{r-1}, P_r)\Big) + \cdots$$

$$+ (-1)^r n(P_1, P_2, \ldots, P_r).$$

Inequalities

THEOREM 4 (Cauchy-Schwarz Inequality). *Let $a_1, a_2, \ldots, a_n$, $b_1, b_2, \ldots, b_n$ be real numbers. Then*

$$\left(\sum_{i=1}^{n} a_i b_i\right)^2 \leq \sum_{i=1}^{n} a_i^2 \sum_{i=1}^{n} b_i^2.$$

THEOREM 5 (Arithmetic-Geometric Means Inequality). *Let $a_1, a_2, \ldots, a_n$ be positive real numbers. Then*

$$(a_1 a_2 \ldots a_n)^{1/n} \leq \frac{a_1 + a_2 + \cdots + a_n}{n},$$

with equality if and only if $a_1 = a_2 = \ldots = a_n$.

Divisibility

DEFINITION 1 (Binomial Coefficients). *Let n be a positive integer and k an integer satisfying $0 \leq k \leq n$. We define the binomial coefficient $\binom{n}{k}$ by*

$$\binom{n}{k} = \frac{n!}{k!(n-k)!},$$

where $0! = 1$ and $n! = n(n-1) \cdots 3 \cdot 2 \cdot 1$.

THEOREM 6 (Binomial Theorem). *Let $a, b \in \mathbb{R}$ and $n \in \mathbb{N}$. Then*

$$(a+b)^n = \sum_{k=0}^{n} \binom{n}{k} a^{n-k} b^k.$$

In particular, it follows from Theorem 6 that

$$\sum_{k=0}^{n} (-1)^k \binom{n}{k} = (1-1)^n = 0 \quad \text{and} \quad \sum_{k=0}^{n} \binom{n}{k} = (1+1)^n = 2^n.$$

DEFINITION 2. *Let $a, b \in \mathbb{Z}$ with $a \neq 0$. We say that a divides b if there exists an integer c such that $b = ac$, in which case we write $a|b$ and say that a is a divisor of b. If a does not divide b, we write $a \nmid b$. In the case where $a|b$ and $1 \leq a < b$, we shall say that a is a proper divisor of b. We write $p^\alpha \| n$ to mean that $p^\alpha | n$ while $p^{\alpha+1}$ does not divide n.*

THEOREM 7 (Euclidean Division). *Let $a, b \in \mathbb{Z}$, $a > 0$. Then, there exist integers q and r such that $b = aq + r$, where $0 \leq r < a$. Moreover, if a does not divide b, then $0 < r < a$.*

DEFINITION 3 (Greatest Common Divisor). *Let $a, b \in \mathbb{Z} \setminus \{0\}$. The greatest common divisor (or GCD) of a and b, denoted by (a, b), is the unique positive integer d satisfying the following two conditions:*

(i) $d|a$ and $d|b$,   (ii) if $c|a$ and $c|b$, then $c \leq d$.

Similarly, if $a_1, a_2, \ldots, a_r \in \mathbb{Z} \setminus \{0\}$, the GCD of $a_1, a_2, \ldots, a_r$, denoted by $(a_1, a_2, \ldots, a_r)$, is the unique positive integer d satisfying the following two conditions:

(i) $d|a_1, d|a_2, \ldots, d|a_r$,   (ii) if $c|a_1, c|a_2, \ldots, c|a_r$, then $c \leq d$.

THEOREM 8. *Let $a_1, a_2, \ldots, a_r \in \mathbb{Z} \setminus \{0\}$. Then there exist integers $x_1, x_2, \ldots, x_r$ such that $(a_1, a_2, \ldots, a_r) = a_1 x_1 + a_2 x_2 + \cdots + a_r x_r$.*

THEOREM 9. *Let $a, b \in \mathbb{Z}$ be such that $ab \neq 0$. Let d be a positive integer. Then*

$$d = (a, b) \iff \begin{cases} d|a \text{ and } d|b, \\ c|a \text{ and } c|b \Rightarrow c|d. \end{cases}$$

THEOREM 10 (Euclid's Algorithm). *Let $a, b \in \mathbb{Z}$, $a > 0$. Applying successively the euclidean division (Theorem 7), we obtain the sequence of equalities*

$$
\begin{aligned}
b &= aq_1 + r_1, & 0 < r_1 < a, \\
a &= r_1 q_2 + r_2, & 0 < r_2 < r_1, \\
r_1 &= r_2 q_3 + r_3, & 0 < r_3 < r_2, \\
&\vdots \\
r_{j-2} &= r_{j-1} q_j + r_j, & 0 < r_j < r_{j-1}, \\
r_{j-1} &= r_j q_{j+1},
\end{aligned}
$$

where $r_j = (a, b)$.

DEFINITION 4. *The integers $a_1, a_2, \ldots, a_r$ are said to be relatively prime if $(a_1, a_2, \ldots, a_r) = 1$, while they are said to be pairwise coprime if $(a_i, a_j) = 1$ when $i \neq j$.*

DEFINITION 5 (Lowest Common Multiple). *Let $a_1, a_2, \ldots, a_r \in \mathbb{Z} \setminus \{0\}$. The lowest common multiple (or LCM) of $a_1, a_2, \ldots, a_r$, denoted by $[a_1, a_2, \ldots, a_r]$, is the smallest positive integer amongst all the common multiples of $a_1, a_2, \ldots, a_r$.*

Prime Numbers

THEOREM 11 (Fundamental Theorem of Arithmetic). *Each integer $n \geq 2$ can be written as a product of prime numbers, and this representation is unique, apart from the order in which the prime factors appear. In particular, n can be written in the form*

$$ n = q_1^{\alpha_1} q_2^{\alpha_2} \cdots q_r^{\alpha_r}, $$

where the q_i's are distinct prime numbers and where the α_i's are positive integers.

THEOREM 12. *If $q_1, q_2, \ldots, q_r$ are prime numbers and if $a = \prod_{i=1}^{r} q_i^{\alpha_i}$ and $b = \prod_{i=1}^{r} q_i^{\beta_i}$, with $\alpha_i \geq 0$ and $\beta_i \geq 0$ for $i = 1, 2, \ldots, r$, then*

$$ (a, b) = \prod_{i=1}^{r} q_i^{\min(\alpha_i, \beta_i)} \quad \text{and} \quad [a, b] = \prod_{i=1}^{r} q_i^{\max(\alpha_i, \beta_i)}. $$

Similarly, if $a = \prod_{i=1}^{r} q_i^{\alpha_i}$, $b = \prod_{i=1}^{r} q_i^{\beta_i}$, $c = \prod_{i=1}^{r} q_i^{\gamma_i}$ with $\alpha_i \geq 0$, $\beta_i \geq 0$ and $\gamma_i \geq 0$ for $i = 1, 2, \ldots, r$, then

$$ (a, b, c) = \prod_{i=1}^{r} q_i^{\min\{\alpha_i, \beta_i, \gamma_i\}} \quad \text{and} \quad [a, b, c] = \prod_{i=1}^{r} q_i^{\max\{\alpha_i, \beta_i, \gamma_i\}}. $$

THEOREM 13 (Euclid's Theorem). *There exist infinitely many prime numbers.*

THEOREM 14 (Dirichlet's Theorem). *Given two positive integers a and b with $(a, b) = 1$, the sequence of numbers $an + b$, $n = 1, 2, \ldots$, contains infinitely many prime numbers.*

THEOREM 15 (Bertrand's Postulate). *For each positive integer n, there exists a prime number p satisfying $n < p \leq 2n$.*

THEOREM 16. *The series $\sum_p 1/p$ and the product $\prod_p \left(1 + \frac{1}{p}\right)$, where in each case p runs through the set of all prime numbers, both diverge.*

THEOREM 17 (Prime Number Theorem). *Let $\pi(x)$ be the number of prime numbers $\leq x$. Then*

$$\lim_{x \to \infty} \frac{\pi(x)}{x/\log x} = 1.$$

THEOREM 18 (Mertens' Theorem). *As $x \to \infty$,*

$$\prod_{p \leq x} \left(1 - \frac{1}{p}\right) \sim \frac{e^{-\gamma}}{\log x},$$

where γ is Euler's constant defined by $\gamma = \lim_{n \to \infty} \left(\sum_{k=1}^{n} \frac{1}{k} - \log n\right)$.

Congruences

DEFINITION 6. *Let $a, b, m \in \mathbb{Z}$, $m \neq 0$. We say that a is congruent to b modulo m, and we write $a \equiv b \pmod{m}$, if $m|a - b$; if a is not congruent to b modulo m, we write $a \not\equiv b \pmod{m}$.*

THEOREM 19. *Let $a, b, c, d, m \in \mathbb{Z}$, $m > 0$. Then*
  (1) *$a \equiv a \pmod{m}$;*
  (2) *$a \equiv b \pmod{m}$ if and only if $b \equiv a \pmod{m}$;*
  (3) *if $a \equiv b \pmod{m}$ and $b \equiv c \pmod{m}$, then $a \equiv c \pmod{m}$;*
  (4) *if $a \equiv b \pmod{m}$ and $c \equiv d \pmod{m}$, then $ac \equiv bd \pmod{m}$ and $ax + cy \equiv bx + dy \pmod{m}$ for all $x, y \in \mathbb{Z}$;*
  (5) *if $a \equiv b \pmod{m}$ and $d|m$, $d > 0$, then $a \equiv b \pmod{d}$.*

THEOREM 20. *Let $a, m, m_1, m_2, \ldots, m_r \in \mathbb{N}$ and $x, y \in \mathbb{Z}$. Then*
  (i) *$ax \equiv ay \pmod{m}$ if and only if $x \equiv y \pmod{m/(a,m)}$;*
  (ii) *if $ax \equiv ay \pmod{m}$ and $(a, m) = 1$, then $x \equiv y \pmod{m}$;*
  (iii) *$x \equiv y \pmod{m_i}$ for $i = 1, 2, \ldots, r$ if and only if*

$$x \equiv y \pmod{[m_1, m_2, \ldots, m_r]}.$$

DEFINITION 7 (Residue modulo m). *If $x \equiv y \pmod{m}$, then y is called a residue of x modulo m. A set of integers $\{y_1, y_2, \ldots, y_m\}$ is called a complete residue system modulo m if for each integer x there exists one and only one y_i such that $x \equiv y_i \pmod{m}$.*

DEFINITION 8 (Reduced residue system). *A reduced residue system modulo m is a set of integers r_i such that $(r_i, m) = 1$, $r_i \not\equiv r_j \pmod{m}$ when $i \neq j$, and such that each integer x relatively prime to m is congruent to a certain r_i modulo m.*

DEFINITION 9 (Euler's function). *The Euler ϕ function is defined by*

$$\phi(n) = \#\{0 < m \leq n \mid (n, m) = 1\}.$$

THEOREM 21 (Fermat's Little Theorem). *Let p be a prime number and a a positive integer such that p does not divide a. Then $a^{p-1} \equiv 1 \pmod{p}$. Moreover, given any integer a, $a^p \equiv a \pmod{p}$.*

THEOREM 22 (Euler's Theorem). *Let $m \in \mathbb{N}$ and $a \in \mathbb{Z}$ be such that $(a, m) = 1$. Then*

$$a^{\phi(m)} \equiv 1 \pmod{m}.$$

THEOREM 23. *Let $m \in \mathbb{N}$, $a, b \in \mathbb{Z}$ and $(a, m) = 1$. Then the congruence $ax \equiv b \pmod{m}$ has a solution given by $x_0 = a^{\phi(m)-1}b$.*

THEOREM 24 (Wilson's Theorem). *Let m be a positive integer. Then*

$$m \text{ is prime} \iff (m-1)! \equiv -1 \pmod{m}.$$

THEOREM 25 (Chinese Remainder Theorem). *Let $m_1, m_2, \ldots, m_r$ be pairwise relatively prime integers. Let $a_1, a_2, \ldots, a_r$ be arbitrary integers. Then the system of congruences*

$$\begin{cases} x \equiv a_1 \pmod{m_1} \\ x \equiv a_2 \pmod{m_2} \\ \quad \vdots \qquad \quad \vdots \\ x \equiv a_r \pmod{m_r} \end{cases}$$

has a solution given by

$$x_0 = \sum_{j=1}^{r} \frac{m}{m_j} b_j a_j,$$

where $m = m_1 m_2 \cdots m_r$ and where each b_j is the solution of the congruence $(m/m_j)b_j \equiv 1 \pmod{m_j}$.

DEFINITION 10. *The sequence of digits which repeat themselves in the decimal expansion of a rational number is called the cycle of the fraction, in which case the number of digits in this cycle is called the period of the fraction.*

DEFINITION 11. *Given an integer $a \geq 2$, a composite number n is called pseudo-prime in basis a if $a^{n-1} \equiv 1 \pmod{n}$.*

The Function $[x]$

DEFINITION 12. *Given a real number x, the expression $[x]$ represents the largest integer $\leq x$. Also, we denote by $\|x\| = \left[x + \frac{1}{2}\right]$ the closest integer to x.*

THEOREM 26. *Let x and y be real numbers. Then*

(i) *$[x + m] = [x] + m$ if m is an integer;*
(ii) *$[x] + [y] \leq [x + y] \leq [x] + [y] + 1$;*
(iii) *the number of positive integers $\leq x$ and divisible by a positive integer a is equal to $[x/a]$.*

THEOREM 27 (Legendre's Theorem). *Let p be a prime number and n a positive integer. Then the largest exponent $\alpha = \alpha_p$ such that $p^\alpha | n!$ is given by*

$$\alpha = \sum_{i=1}^{\infty} \left[\frac{n}{p^i}\right],$$

this last sum being in fact a finite sum since $[n/p^i] = 0$ when $i > \log n / \log p$. It follows that

$$n! = \prod_{p \leq n} p^{\sum_{i=1}^{\infty} [n/p^i]}.$$

Arithmetical Functions

DEFINITION 13. *An arithmetical function is an application from $\mathbb{N}$ to $\mathbb{C}$. An arithmetical function f is said to be multiplicative if $f(1) = 1$ and if $f(mn) = f(m)f(n)$ when $(m, n) = 1$. An arithmetical function f is said to be totally multiplicative (or completely multiplicative) if $f(1) = 1$ and if $f(mn) = f(m)f(n)$ for all positive integers m and n.*

DEFINITION 14. *An arithmetical function f is said to be additive if $f(1) = 0$ and if $f(mn) = f(m) + f(n)$ when $(m, n) = 1$. An arithmetical function f is said to be totally additive (or completely additive) if $f(1) = 0$ and if $f(mn) = f(m) + f(n)$ for all positive integers m and n.*

DEFINITION 15. *Here are some of the arithmetical functions used in this book.*

(i) $\tau(n)$: *the number of divisors[1] of n;*

(ii) $\sigma(n)$: *the sum of the divisors of n; for each real number r, $\sigma_r(n) = \sum_{d|n} d^r$;*

(iii) $\omega(n)$: *the number of distinct prime factors of n, or in other terms $\omega(n) = \sum_{p|n} 1$ and $\omega(1) = 0$;*

(iv) $\Omega(n)$: *the total number of prime factors of n, or in other terms $\Omega(n) = \sum_{p^\alpha \| n} \alpha$ and $\Omega(1) = 0$;*

(v) $\phi(n)$: *the Euler function (see Definition 9);*

(vi) $\mu(n)$: *the Moebius function defined by*

$$\mu(n) = \begin{cases} 0 & \text{if } n \text{ is divisible by a perfect square} > 1, \\ (-1)^{\omega(n)} & \text{otherwise;} \end{cases}$$

(vii) $\lambda(n)$: *the Liouville function defined by $\lambda(n) = (-1)^{\Omega(n)}$;*

(viii) $\gamma(n)$: *the kernel of n defined by $\gamma(1) = 1$ and $\gamma(n) = \prod_{p|n} p$, for $n \geq 2$;*

(ix) $1(n) = 1$ *for each $n \geq 1$;*

(x) $I(n) = n$ *for each $n \geq 1$;*

(xi) $E(n) = \left[\dfrac{1}{n}\right] = \begin{cases} 1 & \text{if } n = 1, \\ 0 & \text{if } n > 1. \end{cases}$

(xii) $\Lambda(n)$: *the von Mangoldt function defined by*

$$\Lambda(n) = \begin{cases} \log p & \text{if } n = p^\alpha \text{ for a prime } p \text{ and some } \alpha \in \mathbb{N}, \\ 0 & \text{otherwise.} \end{cases}$$

DEFINITION 16. *A positive integer n is said to be squarefree if $\mu^2(n) = 1$.*

DEFINITION 17. *Given an arithmetical function f and a positive integer n, then*

(i) $\sum_{d|n} f(d)$ *represents the sum of $f(d)$ as d runs through all the divisors d of n;*

(ii) $\sum_{p|n} f(p)$ *represents the sum of $f(p)$ as p runs through all the prime divisors p of n;*

(iii) $\sum_{n \leq x} f(n)$ *represents the sum $f(1) + f(2) + \cdots + f([x])$;*

(iv) $\sum_{p \leq x} f(p)$ *represents the sum $f(2) + f(3) + f(5) + \cdots + f(p_{\pi(x)})$.*

---

[1]By "divisors", we mean "positive divisors". This remark is valid throughout this book.

THEOREM 28. *Each positive integer n can be written in a unique way as*

$$n = mr^2, \quad where \ \mu^2(m) = 1.$$

THEOREM 29. *For each integer $n \geq 1$,*

$$\sum_{d|n} \mu(d) = E(n).$$

THEOREM 30. *For each integer $n \geq 1$,*

$$\sum_{d|n} \phi(d) = n \quad and \quad \phi(n) = n \sum_{d|n} \frac{\mu(d)}{d} = n \prod_{p|n} \left(1 - \frac{1}{p}\right).$$

THEOREM 31. *For each integer $n \geq 2$ written in its canonical form*

$$n = q_1^{\alpha_1} q_2^{\alpha_2} \cdots q_r^{\alpha_r},$$

we have

$$\tau(n) = (\alpha_1 + 1)(\alpha_2 + 1) \cdots (\alpha_r + 1)$$

and

$$\sigma(n) = \prod_{i=1}^{r} \frac{q_i^{\alpha_i + 1} - 1}{q_i - 1}.$$

DEFINITION 18 (Dirichlet Product). *Given two arithmetical functions f and g, we define their Dirichlet product $f * g$ by $(f * g)(n) = \sum_{d|n} f(d)g(n/d)$.*

THEOREM 32 (Moebius Inversion Formula). *Let F and f be two arithmetical functions. Then, for each integer $n \geq 1$,*

$$F(n) = \sum_{d|n} f(d) \Longleftrightarrow f(n) = \sum_{d|n} \mu(d)F(n/d).$$

DEFINITION 19 (Some Special Numbers).

(1) *A positive integer n is said to be triangular if there exists a positive integer m such that $n = 1 + 2 + \cdots + m$.*
(2) *A positive integer n is said to be perfect if $\sigma(n) = 2n$. More generally, n is said to be k–perfect (k integer ≥ 2), or aliquot, if $\sigma(n) = kn$.*
(3) *The numbers $F_n = 2^{2^n} + 1$ are called Fermat numbers. If F_n is prime, then it is called a Fermat prime.*
(4) *The numbers $M_p = 2^p - 1$, where p is prime, are called Mersenne numbers. If M_p is prime, it is called a Mersenne prime.*
(5) *A positive composite integer n is called a Carmichael number if $b^{n-1} \equiv 1 \pmod{n}$ for each positive integer b such that $(b, n) = 1$.*
(6) *The positive integers m and n are said to be amicable if $\sigma(m) = m + n = \sigma(n)$.*
(7) *An integer $n \geq 2$ is called a powerful number (or squarefull number) if $p|n$ implies that $p^2|n$.*
(8) *Let $k \geq 2$ be an integer. An integer $n \geq 2$ is called a k–powerful number if $p|n$ implies that $p^k|n$.*

Diophantine Equations

THEOREM 33. *Let a, b and c be three integers and let $d = (a, b)$. Then the Diophantine equation*

$$(*) \qquad\qquad ax + by = c$$

has integer solutions if and only if $d | c$, in which case the solutions are given by

$$x = x_0 + \frac{bt}{d}, \qquad y = y_0 - \frac{at}{d} \qquad (t \in \mathbb{Z}),$$

where $x = x_0$, $y = y_0$ is a particular solution of $(*)$.

THEOREM 34. *The primitive positive solutions of $x^2 + y^2 = z^2$ (that is, with x, y, z relatively prime and positive integers), with y even, are $x = r^2 - s^2$, $y = 2rs$, $z = r^2 + s^2$, where r and s are arbitrary integers of opposite parity and satisfying $r > s > 0$ and $(r, s) = 1$.*

Quadratic Reciprocity

DEFINITION 20. *Let a be an integer and p an odd prime number such that $(a, p) = 1$. If the congruence $x^2 \equiv a \pmod{p}$ has a solution, we say that a is a quadratic residue modulo p; otherwise we say that a is a quadratic nonresidue modulo p.*

DEFINITION 21. *Let a be an integer and p an odd prime number such that $(a, p) = 1$. We define the Legendre symbol $\left(\dfrac{a}{p} \right)$ by*

$$\left(\frac{a}{p} \right) = \begin{cases} 1 & \text{if } a \text{ is a quadratic residue modulo } p, \\ -1 & \text{if } a \text{ is a quadratic nonresidue modulo } p. \end{cases}$$

THEOREM 35 (Euler's Criterion). *Let p be an odd prime number. Then, for each integer a such that $(a, p) = 1$,*

$$\left(\frac{a}{p} \right) \equiv a^{(p-1)/2} \pmod{p}.$$

THEOREM 36. *If p is an odd prime number, then*

$$\left(\frac{-1}{p} \right) = (-1)^{(p-1)/2} \quad \text{and} \quad \left(\frac{2}{p} \right) = (-1)^{(p^2-1)/8}.$$

THEOREM 37 (Law of Quadratic Reciprocity). *If p and q are distinct odd prime numbers, then*

$$\left(\frac{p}{q} \right) \left(\frac{q}{p} \right) = (-1)^{\frac{p-1}{2} \frac{q-1}{2}}.$$

Continued Fractions

Given a rational number a/b, with $(a,b) = 1$ and $b > 0$, according to Euclid's algorithm, there exists a positive integer n such that

$$
\begin{aligned}
a &= a_1 b + b_1, & 0 &< b_1 < b, \\
b &= a_2 b_1 + b_2, & 0 &< b_2 < b_1, \\
b_1 &= a_3 b_2 + b_3, & 0 &< b_3 < b_2,
\end{aligned}
$$

$$
\vdots \qquad\qquad\qquad \vdots
$$

$$
\begin{aligned}
b_{n-3} &= a_{n-1} b_{n-2} + b_{n-1}, & 0 &< b_{n-1} < b_{n-2}, \\
b_{n-2} &= a_n b_{n-1}.
\end{aligned}
$$

DEFINITION 22. *Using successively each of these equations, we can write*

$$
\frac{a}{b} = a_1 + \cfrac{1}{(b/b_1)} = a_1 + \cfrac{1}{a_2 + \cfrac{1}{(b_1/b_2)}}
$$

$$
\vdots \qquad\qquad \vdots
$$

$$
= a_1 + \cfrac{1}{a_2 + \cfrac{1}{a_3 + \cfrac{1}{\ddots \cfrac{1}{a_{n-1} + \cfrac{1}{a_n}}}}} =: [a_1, a_2, \ldots, a_n].
$$

This expression is called the development of the number a/b as a finite continued fraction.

THEOREM 38. *A number is rational if and only if it can be expressed as a finite continued fraction.*

DEFINITION 23. *Let $[a_1, a_2, \ldots, a_n]$ be a finite continued fraction. The continued fraction $C_k = [a_1, a_2, \ldots, a_k]$, where $k = 1, 2, \ldots, n$, is called the k–th convergent of the continued fraction $[a_1, a_2, \ldots, a_n]$.*

DEFINITION 24. *Let $[a_1, a_2, \ldots, a_n]$ be a finite continued fraction. We define the corresponding sequences $\{p_n\}$ and $\{q_n\}$ by*

$$
\begin{aligned}
p_0 &= 1, \ p_1 = a_1, & p_n &= a_n p_{n-1} + p_{n-2}, & n &\geq 2, \\
q_0 &= 0, \ q_1 = 1, & q_n &= a_n q_{n-1} + q_{n-2}, & n &\geq 2.
\end{aligned}
$$

We easily obtain the following inequalities:

$$
0 = q_0 \leq q_1 < q_2 < q_3 < \cdots .
$$

THEOREM 39. *Let $[a_1, a_2, \ldots, a_n]$ be a finite continued fraction. The corresponding sequences $\{p_n\}$ and $\{q_n\}$ satisfy*

$$
p_n q_{n-1} - q_n p_{n-1} = (-1)^n.
$$

THEOREM 40. *Each convergent C_k, $k = 1, 2, \ldots, n$, of the finite continued fraction $[a_1, a_2, \ldots, a_n]$ satisfies $C_k = p_k/q_k$.*

DEFINITION 25. *Let $a_1, a_2, \ldots$ be an infinite sequence of integers, all positive except possibly a_1. The expression $[a_1, a_2, \ldots]$ is then called an infinite continued fraction and is defined as the number*

$$\lim_{n \to \infty} [a_1, a_2, \ldots, a_n].$$

(It is indeed possible to prove that this limit always exists (see Niven, Zuckerman and Montgomery [25]).)

THEOREM 41. *If C_n is the n–th convergent of the irrational number α, then*

$$|\alpha - C_n| < \frac{1}{q_n q_{n+1}} < \frac{1}{q_n^2}.$$

THEOREM 42. *A real number is irrational if and only if it can be expressed as an infinite continued fraction.*

THEOREM 43 (Lagrange). *Given an integer $n \geq 1$, let $C_n = p_n/q_n$ be the n-th convergent of the irrational number α. Moreover, let $a, b \in \mathbb{Z}$ with $1 \leq b < q_{n+1}$. Then*

$$|q_n \alpha - p_n| \leq |b\alpha - a|.$$

Classification of Real Numbers

DEFINITION 26. *A complex number x is said to be algebraic if it satisfies an equation of the form*

$$a_n x^n + a_{n-1} x^{n-1} + \cdots + a_1 x + a_0 = 0,$$

where the a_i's are integers not all zero. A complex number which is not algebraic is said to be transcendental.

THEOREM 44. *If x_0 is a root of the equation*

$$x^m + c_1 x^{m-1} + \cdots + c_{m-1} x + c_m = 0,$$

where each $c_i \in \mathbb{Z}$, then x_0 is either an integer or an irrational number.

Two Conjectures

CONJECTURE abc (Masser and Oesterlé, 1985). *Let $\varepsilon > 0$. Then there exists a constant $M = M(\varepsilon) > 0$ such that for each triple of positive integers (a, b, c) pairwise relatively prime and satisfying $a + b = c$, we have*

$$c \leq M \left(\prod_{p \mid abc} p \right)^{1+\varepsilon}.$$

HYPOTHESIS H (A. Schinzel and W. Sierpinski [35]). *Let $k \geq 1$ and $f_1(x), \ldots, f_k(x)$ be irreducible polynomials with integer coefficients whose leading coefficient is positive. Assume that there exists no integer > 1 dividing the products $f_1(n) \ldots f_k(n)$ for all positive integers n. Then, there exist infinitely many positive integers m such that all the numbers $f_1(m), \ldots, f_k(m)$ are primes.*

Part 2

Statements of the Problems

1. Mathematical Induction and Combinatorics

(1) Show that for each positive integer n, we have

$$\sum_{k=1}^{n} k = \frac{n(n+1)}{2}, \quad \sum_{k=1}^{n} k^2 = \frac{n(n+1)(2n+1)}{6}, \quad \sum_{k=1}^{n} k^3 = \left(\frac{n(n+1)}{2}\right)^2.$$

(2) Show that the cube of a positive integer can always be written as the difference of two squares.

(3) Establish a formula for $\displaystyle\sum_{k=2}^{n} \frac{1}{k^2 - 1}$ valid for each positive integer n.

(4) Establish a formula allowing one to obtain the sum of the first n positive even integers.

(5) Show that the formula $\displaystyle\sum_{j=1}^{n} (-1)^j j^2 = (-1)^n \sum_{j=1}^{n} j$ holds for each positive integer n.

(6) Show that $a + b$ is a factor of $a^{2n-1} + b^{2n-1}$ for each integer $n \geq 1$.

(7) Show that $a^2 + b^2$ is a factor of $a^{4n} - b^{4n}$ for each integer $n \geq 1$.

(8) Show that for each positive integer n,

$$a^n - b^n = (a - b) \sum_{k=0}^{n-1} a^k b^{n-1-k}.$$

(9) Show that $\displaystyle\sum_{j=1}^{n} j \cdot j! = (n+1)! - 1$ for each positive integer n.

(10) Prove, using induction, that $(2n)! < 2^{2n}(n!)^2$ for each integer $n \geq 1$.

(11) Use induction in order to prove that $n^3 < n!$ for each integer $n \geq 6$.

(12) Let θ be a real number such that $\theta \geq -1$. Prove, using induction, that for each integer $n \geq 0$, we have $(1 + \theta)^n \geq 1 + n\theta$.

(13) Let θ be a nonnegative real number. Show, using induction, that for each positive integer n, we have $(1 + \theta)^n \geq 1 + n\theta + \frac{n(n-1)}{2}\theta^2$.

(14) Show that for each positive integer n, $\frac{1}{3}(n^3 + 2n)$ is an integer.

(15) Show that $\dfrac{10^n + 3 \cdot 4^{n+2} + 5}{9}$ is an integer for each positive integer n.

(16) Show that if n is a positive integer, then

$$\binom{n}{k} = \binom{n}{k+1} \iff n = 2k + 1.$$

(17) Show that if n is a positive integer, then

(a) $\dbinom{2n}{0} + \dbinom{2n}{2} + \dbinom{2n}{4} + \cdots + \dbinom{2n}{2n} = 2^{2n-1}$;

(b) $\dbinom{2n}{1} + \dbinom{2n}{3} + \cdots + \dbinom{2n}{2n-1} = 2^{2n-1}$.

(18) Prove that for each integer $n \geq 1$, we have
$$n! \leq \left(\frac{n+1}{2}\right)^n.$$

(19) Show that each integer $n > 7$ can be written as a sum containing only the numbers 3 and 5. For example, $8 = 3 + 5$, $9 = 3 + 3 + 3$, $10 = 5 + 5$.

(20) Assume that amongst n points, $n \geq 2$, in a given plane, no three points are on the same line. Show that the number of possible lines passing through these points is $n(n-1)/2$.

(21) Show that for each integer $n \geq 2$,
$$1 + \frac{1}{\sqrt{2}} + \frac{1}{\sqrt{3}} + \cdots + \frac{1}{\sqrt{n}} > \sqrt{n}.$$

(22) Prove that for each positive integer k,
$$1^3 + 3^3 + 5^3 + \cdots + (2k-1)^3 = k^2(2k^2 - 1).$$

(23) We saw in problem 1 that, for each integer $n \geq 1$,
$$
\begin{aligned}
1 + 2 + \cdots + n &= \frac{n(n+1)}{2} = \frac{n^2}{2} + \frac{n}{2}; \\
1^2 + 2^2 + \cdots + n^2 &= \frac{n(n+1)(2n+1)}{6} = \frac{n^3}{3} + \frac{n^2}{2} + \frac{n}{6}; \\
1^3 + 2^3 + \cdots + n^3 &= \frac{n^2(n+1)^2}{4} = \frac{n^4}{4} + \frac{n^3}{2} + \frac{n^2}{4}.
\end{aligned}
$$

Hence, letting $S_k(n) = 1^k + 2^k + \cdots + n^k$ and in light of these three relations, it is normal to conjecture that, for each integer $k \geq 1$, $S_k(n)$ is a polynomial of degree $k + 1$. In fact, in 1654, Blaise Pascal (1623–1662) established that indeed it was the case. His proof used induction and the expansion of the expression $(n+1)^{k+1} - 1$. Provide the details.

(24) Find a formula, valid for each integer $n \geq 2$, for
$$\prod_{i=2}^n \left(1 - \frac{1}{i}\right), \quad \text{and the same for} \quad \prod_{i=2}^n \left(1 - \frac{1}{i^2}\right).$$

(25) Show that, whatever the value of the integer $n \geq 1$, we always have
$$\sum_{i=1}^n \frac{i}{i^4 + i^2 + 1} < \frac{1}{2}.$$

(26) Show that if m, n and r are three positive integers such that
$$S := \frac{1}{m} + \frac{1}{n} + \frac{1}{r} < 1, \quad \text{then} \quad S \leq \frac{41}{42}.$$

(27) Given a positive integer n, let $s(n)$ be the sum of its digits (in basis 10). For each pair of positive integers k, ℓ smaller than 10, let $A_k(\ell)$ be the number of ℓ-digit positive integers n whose sum of digits is equal to k. In other words,
$$A_k(\ell) = \#\{n : 10^{\ell-1} \leq n < 10^\ell, \ s(n) = k\}.$$

Show that

$$A_k(\ell) = \binom{k+\ell-2}{k-1} = \binom{k+\ell-2}{\ell-1},$$

and conclude in particular that $A_k(\ell) = A_\ell(k)$.

(28) Using induction, prove the formulas due to Mariares (1913):

$$1^2 + 3^2 + 5^2 + \cdots + n^2 = \binom{n+2}{3}, \quad \text{if } n \text{ is odd;}$$

$$2^2 + 4^2 + 6^2 + \cdots + n^2 = \binom{n+2}{3}, \quad \text{if } n \text{ is even.}$$

(29) Let S be a set of 10 distinct integers chosen amongst the numbers $1, 2, \ldots, 99$. Show that S must contain two disjoint subsets for which the sum of their respective elements is the same.

(30) Given 51 arbitrary positive integers, show that one can always find two of them whose difference is 50.

(31) In order to acquire problem solving skills, a student decides to solve at least one problem per day and at most 11 per week and to do this for a whole year. Show that there exists a period of consecutive days during which he will solve exactly 20 problems.

(32) On a rectangular table of dimension 120 inches by 150 inches, we set 14 001 marbles. Show that, no matter how these are arranged, one can place a cylindrical glass with a diameter of 5 inches over at least 8 marbles.

(33) Choose n points on a circle and join them pairwise by secants. Taking for granted that no more than two secants can meet at the same point, in how many regions is the circle thus divided?

(34) Say we have three posts and n disks of different diameters placed on one of the posts, ordered by increasing diameters, the largest at the bottom of the post, the smaller at the top. The problem consists in transferring the tower of disks from the first post to the third post, using if need be the second post, but in such a way that, with each move, we do not place the moving disk on a smaller one. Establish the function of n which gives the minimum number of moves. (This problem is known as the "Tower of Hanoi Problem".)

(35) Let $\{F_n : n \in \mathbb{N}\}$ be the sequence of Fibonacci numbers defined by $F_1 = 1, F_2 = 1$ and $F_n = F_{n-1} + F_{n-2}$ for $n \geq 3$. Show that each positive integer can be written as the sum of distinct Fibonacci numbers.

(36) One easily checks that

$$
\begin{aligned}
1 &= 1^2, \\
2 &= -1^2 - 2^2 - 3^2 + 4^2, \\
3 &= -1^2 + 2^2, \\
4 &= -1^2 - 2^2 + 3^2, \\
5 &= 1^2 + 2^2, \\
6 &= 1^2 - 2^2 + 3^2.
\end{aligned}
$$

Hence, we may be tempted to formulate a conjecture, namely that each positive integer n can be written as

$$n = e_1 1^2 + e_2 2^2 + e_3 3^2 + e_4 4^2 + \cdots + e_k k^2,$$

for a certain positive integer k (depending on n), where the $e_i \in \{-1, 1\}$. Prove this conjecture.

2. Divisibility

(37) The mathematician Duro Kurepa defined $!n = 0! + 1! + \cdots + (n-1)!$ for $n \geq 1$ and conjectured that $(!n, n!) = 2$ for all $n \geq 2$. This conjecture has been verified by Ivić and Mijajlović [20] for $n < 10^6$. Using computer software, write a program showing that this conjecture is true up to $n = 1000$.

(38) Consider the situation where the positive integer a is divided by the positive integer b using the euclidian division (see Theorem 7) yielding

$$(*) \qquad\qquad a = 652b + 8634.$$

By how much can we increase both a and b without changing the quotient $q = 652$?

(39) Consider the number $N = 111\ldots11$, here written in basis 2. Write N^2 in basis 2.

(40) Show that $39 | 7^{37} + 13^{37} + 19^{37}$.

(41) Show that, for each integer $n \geq 1$, the number $49^n - 2352n - 1$ is divisible by 2304.

(42) Given any integer $n \geq 1$, show that the number $n^4 + 2n^3 + 2n^2 + 2n + 1$ is never a perfect square.

(43) Let N be a two digit number. Let M be the number obtained from N by interchanging its two digits. Show that 9 divides $M - N$ and then find all the integers N such that $|M - N| = 18$.

(44) Is it true that 3 never divides $n^2 + 1$ for every positive integer n? Explain.

(45) Is it true that 5 never divides $n^2 + 2$ for every positive integer n? Explain. Is the result the same if one replaces the number 5 by the number 7?

(46) Given $s + 1$ integers $a_0, a_1, \ldots, a_s$ and a prime number p, show that p divides the integer

$$N(n) := a_0 + a_1 n + \cdots + a_{s-1}n^{s-1} + a_s n^s$$

if and only if p divides $N(r)$, for an integer r, $0 \leq r \leq p - 1$. Use this to find all integers n such that 7 divides $3n^2 + 6n + 5$.

(47) Compute the value of the expression

$$\frac{(10^4 + 324)(22^4 + 324)(34^4 + 324)(46^4 + 324)(58^4 + 324)}{(4^4 + 324)(16^4 + 324)(28^4 + 324)(40^4 + 324)(52^4 + 324)}.$$

(48) Show that, in any basis, the number 10101 is composite.

(49) Show that the product of four consecutive integers is necessarily divisible by 24.

(50) Show that the number

$$1^{47} + 2^{47} + 3^{47} + 4^{47} + 5^{47} + 6^{47}$$

is a multiple of 7.

(51) Show that the product of any five consecutive positive integers cannot be a perfect square.

(52) Show that $30 | n^5 - n$ for each positive integer n.

(53) Show that $6 | n(n+1)(2n+1)$ for each positive integer n.

(54) Given any integer $n \geq 0$, show that $64^{n+1} - 63n - 64$ is divisible by 3969. More generally, given $a \in \mathbb{N}$, show that for each integer $n \geq 0$, $(a+1)^{n+1} - an - (a+1)$ is divisible by a^2.

(55) Find all positive integers n such that $(n+1)|(n^2+1)$.

(56) Find all positive integers n such that $(n^2+2)|(n^6+206)$.

(57) Identify, if any exist, the positive integers n such that $(n^3+2)|(n^6+216)$.

(58) If a and b are positive integers such that $b|(a^2+1)$, do we necessarily have that $b|(a^4+1)$? Explain.

(59) Let n and k be positive integers.
  (a) For $n \geq k$, show that
$$\frac{n}{(n,k)} \left| \binom{n}{k} \right. .$$

  (b) For $n \geq k$, show that
$$\frac{n+1-k}{(n+1,k)} \left| \binom{n}{k} \right. .$$

  (c) For $n \geq k-1 \geq 1$, show that
$$\frac{(n+1,k-1)}{n+2-k} \binom{n}{k-1} \quad \text{is an integer.}$$

(60) For each integer $n \geq 1$, let $f(n) = 1! + 2! + \cdots + n!$. Find polynomials $P(x)$ and $Q(x)$ such that
$$f(n+2) = P(n)f(n+1) + Q(n)f(n), \quad \text{for each integer } n \geq 1.$$

(61) Show that, for each positive integer n,
$$49 | 2^{3n+3} - 7n - 8.$$

(62) Find all positive integers a for which $a^{10} + 1$ is divisible by 10.

(63) Is it true that $3|2^{2n} - 1$ for each positive integer n? Explain.

(64) Show that if an integer is of the form $6k+5$, then it is necessarily of the form $3k-1$, while the reverse is false.

(65) Can an integer $n > 1$ be of the form $8k+7$ and also of the form $6\ell+5$? Explain.

(66) Let $M_1 = 2+1$, $M_2 = 2 \cdot 3 + 1$, $M_3 = 2 \cdot 3 \cdot 5 + 1$, $M_4 = 2 \cdot 3 \cdot 5 \cdot 7 + 1$, $M_5 = 2 \cdot 3 \cdot 5 \cdot 7 \cdot 11 + 1$, ... Prove none of the numbers M_k is a perfect square.

(67) Verify that if an integer is a square and a cube, then it must be of the form $7k$ or $7k+1$.

(68) If x and y are odd integers, prove that $x^2 + y^2$ cannot be a perfect square.

(69) Show that, for each positive integer n, we have $n^2|(n+1)^n - 1$.

(70) Let $k, n \in \mathbb{N}$, $n \geq 2$. Show that $(n-1)^2|(n^k-1)$ if and only if $(n-1)|k$. More generally, show the following result: Let $a \in \mathbb{Z}$ and $k, n \in \mathbb{N}$ with $n \neq a$; then $(n-a)^2|(n^k - a^k)$ if and only if $(n-a)|ka^{k-1}$.

(71) Let a, b be integers and let n be a positive integer.
  (a) If $a - b \neq 0$, show that
$$\left(\frac{a^n - b^n}{a-b}, a-b \right) = \left(n(a,b)^{n-1}, a-b \right).$$

(b) If $a + b \neq 0$ and if n is odd, show that

$$\left(\frac{a^n + b^n}{a + b}, a + b\right) = \left(n(a, b)^{n-1}, a + b\right).$$

(c) Show that if a and b are relatively prime with $a + b \neq 0$ and if $p > 2$ is a prime number, then

$$\left(\frac{a^p + b^p}{a + b}, a + b\right) = \begin{cases} 1 & \text{if } p \nmid (a + b), \\ p & \text{if } p \mid (a + b). \end{cases}$$

(72) Let k and n be positive integers. Show that the only solutions (k, n) of the equation $(n - 1)! = n^k - 1$ are $(1, 2)$, $(1, 3)$ and $(2, 5)$.

(73) According to Euclid's algorithm, assuming that $b \geq a$ are positive integers, we have

$$\begin{aligned} b &= aq_1 + r_1, & 0 < r_1 < a, \\ a &= r_1 q_2 + r_2, & 0 < r_2 < r_1, \\ r_1 &= r_2 q_3 + r_3, & 0 < r_3 < r_2, \\ &\vdots \\ r_{j-2} &= r_{j-1} q_j + r_j, & 0 < r_j < r_{j-1}, \\ r_{j-1} &= r_j q_{j+1}, \end{aligned}$$

where $r_j = (a, b)$.
  (a) Show that $b > 2r_1$, $a > 2r_2$ and for $k \geq 1$, $r_k > 2r_{k+2}$.
  (b) Deduce that $b > 2^{j/2}$ and therefore that the maximum number of steps in Euclid's algorithm is $[2(\log b / \log 2)]$.

(74) Show that there exist infinitely many positive integers n such that $n \mid 2^n + 1$.

(75) Let a be an integer ≥ 2. Show that for positive integers m and n we have

$$a^n - 1 \mid a^m - 1 \iff n \mid m.$$

(76) Let N_n be an integer formed of n consecutive "1"s. For example, $N_3 = 111$, $N_7 = 1\,111\,111$. Show that $N_n \mid N_m \iff n \mid m$.

(77) Prove that no member of the sequence $11, 111, 1\,111, 11\,111, \ldots$ is a perfect square.

(78) What is the smallest positive integer divisible both by 2 and 3 which is both a perfect square and a sixth power? More generally, what is the smallest positive integer n divisible by both 2 and 3 which is both an n–th power and an m–th power, where $n, m \geq 2$?

(79) Three of the four integers, found between 100 and 1000, with the property of being equal to the sum of the cubes of their digits are 153, 370 and 407. What is the fourth of these integers?

(80) How many positive integers $n \leq 1000$ are not divisible by 2, nor by 3, nor by 5?

(81) Prove the following result obtained in the seventeenth century by Pierre de Fermat (1601–1665): *"Each odd prime number p can be written as the difference of two perfect squares."*

(82) Prove that the representation mentioned in problem 81 is unique.

(83) Is the result of Fermat stated in problem 81 still true if p is simply an odd positive integer?

(84) Let $n = 999\,980\,317$. Observing that $n = 10^9 - 3^9$ and factoring this last expression, conclude that $7 \mid n$.

(85) Show that if an odd integer can be written as the sum of two squares, then it is of the form $4n + 1$.

(86) Let $a, b, c \in \mathbb{Z}$ be such that $abc \neq 0$ and $(a, b, c) = 1$ and such that $a^2 + b^2 = c^2$. Prove that at least one of the integers a and b is even.

(87) For which integer values of k is the number $10^k - 1$ the cube of an integer?

(88) Show that if the positive integer a divides both $42n + 37$ and $7n + 4$ for a certain integer n, then $a = 1$ or $a = 13$.

(89) If a and b are two positive integers and if $\dfrac{1}{a} + \dfrac{1}{b}$ is an integer, prove that $a = b$. Moreover, show that a is then necessarily equal to 1 or 2.

(90) Let $a, b \in \mathbb{N}$ such that $(a, b) = 4$. Find all possible values of (a^2, b^3).

(91) Let $a, b \in \mathbb{N}$ and $d = (a, b)$. Find the value of $(3a + 5b, 5a + 8b)$ in terms of d and more generally that of $(ma + nb, ra + sb)$ knowing that $ms - nr = 1$, where $m, n, r, s \in \mathbb{N}$.

(92) Let $m, n \in \mathbb{N}$. If $d|mn$ where $(m, n) = 1$, show that d can be written as $d = rs$ where $r|m$, $s|n$ and $(r, s) = 1$.

(93) Let a, b, d be nonzero integers, d odd, such that $d|(a + b)$ and $d|(a - b)$. Show that $d|(a, b)$.

(94) Given eight positive composite integers ≤ 360, show that at least two of them have a common factor larger than 1.

(95) If a and b are positive integers such that $(a, b) = 1$ and ab is a perfect square, show that a and b are perfect squares.

(96) Can $n(n+1)$ be a perfect square for a certain positive integer n? Explain.

(97) What are the possible values of the expression $(n, n+14)$ as n runs through the set of positive integers?

(98) Let $n > 1$ an integer. Which of the following statements are true:

$$3|(n^3 - n), \qquad 3|n(n + 1), \qquad 8|(2n + 1)^2 - 1, \qquad 6|n(n + 1)(n + 2).$$

(99) Is it true that if n is an even integer, then $24|n(n + 1)(n + 2)$? Explain.

(100) Let n be an integer such that $(n, 2) = (n, 3) = 1$. Show that $24|n^2 + 47$.

(101) Let $d = (a, b)$, where a and b are positive integers. Show that there are exactly d numbers amongst the integers a, $2a$, $3a, \ldots, ba$ which are divisible by b.

(102) Let a, b be integers such that $(a, b) = d$, and let x_0, y_0 be integers such that $ax_0 + by_0 = d$. Show that:
  (a) $(x_0, y_0) = 1$;
  (b) x_0 and y_0 are not unique.

(103) Let a, m and n be positive integers. If $(m, n) = 1$, show that $(a, mn) = (a, m)(a, n)$.

(104) For all $n \in \mathbb{N}$, show that $(n^2 + 3n + 2, 6n^3 + 15n^2 + 3n - 7) = 1$.

(105) Let $a, b \in \mathbb{Z}$. If $(a, b) = 1$, show that
  (a) $(a + b, a - b) = 1$ or 2;   (b) $(2a + b, a + 2b) = 1$ or 3;
  (c) $(a^2 + b^2, a + b) = 1$ or 2;   (d) $(a + b, a^2 - 3ab + b^2) = 1$ or 5.

(106) Let $a, b \in \mathbb{Z}$. If $(a, b) = 1$, find the possible values of
  (a) $(a^3 + b^3, a^3 - b^3)$;   (b) $(a^2 - b^2, a^3 - b^3)$.

(107) Let a, b and c be integers. For each of the following statements, say if it is true or false. If it is true, give a proof; if it is false, provide a counter-example.
  (a) If $(a, b) = (a, c)$, then $[a, b] = [a, c]$.

(b) If $(a, b) = (a, c)$, then $(a^2, b^2) = (a^2, c^2)$.

(c) If $(a, b) = (a, c)$, then $(a, b) = (a, b, c)$.

(108) Let $a, b \in \mathbb{Z}$ and let $m, n \in \mathbb{N}$. For each of the following statements, say if it is true or false. If it is true, give a proof; if it is false, provide a counter-example.

(a) If $a^n | b^n$, then $a | b$.

(b) If $a^m | b^n$, $m > n$, then $a | b$.

(c) If $a^m | b^n$, $m < n$, then $a | b$.

(109) Let $a, b, c \in \mathbb{Z}$. Show that if $(a, b) = 1$ and $c | a$, then $(c, b) = 1$.

(110) Let $a, b, c \in \mathbb{Z}$. Show that if $(a, bc) = 1$, then $(a, b) = (a, c) = 1$.

(111) Let $a, b \in \mathbb{Z}$. Show that $(a, b) = (a + b, [a, b])$. Using this result, find two positive integers whose sum is 186 and whose LCM is 1440.

(112) Let $a, b, c \in \mathbb{Z}$.

(a) Show that $(a, bc) = (a, (a, b)c)$.

(b) Show that $(a, bc) = (a, (a, b)(a, c))$.

(113) Let $a, b, c \in \mathbb{Z}$. Show that if $(a, c) = 1$, then $(ab, c) = (b, c)$.

(114) Let a, b, m and n be integers. If $(m, n) = 1$, show that $(ma + nb, mn) = (a, n)(b, m)$. Show that this result generalizes the result of problem 103.

(115) Is it possible that $\binom{n}{r}$ is relatively prime with $\binom{n}{s}$, for certain positive integers r, s, n satisfying $0 < r < s \le n/2$? Explain.

(116) Find two positive integers for which the difference between their LCM and their GCD is equal to 143.

(117) Let a, b, c be positive integers. Show that $(a, b, c) = ((a, b), c)$ and $[a, b, c] = [[a, b], c]$. Generalize this result. Use this result to compute $(132, 102, 36)$ and find those integers x, y, z for which $132x + 102y + 36z = (132, 102, 36)$.

(118) Let n be a positive integer. Evaluate $(n, n + 1, n + 2)$ and $[n, n + 1, n + 2]$.

(119) Let a, b, c be positive integers. If $(a, b) = (b, c) = (a, c) = 1$, show that $(a, b, c)[a, b, c] = abc$.

(120) Is it true that if a and b are positive integers such that $(a, b) = 1$, then $(a^2, ab, b^2) = 1$? Explain.

(121) Is it true that if a, b and c are positive integers, then $[a^2, ab, b^2] = [a^2, b^2]$? Explain.

(122) Is it true that if a, b and c are positive integers, then $(a, b, c) = ((a, b), (a, c))$? Explain.

(123) Is it true that $[a, b, c] \cdot (a, b, c) = |abc|$, $\forall a, b, c \in \mathbb{Z} \setminus \{0\}$? Explain.

(124) Let a, b, d, m and n be positive integers such that $a | d^m - 1$, $b | d^n - 1$ and $(a, b) = 1$. Show that $ab | d^{[m,n]} - 1$.

(125) Show that if a is an integer > 1, then, for each pair of positive integers m and n,

$$(a^m - 1, a^n - 1) = a^{(m,n)} - 1.$$

What do we obtain for $(a^m + 1, a^n + 1)$, for $(a^m + 1, a^n - 1)$? More generally, given $a > 1$ and $b > 1$, what are the values of

$$(a^m - b^m, a^n - b^n), \quad (a^m + b^m, a^n + b^n) \text{ and } (a^m + b^m, a^n - b^n)?$$

(126) Show that there exist infinitely many pairs of integers $\{x, y\}$ satisfying $x + y = 40$ and $(x, y) = 5$.

(127) Find all pairs of positive integers $\{a, b\}$ such that $(a, b) = 15$ and $[a, b] = 90$. More generally, if d and m are positive integers, show that there exists a pair of positive integers $\{a, b\}$ for which $(a, b) = d$ and $[a, b] = m$ if and only if $d|m$. Moreover, in this situation, show that the number of such pairs is 2^r, where r is the number of distinct prime factors of m/d.

(128) Prove that one cannot find integers m and n such that $m + n = 101$ and $(m, n) = 3$.

(129) Let $a, m, n \in \mathbb{N}$ with $m \neq n$.
   (a) Show that $(a^{2^n} + 1)|(a^{2^m} - 1)$ if $m > n$.
   (b) Show that $(a^{2^n} + 1, a^{2^m} + 1) = \begin{cases} 1 & \text{if } a \text{ is even,} \\ 2 & \text{if } a \text{ is odd.} \end{cases}$

(130) Let n be a positive integer. Find the greatest common divisor of the numbers
$$\binom{2n}{1}, \binom{2n}{3}, \binom{2n}{5}, \ldots, \binom{2n}{2n-1}.$$

(131) Given $n + 1$ distinct positive integers $a_1, a_2, \ldots, a_{n+1}$ such that $a_i \leq 2n$ for $i = 1, 2, \ldots, n + 1$, show that there exists at least one pair $\{a_j, a_k\}$ with $j \neq k$ such that $a_j | a_k$.

(132) Let $n > 2$. Consider the three n-tuples $(a_1^{(i)}, a_2^{(i)}, \ldots, a_n^{(i)})$, $i = 1, 2, 3$, where each $a_j^{(i)} \in \{+1, -1\}$ and assume that these three n-tuples satisfy $\sum_{j=1}^{n} a_j^{(i)} a_j^{(k)} = 0$ for each pair $\{i, k\}$ such that $1 \leq i < k \leq 3$. Show that $4|n$.

(133) Let A be the set of natural numbers which, in their decimal representation, do not have "7" amongst their digits. Prove that
$$\sum_{n \in A} \frac{1}{n} < +\infty.$$

(134) Let $u_1, u_2, \ldots$ be a strictly increasing sequence of positive integers. Denoting by $[a, b]$ the lowest common multiple of a and b, show that the series
$$\sum_{n=1}^{\infty} \frac{1}{[u_n, u_{n+1}]} \qquad \text{converges.}$$

3. Prime Numbers

(135) Using computer software, write a program
    (a) to generate all Mersenne primes up to $2^{525} - 1$;
    (b) to determine the smallest prime number larger than $10^{100} + 1$.

(136) Write a program that generates prime numbers up to a given number N. One can, of course, use Eratosthenes' sieve.

(137) Use a computer to find four consecutive integers having the same number of prime factors (allowing repetitions).

(138) (a) By reversing the digits of the prime number 1009, we obtain the number 9001, which is also prime. Write a program to find the prime numbers in $[1, 10000]$ verifying this property.
(b) By reversing the digits of the prime number 163, we obtain the number 361, which is a perfect square. Using computer software, write a program to find all prime numbers in $[1, 10000]$ with this property.

(139) Using a computer, find all prime numbers $p \leq 10\,000$ with the property that p, $p + 2$ and $p + 6$ are all primes.

(140) Let p_k be the k-th prime number. Show that $p_k < 2^k$ if $k \geq 2$.

(141) If a prime number $p_k > 5$ is equally isolated from the prime numbers appearing before and after it, that is $p_k - p_{k-1} = p_{k+1} - p_k = d$, say, show that d is a multiple of 6. Then, for each of the cases $d = 6$, 12 and 18, find, by using a computer, the smallest prime number p_k with this property.

(142) Prove that none of the numbers

$$12321, \ 1234321, \ 123454321, \ 12345654321, \ 1234567654321,$$

$$123456787654321, \ 12345678987654321$$

is prime.

(143) For each integer $k \geq 1$, let n_k be the k-th composite number, so that for instance $n_1 = 4$ and $n_{10} = 18$. Use computer software and an appropriate algorithm in order to establish the value of n_k, with $k = 10^\alpha$, for each integer $\alpha \in [2, 10]$.

(144) For each integer $k \geq 1$, let n_k be the k-th number of the form p^α, where p is prime, α a positive integer, so that for instance $n_1 = 2$ and $n_{10} = 16$. Use computer software and an appropriate algorithm in order to establish the value of n_k, with $k = 10^\alpha$, for each integer $\alpha \in [2, 10]$.

(145) Find all positive integers $n < 100$ such that $2^n + n^2$ is prime. To which class of congruence modulo 6 do these numbers n belong?

(146) Show that if the integer $n \geq 4$ is not an odd multiple of 9, then the corresponding number $a_n := 4^n + 2^n + 1$ is necessarily composite. Then, use a computer in order to find all positive integers $n < 1000$ for which a_n is prime.

(147) Consider the sequence (a_n) defined by $a_1 = a_2 = 1$ and, for $n \geq 3$, by $a_n = n! - (n-1)! + \cdots + (-1)^n 2! + (-1)^{n+1} 1!$. Use a computer in order to find the smallest number n such that a_n is a composite number.

(148) The mathematicians Mináč and Willans have obtained a formula for the n-th prime number p_n which is more of a theoretical interest than of a

practical interest:

$$p_n = 1 + \sum_{m=1}^{2^n} \left[\left[\frac{n}{1 + \sum_{j=2}^{m} \left[\frac{(j-1)!+1}{j} - \left[\frac{(j-1)!}{j} \right] \right]} \right]^{1/n} \right],$$

where as usual $[x]$ stands for the largest integer $\leq x$. Prove this formula.

(149) Develop an idea used by Paul Erdős (1913–1996) to show that, for each integer $n \geq 1$,

$$\prod_{p \leq n} p \leq 4^n.$$

His idea was to write

$$\prod_{p \leq n} p = \prod_{p \leq \frac{n+1}{2}} p \cdot \prod_{\frac{n+1}{2} < p \leq n} p$$

and to use the fact that each prime number $p > (n+1)/2$ appears in the factorization of the binomial coefficient $\binom{n}{(n+1)/2}$. Provide the details.

(150) Show that if four positive integers a, b, c, d are such that $ab = cd$, then the number $a^2 + b^2 + c^2 + d^2$ is necessarily composite.

(151) Show that, for each integer $n \geq 1$, the number $4n^3 + 6n^2 + 4n + 1$ is composite.

(152) Show that if p and q are two consecutive odd prime numbers, then $p + q$ is the product of at least three prime numbers (not necessarily distinct).

(153) Does there exist a positive integer n such that $n/2$ is a perfect square, $n/3$ a cube and $n/5$ a fifth power?

(154) Given any integer $n \geq 2$, show that $n^{42} - 27$ is never a prime number.

(155) Let $\theta(x) := \sum_{p \leq x} \log p$. Prove that Bertrand's Postulate follows from the fact that

$$c_1 x < \theta(x) < c_2 x,$$

where $c_1 = 0.73$ and $c_2 = 1.12$.

(156) Use Bertrand's Postulate to show that, for each integer $n \geq 4$,

$$p_{n+1}^2 < p_1 p_2 \cdots p_n,$$

where p_n stands for the n-th prime number.

(157) Certain integers $n \geq 3$ can be written in the form $n = p + m^2$, with p prime and $m \in \mathbb{N}$. This is the case for example for the numbers 3, 4, 6, 7, 8, 9, 11, 12, 14, 15, 16, 17, 18, 19, 20, 21. Let q^r be a prime power, where r is a positive even integer such that $2q^{r/2} - 1$ is composite. Show that q^r cannot be written as $q^r = p + m^2$, with p prime and $m \in \mathbb{N}$.

(158) Show that if p and $8p - 1$ are primes, then $8p + 1$ is composite.

(159) Show that all positive integers of the form $3k + 2$ have a prime factor of the same form, that all positive integers of the form $4k + 3$ have a prime factor of the same form, and finally that all positive integers of the form $6k + 5$ have a prime factor of the same form.

(160) A positive integer n has a *Cantor expansion* if it can be written as

$$n = a_m m! + a_{m-1}(m - 1)! + \cdots + a_2 2! + a_1 1!,$$

where the a_j's are integers satisfying $0 \leq a_j \leq j$.

(a) Find the Cantor expansion of 23 and of 57.

(b) Show that all positive integers n have a Cantor expansion and more-over that this expansion is unique.

(161) If $p > 1$ and $d > 0$ are integers, show that p and $p + d$ are both primes if and only if

$$(p-1)!\left(\frac{1}{p} + \frac{(-1)^d d!}{p+d}\right) + \frac{1}{p} + \frac{1}{p+d}$$

is an integer.

(162) Find all prime numbers p such that $p + 2$ and $p^2 + 2p - 8$ are primes.

(163) Is it true that if p and $p^2 + 8$ are primes, then $p^3 + 4$ is prime? Explain.

(164) Let $n \geq 2$. Show that the integers n and $n + 2$ form a pair of twin primes if and only if

$$4\left((n-1)! + 1\right) + n \equiv 0 \pmod{n(n+2)}.$$

(165) Identify each prime number p such that $2^p + p^2$ is also prime.

(166) For which prime number(s) p is $17p + 1$ a perfect square?

(167) Given two integers a and b such that $(a, b) = p$, where p is prime, find all possible values of:

(a) (a^2, b); (b) (a^2, b^2); (c) (a^3, b); (d) (a^3, b^2).

(168) Given two integers a and b such that $(a, p^2) = p$ and $(b, p^4) = p^2$, where p is prime, find all possible values of:

(a) (ab, p^5); (b) $(a + b, p^4)$; (c) $(a - b, p^5)$; (d) $(pa - b, p^5)$.

(169) Given two integers a and b such that $(a, p^2) = p$ and $(b, p^3) = p^2$, where p is a prime number, evaluate the expressions $(a^2 b^2, p^4)$ and $(a^2 + b^2, p^4)$.

(170) Let p be a prime number and a, b, c be positive integers. For each of the following statements, say if is true or false. If it is true, give a proof; if it is false, provide a counter-example.

(a) If $p|a$ and $p|(a^2 + b^2)$, then $p|b$.

(b) If $p|a^n$, $n \geq 1$, then $p|a$.

(c) If $p|(a^2 + b^2)$ and $p|(b^2 + c^2)$, then $p|(a^2 - c^2)$.

(d) If $p|(a^2 + b^2)$ and $p|(b^2 + c^2)$, then $p|(a^2 + c^2)$.

(171) Let a, b and c be positive integers. Show that $abc = (a, b, c)[ab, bc, ac] = (ab, bc, ac)[a, b, c]$.

(172) Let a, b and c be positive integers and assume that $abc = (a, b, c)[a, b, c]$. Show that this necessarily implies that $(a, b) = (b, c) = (a, c) = 1$.

(173) Let a, b and c be positive integers. Show that $(a, b, c) = \dfrac{(a, b)(b, c)(a, c)}{(ab, bc, ac)}$

and that $[a, b, c] = \dfrac{abc\,(a, b, c)}{(a, b)(b, c)(a, c)}$.

(174) Let a, b and c be positive integers. Show that

$$\frac{[a, b, c]^2}{[a, b][b, c][c, a]} = \frac{(a, b, c)^2}{(a, b)(b, c)(c, a)}.$$

(175) Find three positive integers a, b, c such that

$$[a, b, c] \cdot (a, b, c) = \sqrt{abc}.$$

(176) Let $\#n = [1, 2, 3, \ldots, n]$ be the lowest common multiple of the numbers $1, 2, \ldots, n$. Show that
$$\prod_{p \leq n} p \leq \#n = \prod_{p \leq n} p^{[\log n / \log p]}.$$

(177) Let p be a prime number and r a positive integer. What are the possible values of $(p, p + r)$ and of $[p, p + r]$?

(178) Let $p > 2$ be a prime number such that $p | 8a - b$ and $p | 8c - d$, where $a, b, c, d \in \mathbb{Z}$. Show that $p | (ad - bc)$.

(179) Show that, if $\{p, p+2\}$ is a pair of twin primes with $p > 3$, then 12 divides the sum of these two numbers.

(180) Let n be a positive integer. Show that if n is a composite integer, then $n | (n - 1)!$ except when $n = 4$.

(181) For which positive integers n is it true that
$$\sum_{j=1}^{n} j \ \Bigg| \ \prod_{j=1}^{n} j?$$

(182) Let $\pi = 3.141592\ldots$ be Archimede's constant, and for each positive real number x, let $\pi_2(x)$ be the function that counts the number of pairs of twin primes $\{p, p+2\}$ such that $p \leq x$. Show that
$$\pi_2(x) = 2 + \sum_{7 \leq n \leq x} \sin\left(\frac{\pi}{2}(n+2)\left[\frac{n!}{n+2}\right]\right) \cdot \sin\left(\frac{\pi}{2}n\left[\frac{(n-2)!}{n}\right]\right),$$
where $[y]$ stands for the largest integer $\leq y$.

(183) Given an integer $n \geq 2$, show, without using Bertrand's Postulate, that there exists a prime number p such that $n < p < n!$.

(184) In 1556, Niccòlo Tartaglia (1500–1557) claimed that the sums
$$1 + 2 + 4, \ 1 + 2 + 4 + 8, \ 1 + 2 + 4 + 8 + 16, \ \ldots$$
stood successively for a prime number and a composite number. Was he right?

(185) Show that if $a^n - 1$ is prime for certain integers $a > 1$ and $n > 1$, then $a = 2$ and n is prime.

REMARK: *The integers of the form $2^p - 1$, where p is prime, are called Mersenne numbers. We denote them by M_p in memory of Marin Mersenne (1588–1648), who had stated that M_p is prime for*
$$p = 2, 3, 5, 7, 13, 17, 19, 31, 67, 127, 257$$
and composite for all the other primes $p < 257$. This assertion of Mersenne can be found in the preface of his book Cogita Physico-mathematica, *published in Paris in 1644. Since then, we have found a few errors in the computations of Mersenne: indeed M_p is not prime for $p = 67$ and $p = 257$, while M_p is prime for $p = 61$, $p = 89$ and $p = 109$. One can find in the appendix C of the book of J.M. De Koninck and A. Mercier [8] the list of Mersenne primes M_p corresponding to the prime numbers p satisfying $2 \leq p \leq 44\,497$. Note on the other hand that it has recently been discovered that $2^{32\,582\,657} - 1$ is prime (in September 2006), which brings to 44 the total number of known Mersenne primes. It is also known that the primes*

M_p *are closely related to the* PERFECT NUMBERS, *in the sense that, as was shown by Leonhard Euler (1707–1783), n is an even perfect number if and only if* $n = 2^{p-1}(2^p - 1)$, *where* $2^p - 1$ *is a Mersenne prime.*

(186) Show that if there exists a positive integer n and an integer $a \geq 2$ such that $a^n + 1$ is prime, then a is even and $n = 2^r$ for a certain positive integer r.

REMARK: *The prime numbers of the form* $2^{2^k} + 1$, $k = 0, 1, 2, \ldots$, *are called "Fermat primes". The reason is that Pierre de Fermat claimed in 1640 (although saying he could not prove it) that all the numbers of the form* $2^{2^k} + 1$ *are prime. One hundred years later, Euler proved that*

$$2^{2^5} + 1 = 4294967297 = 641 \cdot 6700417.$$

As of today, we still do not know if, besides the cases $k = 0, 1, 2, 3, 4$, *primes of the form* $2^{2^k} + 1$ *exist. Nevertheless, it is known that* $2^{2^k} + 1$ *is composite for* $5 \leq k \leq 32$; *see H.C. Williams* [41] *and the site* www.prothsearch.net/fermat.html.

(187) Show that the equation $(2^x - 1)(2^y - 1) = 2^{2^z} + 1$ is impossible for positive integers x, y and z. (This implies in particular that a Fermat number, that is a number of the form $2^{2^k} + 1$, cannot be the product of two Mersenne numbers.)

(188) Prove by induction that, for each integer $n \geq 1$,

$$F_0 F_1 F_2 \cdots F_{n-1} = F_n - 2,$$

where $F_i = 2^{2^i} + 1$, $i = 0, 1, 2, \ldots$.

(189) Use the result of problem 188 in order to prove that if m and n are distinct positive integers, then $(F_m, F_n) = 1$.

(190) A positive integer n is said to be *pseudoprime in basis* $a \geq 2$ if it is composite and if $a^{n-1} \equiv 1 \pmod{n}$. Find the smallest number which is pseudoprime in each of the bases 2, 3, 5 and 7.

(191) Use Problem 189 to prove that there exist infinitely many primes.

(192) Consider the numbers $f_n = 2^{3^n} + 1$, $n = 1, 2, \ldots$, and show they are all composite and in particular that, for each positive integer n,
    (a) $3^{n+1} | f_n$;   (b) $p | f_n \Rightarrow p | f_{n+1}$.

(193) Show that there exist infinitely many prime numbers p such that the numbers $p - 2$ and $p + 2$ are both composite.

(194) Show that 641 divides $F_5 = 2^{2^5} + 1$ without doing the explicit division.

(195) Use an induction argument in order to prove that each Fermat number $F_n = 2^{2^n} + 1$, where $n \geq 2$, ends with the digit 7.

(196) Let n be a positive integer and consider the set $E = \{1, 2, \ldots, n\}$. Let 2^k be the largest power of 2 which belongs to E. Show that for all $m \in E \setminus \{2^k\}$, we have $2^k \nmid m$. Using this result, show that $\sum_{j=1}^n 1/j$ is not an integer if $n > 1$.

(197) Show that, for each positive integer n, one can find a prime number $p < 50$ such that $p | (2^{5n} - 1)$.

(198) Show that the integers defined by the sequence of numbers

$$M_k = p_1 p_2 \cdots p_k + 1 \qquad (k = 1, 2, \ldots),$$

where p_j stands for the j–th prime number, are prime numbers for $1 \leq k \leq 5$ and composite numbers for $k = 6, 7$. What about M_8, M_9 and M_{10}?

(199) Use the proof of Euclid's Theorem on the infinitude of primes to show that, if we denote by p_r the r-th prime number, then $p_r \leq 2^{2^{r-1}}$ for each $r \in \mathbb{N}$.

(200) In Problem 199, we obtained an upper bound for p_r, the r-th prime number, namely $p_r \leq 2^{2^{r-1}}$. Use this inequality to obtain a lower bound for $\pi(x)$, the number of prime numbers $\leq x$. More precisely, show that, for $x \geq 3$, $\pi(x) \geq \log \log x$.

(201) Show that there exist infinitely many prime numbers of the form $4n + 3$.

(202) Show that there exist infinitely many prime numbers of the form $6n + 5$.

(203) Let $f \colon \mathbb{N} \to \mathbb{R}$ be the function defined by

$$f(x) = a_r x^r + a_{r-1} x^{r-1} + \cdots + a_1 x + a_0,$$

where $a_r \neq 0$ and where each a_i, $0 \leq i \leq r$, is an integer. Show that, by an appropriate choice of a_i, $0 \leq i \leq r$, the set $\{f(n) : n \in \mathbb{N}\}$ contains at least r prime numbers.

(204) Consider the positive integers which can be written as an alternating sequence of 0's and 1's. The number $101\,010\,101$ is such a number and observe that $101\,010\,101 = 41 \cdot 271 \cdot 9091$. Besides 101, do there exist other prime numbers of this form?

(205) Find all prime numbers of the form $2^{2^n} + 5$, where $n \in \mathbb{N}$. Would the question be more difficult if one replaces the number 5 by another number of the form $3k + 2$? Explain.

(206) The largest gaps between two consecutive prime numbers $p_r < p_{r+1} < 100$ occur successively when

$$p_{r+1} - p_r = 5 - 3 = 2,$$
$$p_{r+1} - p_r = 11 - 7 = 4,$$
$$p_{r+1} - p_r = 29 - 23 = 6,$$
$$p_{r+1} - p_r = 97 - 89 = 8.$$

Is it true that these constantly increasing gaps always occur by jumps of length 2? In other words, does the first gap of length $2k$ always occur before the first gap of length $2k + 2$?

(207) Show that $\displaystyle\sum_{\alpha=2}^{\infty} \sum_p \frac{1}{p^\alpha} < 1$, where the inner sum runs over all the prime numbers p.

(208) Let

$$f(x) = \pi(x) + \frac{1}{2}\pi(x^{1/2}) + \frac{1}{3}\pi(x^{1/3}) + \frac{1}{4}\pi(x^{1/4}) + \cdots,$$

be a series which is in fact a finite sum for each real number $x \geq 1$ since $\pi(x^{1/n}) = 0$ as soon as $n > \log x / \log 2$. Show that

$$\pi(x) = \sum_{n=1}^{\infty} \frac{\mu(n)}{n} f(x^{1/n}).$$

REMARK: It is possible to show that $f(x)$ is a better approximation of $\pi(x)$ than $Li(x) := \displaystyle\int_2^x \frac{dt}{\log t}$ (see H. Riesel [**31**]).

(209) Let $n \geq 2$ be an integer. Show that the interval $[n, 2n]$ contains at least one perfect square.

(210) If n is a positive integer such that $3n^2 - 3n + 1$ is composite, show that n^3 cannot be written as $n^3 = p + m^3$, with p prime and m a positive integer.

(211) It is conjectured that there exist infinitely many prime numbers p of the form $p = n^2 + 1$. Identify the primes $p < 10\,000$ of this particular form. Why is the last digit of such a prime number p always 1 or 7? Is there any reasonable explanation for the fact that the digit 7 appears essentially twice as often?

(212) Show that, for each integer $n \geq 2$,

$$(n!)^{1/n} \leq \prod_{p \leq n} p^{\frac{1}{p-1}}.$$

(213) For each integer $N \geq 1$, let $S_N = \{n^2 + 2 : 6 \leq n \leq 6N\}$. Show that no more than $\frac{1}{6}$ of the elements of S_N are primes.

(214) Let p be a prime number and consider the integer $N = 2 \cdot 3 \cdot 5 \cdots p$. Show that the $(p-1)$ consecutive integers

$$N + 2, N + 3, N + 4, \ldots, N + p$$

are composite.

(215) Let $n > 1$ be an integer with at least 3 digits. Show that
   (a) $2 | n$ if and only if the last digit of n is divisible by 2;
   (b) $2^2 | n$ if and only if the number formed with the last two digits of n is divisible by 4;
   (c) $2^3 | n$ if and only if the number formed with the last three digits of n is divisible by 8.
Can one generalize?

(216) For each integer $n \geq 2$, let

$$P(n) = \prod_{\substack{p | n \\ p > \log n}} \left(1 - \frac{1}{p}\right).$$

Show that $\displaystyle\lim_{n \to \infty} P(n) = 1$.

(217) Prove that there exists an interval of the form $[n^2, (n+1)^2]$ containing at least 1000 prime numbers.

(218) Use the Prime Number Theorem (see Theorem 17) in order to prove that the set of numbers of the form p/q (where p and q are primes) is dense in the set of positive real numbers.

(219) Show that the sum of the reciprocals of a finite number of distinct prime numbers cannot be an integer.

(220) Use the fact that there exists a positive constant c such that if $x \geq 100$,

$$(1) \qquad \sum_{p \leq x} \frac{1}{p} = \log \log x + c + R(x) \quad \text{with } |R(x)| < \frac{1}{\log x}$$

and moreover that, for $x \geq 2$,

(2) $$\pi(x) := \sum_{p \leq x} 1 < \frac{3}{2} \frac{x}{\log x}$$

in order to prove that if $P(n)$ stands for the largest prime factor of n, then

(3) $$\frac{1}{x} \#\{n \leq x : P(n) > \sqrt{x}\} = \log 2 + T(x) \quad \text{with } |T(x)| < \frac{9}{2} \frac{1}{\log x}.$$

Use this result to show that more than $\frac{2}{3}$ of the integers have their largest prime factor larger than their square root, or in other words that the density of the set of integers n such that $P(n) > \sqrt{n}$ is larger than $\frac{2}{3}$.

(221) Prove the following formula (due to Adrien-Marie Legendre (1752–1833)):

$$\pi(x) = \pi(\sqrt{x}) + \sum_{n | p_1 \cdots p_r} \mu(n) \left[\frac{x}{n}\right] - 1,$$

where $r = \pi(\sqrt{x})$.

(222) Consider the following two conjectures:

A. *(Goldbach Conjecture) Each even integer ≥ 4 can be written as the sum of two primes.*

B. *Each integer > 5 can be written as the sum of three prime numbers.*

Show that these two conjectures are equivalent.

(223) Show that $\pi(m)$, the number of prime numbers not exceeding the positive integer m, satisfies the relation

$$\pi(m) = \sum_{j=2}^{m} \left[\frac{(j-1)! + 1}{j} - \left[\frac{(j-1)!}{j} \right] \right],$$

where $[y]$ stands for the largest integer $\leq y$.

(224) Given a sequence of natural numbers $\mathcal{A}$, let $A(n) = \#\{m \leq n : m \in \mathcal{A}\}$, and let us denote respectively by

$$\underline{\mathbf{d}}\,\mathcal{A} = \liminf_{n \to \infty} \frac{A(n)}{n} \quad \text{and} \quad \overline{\mathbf{d}}\,\mathcal{A} = \limsup_{n \to \infty} \frac{A(n)}{n}$$

the *asymptotic lower density* and *asymptotic upper density* of the sequence $\mathcal{A}$. On the other hand, if both these densities are equal, we say that the sequence $\mathcal{A}$ has density $\mathbf{d}\,\mathcal{A} = \underline{\mathbf{d}}\mathcal{A} = \overline{\mathbf{d}}\mathcal{A}$. Prove that:

(a) the density of the sequence made up of all the multiples of a natural number a is equal to $1/a$;

(b) the density of the sequence made up of all the multiples of a natural number a which are not divisible by the natural number a_0 is equal to $\frac{1}{a} - \frac{1}{[a, a_0]}$;

(c) the density of the sequence made up of all natural numbers which are not divisible by any of the prime numbers $q_1, q_2, \ldots, q_r$ is equal to $\prod_{i=1}^{r} \left(1 - \frac{1}{q_i}\right)$.

(225) Let $\mathcal{A}$ be the set of natural numbers n such that $2^{2k} \leq n < 2^{2k+1}$ for a certain integer $k \geq 0$, so that

$$\mathcal{A} = \{1, 4, 5, 6, 7, 16, 17, \ldots, 31, 64, 65, \ldots, 127, 256, 257, \ldots\}.$$

Show that

$$\underline{\mathbf{d}}\,\mathcal{A} \neq \overline{\mathbf{d}}\,\mathcal{A}.$$

(226) We say that a sequence of natural numbers $\mathcal{A}$ is *primitive* if no element of $\mathcal{A}$ divides another one. Examples of such sequences are: the sequence of prime numbers, the sequence of natural numbers having exactly k prime factors (k fixed), and finally the sequence of integers n belonging to the interval $]k, 2k]$ (k fixed). Show that if $\mathcal{A}$ is a primitive sequence, then $\overline{\mathbf{d}}\,\mathcal{A} \leq \frac{1}{2}$.

(227) Let $\mathcal{A}$ be a primitive sequence (see Problem 226). Show that

$$\sum_{a \in \mathcal{A}} \frac{1}{a \log a} < +\infty.$$

(228) Let $E = \{a + b\sqrt{-5} \mid a, b \in \mathbb{Z}\}$.
  (a) Show that the sum and the product of elements of E are in E.
  (b) Define the norm of an element $z \in E$ by $\|z\| = \|a + b\sqrt{-5}\| = a^2 + 5b^2$. We say that an element $p \in E$ is *prime* if it is impossible to write $p = n_1 n_2$, with $n_1, n_2 \in E$, $\|n_1\| > 1$, $\|n_2\| > 1$; we say that it is *composite* if it is not prime. Show that, in E, 3 is a prime number and 29 is a composite number.
  (c) Show that the factorization of 9 in E is not unique.

(229) Let A be a set of natural numbers and let $A(x) = \#\{n \leq x : n \in A\}$. Show that, for all $x \geq 1$,

$$\sum_{\substack{n \leq x \\ n \in A}} \frac{1}{n} = \sum_{n \leq x} \frac{A(n)}{n(n+1)} + \frac{A(x)}{[x]+1}.$$

4. Representations of Numbers

(230) A number $n = d_1 d_2 \cdots d_r$, where $d_1, d_2, \ldots, d_r$ are the digits of n, is called a *palindrome* if it remains unchanged when its digits are reversed, that is if $n = d_r d_{r-1} \cdots d_1$. Hence the numbers 36763 and 437734 are both palindromes. Show that each palindrome having an even number of digits is divisible by 11.

(231) The smallest number $n > 2$ which is equal to the sum of the factorials of its digits in basis 15 is $1\,441$ (here, $1\,441 = [6, 6, 1]_{15} = 6! + 6! + 1!$). How can one find another such number $n > 2$ without using a computer?

(232) Let r be a positive integer and let n be a number which can be written as a sum of r distinct factorials, that is for which there exist positive integers $d_1 < d_2 < \ldots < d_r$ such that

$$n = d_1! + d_2! + \cdots + d_r!.$$

Prove that such a representation is unique.

(233) Let c be a positive odd integer. Show that the equation $x^2 - y^3 = 8c^3 - 1$ has no solutions in positive integers x and y, and use this to show that there exist infinitely many positive integers which are not of the form $x^2 - y^3$.

(234) Show that the last four digits of the decimal representation of 5^n, for $n = 4, 5, 6, \ldots$, form a periodic sequence. What is this period?

(235) Show that there exist infinitely many natural numbers which cannot be written as the sum of one, two or three cubes.

(236) Show that every integer can be written as the sum of five cubes.

(237) Given a positive integer n, let $s(n)$ be the sum of its digits, so that for example $s(12) = 3$ and $s(924) = 15$. Find all pairs of integers $m < n$ such that $s(m)^2 = n$ and $s(n)^2 = m$.

(238) The Egyptians used to express each fraction (except $\frac{2}{3}$) as a sum of unitary fractions (that is, fractions of the form $1/n$, where n is a positive integer).
   (a) Prove the result etablished by James Joseph Sylvester (1814–1897) to the effect that each fraction n/m, $n < m$, $(n, m) = 1$, can be written as a sum of unitary fractions.
   (b) Show that such a representation is not necessarily unique.
   (c) Show that if n is of the form $n = 4m + 3$, then $4/n$ is the sum of three unitary distinct fractions.

(239) A positive integer is said to be *complete* if its square uses each of the 10 digits exactly once. Use a computer to find the smallest complete number and then show that a complete number cannot be a prime number.

(240) Show that 2 is the only prime number p which can be written as $p = x^3 + y^3$ with $x, y \in \mathbb{N}$.

(241) Prove that a prime number p can be written as the difference of two positive cubes if and only if $p = 3k(k+1) + 1$ for a certain positive integer k. Find the ten smallest prime numbers of this form.

(242) Prove that an integer n can be written as the difference of two squares if and only if n is not of the form $4k + 2$.

(243) We say that an integer $n > 1$ is *automorphic* if the number n^2 ends with the same digits as n. Hence 5, 25 and 625 are automorphic numbers. Show that there exist infinitely many automorphic numbers.

(244) Show that for each number $n = d_1 d_2 \cdots d_r > 9$, where $d_1, d_2, \ldots, d_r$ stand for the digits of n in basis 10, we have

$$n - d_1 \cdot d_2 \cdots d_r \geq 10^{r-1}.$$

(245) Find the largest positive integer which is equal to the sum of the fifth powers of its digits added to the product of its digits.

(246) Show how one can construct the only sequence (a_k) of positive integers having the following properties:
   (i) a_k is made of k digits;
   (ii) 2^k divides a_k;
   (iii) a_k contains only the digits 1 and 2.
Generate the first 14 terms of this sequence.

(247) For each positive integer n, let $s(n)$ be the sum of its digits. Given an integer $k \geq 2$ which is not a multiple of 3, let $\rho(k)$ be the smallest prime number p such that $s(p) = k$, if such a prime number p exists. In the particular case $k = 2$, it seems that there are only three prime numbers p such that $s(p) = 2$, namely 2, 11 and 101, and we have in particular that $\rho(2) = 2$. We also have that $\rho(4) = 13$, $\rho(5) = 5$, $\rho(7) = 7$, $\rho(8) = 17$, $\rho(10) = 19$, $\rho(11) = 29$, and so on; the function $\rho(k)$ increases quite fast; for instance, $\rho(80) = 998\,999\,999$. For each integer $k > 2$ which is not a multiple of 3, the candidates p such that $s(p) = k$ appear to be numerous, and in fact there seems to be infinitely many of them. It therefore appears that the function $\rho(k)$ is well defined. However, it is not at all obvious that given a particular integer $k > 2$, one can always find at least one prime number p such that $s(p) = k$. Nevertheless, prove that if $\rho(k)$ exists, then

$$\rho(k) \geq (a+1)10^b - 1, \quad \text{where } b = [k/9] \text{ and } a = k - 9b.$$

(248) Given a positive integer m, set

$$P(m) = \prod_{n \neq m} \frac{n^3 - m^3}{n^3 + m^3},$$

where the infinite product runs over all positive integers $n \neq m$. Show that

$$P(m) = (-1)^{m+1} \frac{2}{3}(m!)^2 \prod_{n=1}^{m} \frac{n+m}{n^3 + m^3}.$$

(249) Find all positive integers n which can be written as the sum of the factorials of their digits.

(250) Find the only positive integer $n = d_1 d_2 \cdots d_{2k}$, where $d_1, d_2, \ldots, d_{2k}$ stand for the digits (an even number of them) of n, such that

$$n = d_1^{d_2} \cdot d_3^{d_4} \cdots d_{2k-1}^{d_{2k}}.$$

More precisely, proceed in two steps. First, show that there exists only a finite number of positive integers with this property. Afterwards, use a computer to find this number, thereby elaborating a process which allows one to minimize the number of candidates.

(251) Show that there exist exactly six positive integers n with the property that the sum of their digits added to the sum of the cubes of their digits is equal to the number n itself, that is such that

$$n = d_1 d_2 \cdots d_r = d_1 + d_2 + \cdots + d_r + d_1^3 + d_2^3 + \cdots + d_r^3,$$

where $d_1, d_2, \ldots, d_r$ stand for the digits of n.

(252) Given a positive integer k, let $g(k)$ be the smallest number r which has the property that each positive integer can be written as $x_1^k + x_2^k + \cdots + x_r^k$, where the x_i's are nonnegative integers. In 1770, Joseph-Louis Lagrange (1736–1813) showed that $g(2) = 4$. The problem of calculating the value of $g(k)$ is known as the *Waring problem*. The mere fact that the function $g(k)$ is well defined is not at all obvious; as a matter of fact, it is only in 1909 that David Hilbert (1862–1943) finally showed that $g(k)$ exists for each positive integer k. It is conjectured that

$$g(k) = 2^k - 2 + \left[\left(\frac{3}{2} \right)^k \right] \qquad (k \geq 1).$$

Around 1772, Johannes Albert Euler (the son of the famous Leonhard Euler) proved that this last quantity is actually a lower bound for $g(k)$. Reconstruct this proof by considering the integer $n = q2^k - 1$, where the number q is defined implicitly by

$$3^k = q2^k + r \qquad (1 \leq r < 2^k).$$

(253) Find the only positive integer whose square and cube, taken together, use all the digits from 0 to 9 exactly once.

(254) Find the only positive integer whose square and cube, taken together, use all the digits from 0 to 9 exactly twice.

(255) Show that there are only a finite number of positive integers whose square and cube, taken together, uses all the digits from 0 to 9 exactly three times, and find these numbers.

(256) A positive integer n having $2r$ digits, $r \geq 1$, is called a *vampire number* if it can be written as the product of two positive integers, each of r digits, the union of their digits giving all the digits appearing in n. Hence $1260 = 21 \times 60$ is the smallest vampire number. Use a computer to find the seven vampire numbers made up of four digits.

(257) Given a positive integer $n = d_1 d_2 \cdots d_r$, where $d_1, d_2, \ldots, d_r$ stand for the digits of n, we let $g_3(n) = d_1^3 + d_2^3 + \cdots + d_r^3$. Find all positive integers n such that $g_3(g_3(n)) = n$.

(258) Given a positive integer $n = d_1 d_2 \cdots d_r$, where $d_1, d_2, \ldots, d_r$ stand for the digits of n, let $f(n) = d_1! + d_2! + \cdots + d_r!$. For each positive integer k, let f_k stand for the k-th iteration of the function f, that is $f_1(n) = f(n)$, $f_2(n) = f(f(n))$, and so on. Using a computer, show that, for every positive integer n, the iteration

$$(*) \qquad f_1(n), f_2(n), f_3(n), \ldots, f_k(n), \ldots$$

always ends up in an infinite loop. If $n = 1$, this loop is $1, 1, 1, \ldots$; establish that if $n > 1$, then the iteration $(*)$ eventually enters one of the following six loops:

$$2, 2, 2, \ldots$$
$$145, 145, 145, \ldots$$
$$169, 363601, 1454, 169, \ldots$$
$$871, 45361, 871, \ldots$$
$$872, 45362, 872, \ldots$$
$$40585, 40585, 40585, \ldots$$

(259) Let n, k be arbitrary integers larger than 1. Show that there exists a polynomial $p(x)$ of degree k with integer coefficients and a positive integer m such that $n = p(m)$.

(260) If the number $111\ldots1$, made of k times the digit 1, is prime, show that k is prime.

(261) Prove that it is impossible to find three prime numbers $q_1 < q_2 < q_3$ such that

(1) $$q_1 q_2 q_3 = q_1^3 + q_2^3 + q_3^3.$$

What if, instead of (1), we have

(2) $$q_1 q_2 q_3 = q_1^2 + q_2^2 + q_3^2 ?$$

(262) Find all positive integers n such that $\frac{1}{n} = 0.\overline{n}$.

(263) Use a computer to find the eight positive composite integers $n < 10^6$ such that

(*) $$n = q_1^{\alpha_1} q_2^{\alpha_2} \cdots q_r^{\alpha_r} = q_1^a + q_2 + \cdots + q_r,$$

for a certain positive integer a, where $q_1 < q_2 < \ldots < q_r$ are the prime factors of n.

(264) Show that $\sigma(n)$ is a power of 2 if and only if n is a product of Mersenne primes.

(265) What are the positive integers which can be represented in the form

$$\binom{k}{2} + kn, \qquad k > 1, \ n \geq 1?$$

5. Congruences

(266) For which positive integers n is the number $3^n + 1$ a multiple of 10?

(267) Find the smallest positive residue modulo 7 of $1! + 2! + \cdots + 50!$.

(268) What is the remainder of the division of $\sum_{i=1}^{111} i!$ by 12?

(269) Show that for each positive integer n, $10 \cdot 32^n + 1$ is a composite number.

(270) Is it true that 36 divides $n^6 + n^2 + 4$ for infinitely many positive integers n? Explain.

(271) In a letter sent to Christian Huygens (1629–1695) in 1659, Fermat wrote that using his method of infinite descent, he was successful in showing that no integer of the form $3k - 1$ can be written as $x^2 + 3y^2$ (with x and y integers). Is it possible to prove this result in a very simple manner? Explain.

(272) Let m and n be positive integers such that $p^m \| n$ for a certain prime number p. Show that

$$\frac{n!}{p^m} \equiv (-1)^m \prod_{k=0}^{\left[\frac{\log n}{\log p}\right]} \left(\left[\frac{n}{p^k}\right] - p\left[\frac{n}{p^{k+1}}\right]\right)! \pmod{p}.$$

(273) Let n be a positive integer. Show that the last digit of n^{13} is the same as the last digit of n.

(274) Find the smallest positive integer n such that $\sqrt[7]{n/7}$ and $\sqrt[11]{n/11}$ are both integers.

(275) Show that there exists an arbitrarily long sequence of consecutive integers, each divisible by a perfect square.

(276) Let a and b be integers and let m and n be positive integers. Show that the system of congruences

$$x \equiv a \pmod{m},$$
$$x \equiv b \pmod{n}$$

has solutions if and only if $(m, n) | (a - b)$.

(277) Let p be a prime number. Show that if k is an integer, $1 \le k < p$, then $\binom{p}{k} \equiv 0 \pmod{p}$.

(278) (a) Let $x_1, x_2, \ldots, x_n$ be integers. Show that

$$(x_1 + x_2 + \cdots + x_n)^p \equiv x_1^p + x_2^p + \cdots + x_n^p \pmod{p}.$$

(b) Show that if a and b are integers such that $a^p \equiv b^p \pmod{p}$, then $a^p \equiv b^p \pmod{p^2}$.

(279) Let p be an odd prime number and let k be an integer such that $1 \le k < p$. Show that

$$\binom{p-1}{k} \equiv (-1)^k \pmod{p}.$$

(280) Let p be a prime number and let r be an integer such that $1 \le r < p$. If $(-1)^r r! \equiv 1 \pmod{p}$, show that

$$(p - r - 1)! \equiv -1 \pmod{p}.$$

Use this result to show that $259! \equiv -1 \pmod{269}$ and $463! \equiv -1 \pmod{479}$.

(281) Let $\alpha \geq 3$ and $\beta \geq 6$ be two integers. Show that the equation $2^\beta - 1 = 3p^\alpha$ has no solutions for p prime.

(282) Let p be a prime number and let $n = 2p + 1$. Show that if n is not a multiple of 3 and if $2^{n-1} \equiv 1 \pmod{n}$, then n is prime.

(283) Let p be a prime number and k a positive integer. Show that

$$(*) \qquad a \equiv b \pmod{p^k} \implies a^p \equiv b^p \pmod{p^{k+1}}.$$

Then, prove that if $p > 2$, $p \nmid a$ and $p^k \| a - b$, then $p^{k+1} \| a^p - b^p$.

(284) If p is a prime number, can the equation $p^\delta + 1 = 2^\nu$ have solutions with integers $\delta \geq 2$ and $\nu \geq 2s$?

(285) Show that the equation $1 + n + n^2 = m^2$, where m and n are positive integers, is impossible.

(286) Show that the only solution of the equation $1 + p + p^2 + p^3 + p^4 = q^2$, where p and q are primes, is $\{p, q\} = \{3, 11\}$.

(287) Let x_1, x_2, x_3, x_4 and x_5 be integers such that

$$x_1^3 + x_2^3 + x_3^3 + x_4^3 + x_5^3 = 0.$$

Show that necessarily one of the x_i's is a multiple of 7.

(288) Show that $2^p + 3^p$ is not a power (> 1) of an integer if p is prime.

(289) Show that for each positive integer n,

$$1^n + 2^n + 3^n + 4^n + 5^n + 6^n$$

is divisible by 7 if and only if n is not divisible by 6.

(290) Is it true that if n is a positive odd integer whose last digit in decimal representation is different from 5, then the last two digits of the decimal representation of n^{400} are 0 and 1? Explain.

(291) What are the possible values of the last digit of 4^m for each $m \in \mathbb{N}$?

(292) Show that the difference of two consecutive cubes is never divisible by 3, nor by 5.

(293) Is it true that $27 | (2^{5n+1} + 5^{n+2})$ for each integer $n \geq 0$? Explain.

(294) Show that for each positive integer k, the number $(13^2)^{2k+1} + (98^2)^{2k+1}$ is divisible by 337.

(295) Find the last two digits of the decimal representation of $19^{19^{19}}$.

(296) If a and b are positive integers such that $(ab, 70) = 1$, show that $a^{12} - b^{12} \equiv 0 \pmod{280}$.

(297) Show that for each integer $n \geq 2$, $n^{13} - n$ is divisible by 2730.

(298) Find the smallest positive integer which divided by 12, by 17, by 45 or by 70 gives in each case a remainder of 4.

(299) If n is an arbitrary positive integer, is the number

$$3n^{13} + 4n^{11} + n^7 + 3n^5 + 3n$$

divisible by 7?

(300) Let p be a prime number; show that $\binom{2p}{p} \equiv 2 \pmod{p}$.

(301) Show that a 3-digit positive integer whose decimal representation is of the form "abc" (for three digits a, b and c) is divisible by 7 if and only if $2a + 3b + c$ is divisible by 7.

(302) Show that a 6-digit positive integer whose decimal representation is of the form "*abcabc*" (for three digits a, b and c) is necessarily divisible by 13.

(303) Show that $561|2^{561} - 2$ and that $561|3^{561} - 3$.

(304) Given a positive integer n, show that

$$\frac{12}{35}n^{13} + \frac{23}{35}n$$

is an integer.

(305) Does there exist a rational number r such that for each positive integer n relatively prime with 481,

$$\frac{50}{481}n^{36} + r$$

is a positive integer?

(306) Let p be an odd prime number, $p \neq 5$. Show that p divides infinitely many integers amongst $1, 11, 111, 1\,111, \ldots$.

(307) According to Fermat's Little Theorem, if n is an odd prime number and if a is a positive integer such that $(a, n) = 1$, then $a^{n-1} \equiv 1 \pmod{n}$. Show that the reverse of this result is false.

(308) Let $p > 3$ be a prime number. Show that $ab^p - ba^p \equiv 0 \pmod{6p}$ for any integers a and b.

(309) If n is a positive integer, is it true that

$$1 + 2 + 3 + \cdots + (n - 1) \equiv 0 \pmod{n}?$$

Explain.

(310) For which positive integers n do we have

$$1^2 + 2^2 + 3^2 + \cdots + (n - 1)^2 \equiv 0 \pmod{n}?$$

(311) Is it true that if n is a positive integer divisible by 4, then

$$1^3 + 2^3 + 3^3 + \cdots + (n - 1)^3 \equiv 0 \pmod{n}?$$

(312) Prove that for each positive integer n, we have

$$5^n \equiv 1 + 4n \pmod{16} \quad \text{and} \quad 5^n \equiv 1 + 4n + 8n(n - 1) \pmod{64}.$$

(313) Show that for each positive integer $k \geq 3$,

$$5^{2^{k-3}} \not\equiv 1 \pmod{2^k} \quad \text{while} \quad 5^{2^{k-2}} \equiv 1 \pmod{2^k}.$$

More generally, show that for $k > 2$ and a given odd integer a, we have

$$a^{2^{k-2}} \equiv 1 \pmod{2^k}.$$

(314) Show that

$$\frac{n^5}{5} + \frac{n^3}{3} + \frac{7n}{15}$$

is an integer for all $n \in \mathbb{N}$. More generally, show that if p and q are prime numbers, then

$$\frac{n^p}{p} + \frac{n^q}{q} + \frac{(pq - p - q)n}{pq}$$

is an integer for all $n \in \mathbb{N}$.

(315) Find the solution of the congruence $x^{24} + 7x \equiv 2 \pmod{13}$.

(316) Because of Wilson's Theorem, the numbers $2, 3, 4 \ldots, 15$ can be arranged in seven pairs $\{x, y\}$ such that $xy \equiv 1 \pmod{17}$. Find these seven pairs.

(317) Let $m = m_1 m_2 \cdots m_r$, where the m_i's are integers > 1 and pairwise coprime. Show that

$$m_1^{\phi(m)/\phi(m_1)} + m_2^{\phi(m)/\phi(m_2)} + \cdots + m_r^{\phi(m)/\phi(m_r)} \equiv r - 1 \pmod{m}.$$

(318) Let p be a prime number and k an integer, $0 < k < p$. Show that

$$(k-1)!(p-k)! \equiv (-1)^k \pmod{p}.$$

(319) If p and q are distinct prime numbers, is it true that we always have

$$p^{q-1} + q^{p-1} \equiv 1 \pmod{pq}?$$

More generally, if m and n are positive integers such that $(m,n) = 1$, is it true that

$$n^{\phi(m)} + m^{\phi(n)} \equiv 1 \pmod{mn}?$$

(320) Show that for each positive integer n,

$$3^{2n+2} \equiv 8n + 9 \pmod{64}.$$

(321) Let $p \geq 5$ be a prime number. Find the value of $(p!, (p-2)! - 1)$.
(322) Show that

$$5^{6614} - 12^{857} \equiv 1 \pmod{7}.$$

(323) *Divisibility tests.* Let N be a positive integer whose decimal representation is $N = a_n 10^n + \cdots + a_2 10^2 + a_1 10 + a_0$, where $0 < a_n \leq 9$ and for $k = 0, \ldots, n-1$, $0 \leq a_k \leq 9$. Show that
(a) N is divisible by 3 $\iff$ $a_n + a_{n-1} + \cdots + a_1 + a_0 \equiv 0 \pmod{3}$.
(b) N is divisible by 4 $\iff$ $10a_1 + a_0 \equiv 0 \pmod{4}$.
(c) N is divisible by 6 $\iff$ $4(a_n + \cdots + a_1 + a_0) \equiv 3a_0 \pmod{6}$.
(d) N is divisible by 7 $\iff$ $(100a_2 + 10a_1 + a_0) - (100a_5 + 10a_4 + a_3) + (100a_8 + 10a_7 + a_6) - \cdots \equiv 0 \pmod{7}$.
(e) N is divisible by 8 $\iff$ $100a_2 + 10a_1 + a_0 \equiv 0 \pmod{8}$.
(f) N is divisible by 9 $\iff$ $a_n + a_{n-1} + \cdots + a_0 \equiv 0 \pmod{9}$.
(g) N is divisible by 11 $\iff$ $a_n - a_{n-1} + \cdots + (-1)^n a_0 \equiv 0 \pmod{11}$.
(324) Assume that 168 divides the integer whose decimal representation is "$770ab45c$". Find the digits a, b and c.
(325) Let a be an integer ≥ 2 and let $m \in \mathbb{N}$. If $(a,m) = (a-1,m) = 1$, show that

$$1 + a + a^2 + \cdots + a^{\phi(m)-1} \equiv 0 \pmod{m}.$$

(326) Let p be a prime number. Show that for each $a \in \mathbb{N}$, we have

$$a^{(p-1)!+1} \equiv a \pmod{p}.$$

(327) Show that if p is a prime number, then $1^{p-1} + 2^{p-1} + \cdots + (p-1)^{p-1} \equiv -1 \pmod{p}$.
(328) Show that if p is an odd prime number, then $1^p + 2^p + \cdots + (p-1)^p \equiv 0 \pmod{p}$.
(329) Let p be an odd prime number. Show that

$$\sum_{k=1}^{p-1} (k-1)!(p-k)!k^{p-1} \equiv 0 \pmod{p}.$$

(330) Letting p be a prime number of the form $4n+1$, show that $((2n)!)^2 \equiv -1$ (mod p). More generally, if p is a prime number and if $m + n = p - 1$, $m \geq 0$, $n \geq 0$, show that

$$m!\, n! \equiv (-1)^{m+1} \pmod{p}.$$

(A similar result was obtained in Problem 318.) Use this last formula to prove that

$$\left\{ \left(\frac{p-1}{2} \right)! \right\}^2 \equiv (-1)^{(p+1)/2} \pmod{p}.$$

(331) Show that an integer $n > 2$ is prime if and only if n divides the number $2(n-3)! + 1$.

(332) Show that if p is a prime number and a an arbitrary integer, then p divides the expression $a^p + a(p-1)!$.

(333) Show that if $\pi = 3.141592\ldots$ stands for Archimede's constant and $\pi(x)$ stands for the number of prime numbers $p \leq x$, then

$$\pi(x) = \sum_{2 \leq n \leq x} \left[\cos^2 \left(\pi \frac{(n-1)! + 1}{n} \right) \right],$$

where $[y]$ stands for the largest integer smaller or equal to y.

(334) Let $m_1, m_2 \in \mathbb{N}$ be such that $(m_1, m_2) = 1$. If a, r and s are positive integers such that $a^r \equiv 1$ (mod m_1) and $a^s \equiv 1$ (mod m_2). Show that

$$a^{[r,s]} \equiv 1 \pmod{m_1 m_2}.$$

(335) Let m be a positive integer. Show that for each $a \in \mathbb{N}$,

$$a^m \equiv a^{m-\phi(m)} \pmod{m}.$$

(336) Let m be a positive odd integer. Show that the sum of the elements of a complete residue system modulo m is congruent to 0 (mod m).

(337) Let $a, b \in \mathbb{Z}$, $m \in \mathbb{N}$. If E is a complete residue system modulo m and if $(a, m) = 1$, show that

$$E' = \{ax + b \mid x \in E\}$$

is also a complete residue system modulo m.

(338) Is it possible to construct a reduced residue system modulo 7 made up entirely of multiples of 6? Explain.

(339) Let $m > 2$ be an integer. Show that the sum of the elements of a reduced residue system modulo m is congruent to 0 (mod m).

(340) If $\{r_1, r_2, \ldots, r_{p-1}\}$ is a reduced residue system modulo a prime number p, show that

$$\prod_{j=1}^{p-1} r_j \equiv -1 \pmod{p}.$$

(341) Let $a, b \in \mathbb{Z}$, $m \in \mathbb{N}$. Using a counter-example, show that if E is a reduced residue system modulo m and if $(a, m) = 1$, then the set $\{ax + b \mid x \in E\}$ is not necessarily a reduced residue system modulo m.

(342) Find all integers x, y and z with $2 \leq x \leq y \leq z$ such that

$$xy \equiv 1 \pmod{z}, \quad xz \equiv 1 \pmod{y}, \quad yz \equiv 1 \pmod{x}.$$

(343) Let n and k be positive integers. Show that there exists a sequence of n consecutive composite integers such that each is divisible by at least k distinct prime numbers. Using this result, find the smallest sequence of four consecutive integers divisible by 3, 5, 7 and 11 respectively.

(344) Find all positive integers which give the remainder 1, 2 and 3 when divided respectively by 3, 4 and 5.

(345) Find the smallest integer $a > 2$ such that
$$2|a, \quad 3|a+1, \quad 4|a+2, \quad 5|a+3, \quad 6|a+4.$$

(346) Find the cycle and the period of $1/3, 1/3^2, 1/3^3, 1/3^4, 1/7, 1/7^2, 1/7^3$. Let p be an arbitrary prime number for which the period of $1/p$ is m. Using these computations, what should one conjecture regarding the periods of $1/p^2, 1/p^3, \ldots, 1/p^n$?

(347) The decimal expansion of $2/3 = 0.666\ldots$ consists in a repetition of $6 = 2 \cdot 3$. The same phenomenon occurs with the decimal expansion of $1/3 = 0.333\ldots$. Find all positive rational numbers a/b with $(a, b) = 1$, whose decimal expansion is formed by an infinite repetition of the product of its numerator and of its denominator.

(348) Show that the period of a fraction m/n with $m < n$, $(m, n) = 1$, $(n, 10) = 1$ is the smallest positive integer h such that $10^h \equiv 1 \pmod{n}$.

(349) If m/n has the cycle $a_1 a_2 \cdots a_h$, show that $m | a_1 a_2 \cdots a_h$.

(350) If $m/n = 0.\overline{a_1 a_2 \ldots a_r}$, show that
$$\frac{m}{n} = \frac{a_1 a_2 \ldots a_r}{10^r - 1},$$
where the numerator is the number made up of the r digits $a_1, a_2, \ldots, a_r$ (and not of their product).

6. Primality Tests and Factorization Algorithms

(351) Let $d > 1$ be a proper divisor of the positive integer n. Prove that $2^{n-1} + 2^{d-1} - 1$ is a composite number.

(352) Prove that if a Mersenne number, that is a number of the form $2^q - 1$ where q is prime, is not squarefree, then it must be divisible by a Wieferich prime, that is a prime number p such that $2^{p-1} \equiv 1 \pmod{p^2}$.

(353) Find the three smallest prime factors of the number $n = 5^{96} - 7^{112}$.

(354) Let $m \geq 4$ be an even integer and let $a \geq 2$ be an integer. Show that $\dfrac{m^a}{2} + \dfrac{m}{2} - 1$ is a composite number.

(355) Show that the sequence $2^{2^n} + 3$, $n = 1, 2, \ldots$, contains infinitely many composite numbers.

(356) Use Problem 354 to prove that $2^{2^6} + 15$ is a composite number.

(357) Is it true that $2^{2^n} + 15$ is a prime number for each integer $n \geq 0$? If it is true, prove it. If it is false, provide a counter-example.

(358) By a close examination of the representation of the number n given in Problem 84, obtain that $973 | n$ and therefore that 139 is a prime factor of n.

(359) Knowing that the number $n = 999\,951$ has a prime factor p such that $300 < p < 400$ and observing that $n + 49 = 10^6$, find this number p.

(360) Show that 127 is a prime divisor of $2^{21} - 1$.

(361) Find four prime factors of $2^{2^6} - 1$.

(362) Prove that at least one third of the integers of the form $n10^n + 1$ are composite.

(363) Use Problem 75 to show that 3, 7 and 31 are prime factors of $2^{30} - 1$ and that 31 and 127 are prime factors of $2^{35} - 1$.

(364) Let $n = 2^{30} - 1$. Show that $11 | n$ without computing explicitly the value of n.

(365) Use Problem 75 to show that 2, 5, 7 and 13 are prime factors of $3^{12} - 1$ and that 2, 5, 7, 13, 41 and 73 are prime factors of $3^{24} - 1$.

(366) Given two integers a and m larger than 1, show that, if m is odd, then $a + 1$ is a divisor of $a^m + 1$. Use this result to obtain the factorization of 1001.

(367) Generalize the result of Problem 366 to obtain that if a and m are two integers larger than 1 and if $d \geq 1$ is an odd divisor of m, then $a^{m/d} + 1$ is a divisor of $a^m + 1$. Use this result to show that 101 is a factor of 1000001.

(368) Show that 7, 11 and 13 are factors of $10^{15} + 1$.

(369) Show that $n^4 + 4$ is a composite number for each integer $n \geq 2$. More generally show that if a is a positive integer such that $2a$ is a perfect square, then $n^4 + a^2$ is a composite number provided $n \geq \sqrt{2a}$.

(370) Show that there exist infinitely many composite numbers of the form $k10^k + 1$.

(371) Show that if the number $k + 2$ is prime, then it is a prime divisor of the number $2k^k + 1$.

(372) Find three factors of $2^{58} + 1$.

(373) Let $M_p = 2^p - 1$, where p is an odd prime number. Show that all the factors of M_p are of the form $2kp + 1$, where $k \in \mathbb{N}$.

(374) The primality test of Lucas-Lehmer may be read as follows: "Let p be an odd prime number. The Mersenne number $M_p = 2^p - 1$ is prime if and only if $M_p | S_{p-1}$, where $S_1 = 4$ and $S_{n+1} \equiv S_n^2 - 2 \pmod{M_p}$, $n \geq 1$." Use this test (and a computer) to prove that M_{61} is prime.

(375) Factor the number $n = 10^{48} - 1$. A computer may prove handy to obtain certain factors of n smaller than 10^9.

(376) In 1960, Waclaw Sierpinski (1882–1969) proved that there exist infinitely many integers k such that each of the numbers $N = k \cdot 2^n + 1$ $(n = 1, 2, 3, \ldots)$ is composite. Three years later, Selfridge proved that the number $k = 78\,557$ is such a number. Prove this last result of Selfridge by establishing that, in this case, N is always divisible by 3, 5, 7, 13, 19, 37 or 73.

(377) Find three prime factors of $10^{27} + 1$.

(378) In order to obtain the factorization of the odd integer $n > 1$, it certainly helps to notice that, if n is composite, it is always possible to write n as

$$(*) \quad n = x^2 - y^2 = (x+y)(x-y) \quad \text{with } x, y \text{ positive integers, } x - y > 1,$$

thus revealing the factors $x + y$ and $x - y$ of n (see Problems 81 and 82). To obtain a representation of type $(*)$, we may proceed as follows. We look for an integer x such that $x^2 - n$ is a perfect square, that is such that

$$x^2 - n = y^2.$$

As a first value for x, we choose the smallest integer k such that $k^2 \geq n$, and then we try with $k + 1$, and so on. By proceeding in this manner, it is clear that we will eventually find an integer x such that $x^2 - n$ is a perfect square, the reason being that n is odd and composite. This factorization method is called FERMAT'S FACTORIZATION METHOD.

To show the method, we take $n = 2001$. Since $\sqrt{n} = 44.7325\ldots$, we shall successively try several values of x starting with $x = 45$; we then obtain the following table:

x	$x^2 - n = ?$	Perfect square ?
45	$45^2 - 2001 = 24$	NO
46	$46^2 - 2001 = 115$	NO
47	$47^2 - 2001 = 208$	NO
48	$48^2 - 2001 = 303$	NO
49	$49^2 - 2001 = 400$	YES

Hence, $2001 = 49^2 - 20^2 = (49 + 20)(49 - 20) = 69 \cdot 29$, thus providing a factorization of 2001.

Proceed as above in order to factorize 2009, and then use Fermat's factorization method to find two proper divisors of $n = 289\,751$.

(379) Fermat's factorization method works very well when the odd integer n which is to be factored has two divisors of roughly the same size. But if $n = pq$, where $p < q$ are far apart, the number of steps to reach a factorization may be very large. But this difficulty may be overcome. For instance, take the number $n = 1\,254\,713$. Multiply this number by a small prime number p_0, the goal being to obtain a number $m = p_0 n = d_1 d_2$,

where d_1 and d_2 are two positive integers whose quotient d_2/d_1 is close to 1. Use this strategy to obtain the factorization of n.

(380) Assume that $n = pq$, where p and q are two prime numbers satisfying $p < q < 2p$. Let δ be the number defined by

$$\frac{q}{p} = 1 + \delta,$$

so that $0 < \delta < 1$. Show that the number of steps necessary to factorize n by using Fermat's factorization method is approximately $\dfrac{p\delta^2}{8}$.

(381) Knowing that the number $n = 188\,686\,013$ is the product of two prime numbers p and q such that

$$\left| \frac{p}{q} - 3 \right| < \frac{1}{100},$$

find the factorization of n.

(382) Given an integer $r \geq 2$ and an odd integer $k \geq 5$, consider the number

$$n = r^k + r^{k-1} + \cdots + r^2 + r + 1.$$

Prove that the number n has at least three prime factors and moreover that they are distinct if $r \geq 3$ or if $r = 2$ and $k \geq 7$.

(383) Let k be a positive integer. Show that $\{2^k \pm 2^{k-1} \pm 2^{k-2} \pm \cdots \pm 2^1 \pm 1\}$ represents the set of all positive odd numbers $\leq 2^{k+1} - 1$.

(384) The number 11 is prime, while it is easy to check that the numbers 111, 1 111 and 11 111 are composite.

   (i) Show that if a number of the form $\underbrace{111\ldots1}_{k} = (10^k - 1)/9$ is prime, then the number k is necessarily a prime.

  (ii) Show that, if p is a prime number, then each prime factor of $(10^p - 1)/9$ is of the form $2jp + 1$ for a certain positive integer j.

 (iii) Use a computer to find the five smallest prime numbers p such that the number corresponding to $(10^p - 1)/9$ is prime.

 (iv) Use a computer to obtain the factorization of the numbers $(10^p - 1)/9$ for each prime number $p < 50$.

(385) Show that each positive integer n for which there exist positive integers k, x and y such that

(∗) $$n = x^{2k+1} + y^{2k+1}$$

is composite.

(386) Let n be a positive odd integer for which there exists a prime number $p_0 < \sqrt{n}$ such that $p_0 \cdot n$ can be written as the sum of two positive cubes. Show that n must be a composite number.

(387) Consider the number $n = 52\,657\,403$. Show that $7n$ can be written as the sum of two cubes (one of which is rather small!) and conclude that n is composite and divisible by 719.

(388) Consider the number $n = 237\,749\,938\,896\,803$. Show that $11n$ can be written as the sum of two fifth powers (one of which is rather small!) and conclude that n is composite and divisible by 1213.

(389) Let $n \geq 3$ be a squarefree odd composite number. Show that if for each prime divisor p of n, we have $p - 1 | n - 1$, then n is a Carmichael number.

(390) Let $p \geq 5$ be a prime number such that $2p - 1$ and $3p - 2$ are primes. Show that the number $n = p(2p - 1)(3p - 2)$ is a Carmichael number.

(391) Use Korselt's Criterion (mentioned in the remark on the solution of Problem 389) in order to prove that each Carmichael number must have at least three distinct prime factors.

(392) In the remark attached to the solution of Problem 389, we observed that an integer $n = q_1 q_2 \cdots q_k$, where $k \geq 3$ and $2 < q_1 < q_2 < \ldots < q_k$ are prime numbers, is a Carmichael number if and only if

$$(*) \qquad q_j - 1 \Big| \prod_{i=1}^{k} q_i - 1 \qquad (j = 1, 2, \ldots, k).$$

Show that condition $(*)$ can be replaced by the condition

$$q_j - 1 \Big| \prod_{\substack{i=1 \\ i \neq j}}^{k} q_i - 1 \qquad (j = 1, 2, \ldots, k).$$

(393) Observing that

$$(*) \qquad 327\,763 = 30^3 + 67^3 = 51^3 + 58^3,$$

find the factorization of $327\,763$.

(394) Searching for a prime factor of $n = 48\,790\,373$, we observe that

$$7 \cdot n = 341\,532\,611 = 699^3 + 8^3.$$

Use this information to obtain the factorization of n.

(395) In 1956, Paul Erdős raised the question of obtaining the value of the smallest integer $n > 3$ such that $2^n - 7$ is prime. Use a computer to find this number n as well as the five next numbers n with the same property. Show that, in this search, one may ignore even integers n, the integers $n \equiv 1 \pmod 4$, the integers $n \equiv 7 \pmod{10}$ as well as the integers $n \equiv 11 \pmod{12}$.

(396) Let $a \geq 2$ be an integer and let p be a prime number such that p does not divide $a(a^2 - 1)$. Show that the number

$$n = \frac{a^{2p} - 1}{a^2 - 1}$$

is pseudoprime in basis a. Use this method to find pseudoprimes in basis 2 and 3.

(397) Show that there exist infinitely many pseudoprimes in basis 2.

(398) Let a and m be two positive integers such that $(a, m) = 1$. We say that s is the order of a modulo m if s is the smallest positive integer such that $a^s \equiv 1 \pmod m$. Show that if $a^n \equiv 1 \pmod m$, then $s | n$.

(399) (LUCAS' TEST) Let $n \geq 3$ be an integer such that for each prime factor q of $n - 1$ there exists an integer $a > 1$ such that $a^{n-1} \equiv 1 \pmod n$ and $a^{(n-1)/q} \not\equiv 1 \pmod n$. Show that n is prime.

(400) Let $n = 10^{12} + 61$. First verify that $2^2 \cdot 5 \cdot 3947 \cdot 12667849$ is indeed the factorization of $n - 1$, and then use Lucas' Test, explained in Problem 399 (with an appropriate choice of a), to show that n is prime.

(401) Use the primality test of Lucas, explained in Problem 399, to prove that the numbers $n = r^4 + 1$, where r takes successively the values 1910, 1916 and 1926, are all primes.

(402) Let $n = 10^{12} + 63$. Verify that $n - 1 = 2 \cdot 3^2 \cdot 7 \cdot 47 \cdot 168861871$, and then use Lucas' Test, explained in Problem 399 (with an appropriate choice of a), to show that n is prime.

(403) (POLLARD $p - 1$ FACTORIZATION METHOD) Let n be a positive integer. Assume that n has an odd prime factor p such that $p - 1$ has all its prime factors $\leq k$, where k is a relatively small positive integer (such as $k = 100$ or 1000 or 10000), so that $(p - 1) | k!$. Let m be the residue modulo n of $2^{k!}$ and let $g = (m - 1, n)$. Show that $g > 1$, thus identifying a factor of n.

(404) Use the Pollard $p - 1$ factorization method to find the smallest prime factor of the Fermat number $F_9 = 2^{2^9} + 1$.

(405) Use the Pollard $p - 1$ factorization method and a computer to factor the number 252123019542987435093029.

(406) Use the Pollard $p - 1$ factorization method and a computer to obtain the three prime factors of the Mersenne number

$$2^{71} - 1 = 2361183241434822606847.$$

(407) Use the Pollard $p - 1$ factorization method and a computer to factor the number 136258390321.

(408) Let $n = 302\,446\,877$. Let m be the quantity $2^{25!}$ modulo n. Show that $g = (m - 1, n) = 17\,389$. Use the Pollard $p - 1$ factorization method to conclude that $17\,389$ is a (prime) divisor of $302\,446\,877$.

(409) Show that each prime factor p of the Fermat number $F_n = 2^{2^n} + 1$ with $n \geq 2$ is of the form $p = k \cdot 2^{n+2} + 1$, $k \in \mathbb{N}$.

(410) Use the result of Problem 409 in order to prove that 641 is a prime factor of $F_5 = 2^{2^5} + 1 = 4\,294\,967\,297$.

(411) Use the result of Problem 409 in order to prove that $274\,177$ is a prime factor of

$$F_6 = 2^{2^6} + 1 = 18\,446\,744\,073\,709\,551\,617.$$

(412) (PEPIN'S TEST) Let $F_n = 2^{2^n} + 1$ be a Fermat number and let $k > 2$ be an integer. Show that, for $n \geq 2$,

$$F_n \text{ is prime and } \left(\frac{k}{F_n} \right) = -1 \quad \Longleftrightarrow \quad k^{\frac{F_k - 1}{2}} \equiv -1 \pmod{F_n}.$$

7. Integer Parts

(413) Let $\alpha, \beta \in \mathbb{R}$. Show that
  (a) $[\alpha] + [\beta] + [\alpha + \beta] \leq [2\alpha] + [2\beta]$;
  (b) $[\alpha] + [\beta] + 2[\alpha + \beta] \leq [3\alpha] + [3\beta]$;
  (c) $[\alpha] + [\beta] + 3[\alpha + \beta] \leq [4\alpha] + [4\beta]$;
  (d) $2[\alpha] + 2[\beta] + 2[\alpha + \beta] \leq [4\alpha] + [4\beta]$;
  (e) $3[\alpha] + 3[\beta] + [\alpha + \beta] \leq [4\alpha] + [4\beta]$.

(414) Show that $\dfrac{(2n)!}{(n!)^2}$ is an even integer for each $n \in \mathbb{N}$.

(415) Let $m, n \in \mathbb{N}$. Show that:
  (a) $\dfrac{(2m)!(2n)!}{m!n!(m + n)!}$ is an integer;   (b) $\dfrac{(4m)!(4n)!}{n!m!\big((m + n)!\big)^3}$ is an integer.

(416) Let $a_i \geq 0$, $i = 1, 2, \ldots, r$, be integers such that $a_1 + a_2 + \cdots + a_r = n$. Show that $\dfrac{n!}{a_1! a_2! \cdots a_r!}$ is an integer.

(417) How many zeros appear at the end of the decimal representation of $23!$?

(418) Show that the last digit of $n!$ which is different from 0 is always an even number provided $n \geq 5$.

(419) Find all positive integers n for which the number of zeros appearing at the end of the decimal representation of $n!$ is 57. What happens when the number of zeros is 60 or 61?

(420) Let n be a positive integer.
  (a) Show that the largest integer α such that 5^α divides $(5^n - 3)!$ is $\dfrac{5^n - 4n - 1}{4}$.

  (b) Let p be a prime number and i a positive integer smaller than p. Show that the largest integer α such that p^α divides $(p^n - i)!$ is
  $$\frac{p^n - (p - 1)n - 1}{p - 1}.$$

(421) Let n be a positive integer. Find a formula which reveals explicitly, for a given prime number p, the unique value of α such that
$$p^\alpha \,\Big\|\, \prod_{i=1}^{n}(2i).$$

(422) Let n be a positive integer. Find a formula which reveals explicitly, for a given prime number p, the unique value of α such that
$$p^\alpha \,\Big\|\, \prod_{i=0}^{n}(2i + 1)$$

and use this to show that
$$n = \sum_{k=1}^{\infty} \left(\left[\frac{2n + 1}{2^k} \right] - \left[\frac{n}{2^k} \right] \right).$$

(423) Find all natural numbers n having the property that $[\sqrt{n}\,]$ is a divisor of n.

(424) Prove that for each integer $n \geq 1$,
$$\left[\sqrt{n} + \sqrt{n+1}\right] = \left[\sqrt{4n+1}\right] = \left[\sqrt{4n+2}\right] = \left[\sqrt{4n+3}\right].$$

(425) Prove that for each integer $n \geq 0$,
$$\left[\sqrt{n} + \sqrt{n+1} + \sqrt{n+2}\right] = \left[\sqrt{9n+8}\right].$$

(426) Let m and k be positive integers. Show that
$$\left[\frac{m-k}{k}\right] + \left[-\left(\frac{m+1}{k}\right)\right] + 2 = 0.$$

(427) Show that
$$\lim_{m \to \infty} \left[\cos^2(m!\pi x)\right] = \begin{cases} 0 & \text{if } x \in \mathbb{R} \setminus \mathbb{Q}, \\ 1 & \text{if } x \in \mathbb{Q}, \end{cases}$$

where $[y]$ stands for the largest integer smaller or equal to y, and thus establish that the function $f : \mathbb{R} \to \{0,1\}$ defined by
$$f(x) = \lim_{m \to \infty} \left[\cos^2(m!\pi x)\right]$$

represents the characteristic function of the rational numbers.

(428) Show that, for each positive integer n,
$$\left[\frac{n+1}{2}\right] + \left[\frac{n+2}{4}\right] + \left[\frac{n+4}{8}\right] + \left[\frac{n+8}{16}\right] + \cdots = n.$$

(429) Show that for each $m \in \mathbb{Z}$,
$$\left[\frac{m - \left[\frac{m-17}{25}\right]}{3}\right] = \left[\frac{8m+13}{25}\right].$$

(430) Let $m \in \mathbb{Z}$. Show that the expression
$$\left[\frac{3m+4}{13}\right] - \left[\frac{m - 28 - \left[\frac{m-7}{13}\right]}{4}\right]$$

does not depend on m.

(431) Given an integer $n \geq 2$, show that, for each positive integer $k < n$,
$$(*) \qquad \sum_{j=1}^{k} \left[\frac{n-j}{k}\right] = n - k.$$

(432) Show that if $\alpha \in \mathbb{R}$, we have $[\alpha] + \left[\alpha + \frac{1}{n}\right] + \cdots + \left[\alpha + \frac{n-1}{n}\right] = [n\alpha]$
for each positive integer n.

(433) Show that if $\alpha \in \mathbb{R}$, we have $\left[\frac{\alpha}{n}\right] + \left[\frac{\alpha+1}{n}\right] + \cdots + \left[\frac{\alpha+n-1}{n}\right] = [\alpha]$
for each positive integer n.

(434) Let m and n be positive integers such that $(m, n) = 1$. Show that

$$\sum_{k=1}^{n-1} \left[\frac{mk}{n} \right] = \frac{(m-1)(n-1)}{2}.$$

(435) Let m and n be positive integers such that $(m, n) = d$. Show that

$$\sum_{k=1}^{n-1} \left[\frac{mk}{n} \right] = \frac{(m-1)(n-1)}{2} + \frac{d-1}{2}.$$

(436) Establish the formula obtained in 1997 by Marcelo Polezzi that provides the value of the greatest common divisor of two positive integers m and n:

$$(m, n) = 2 \sum_{j=1}^{m-1} \left[\frac{jn}{m} \right] + m + n - mn.$$

(437) Consider the arithmetical function f defined by

$$f(n) = (n+1)^2 + n - \left[\sqrt{(n+1)^2 + n + 1} \right]^2 \qquad (n = 1, 2, 3, \ldots).$$

Evaluate the quantity $f(n) - n$.

(438) Given an integer $n \geq 2$, show that

$$n! = \prod_{p \leq n} p^{\alpha_p}, \qquad \text{where} \qquad \alpha_p = \frac{n - s_p(n)}{p - 1},$$

where $s_p(n)$ stands for the sum of the digits of n in basis p.

(439) Evaluate the series

$$\sum_{n=1}^{\infty} \frac{2^{\|\sqrt{n}\|} + 2^{-\|\sqrt{n}\|}}{2^n},$$

where $\|x\|$ stands for the closest integer to x.

(440) The characteristic function of the odd numbers defined, for each integer $n \geq 1$, by

$$\chi(n) = \begin{cases} 1 & \text{if } n \text{ is odd,} \\ 0 & \text{if } n \text{ is even,} \end{cases}$$

can be written in a single expression by using the function $[x]$; indeed,

$$\chi(n) = n - 2 \left[\frac{n}{2} \right] \qquad (n \geq 1).$$

Find a similar simple formula for the function

$$f(n) = \begin{cases} 1 & \text{if } n \text{ is odd,} \\ 2 & \text{if } n \text{ is even.} \end{cases}$$

(441) In Problem 1, we established the two formulas

$$1^2 + 2^2 + 3^2 + \cdots + n^2 = \frac{n(n+1)(2n+1)}{6} \qquad (n = 1, 2, 3, \ldots),$$

$$1^3 + 2^3 + 3^3 + \cdots + n^3 = \frac{n^2(n+1)^2}{4} \qquad (n = 1, 2, 3, \cdots).$$

Establish similar formulas for the two sums

$$A_n = [1^{1/2}] + [2^{1/2}] + [3^{1/2}] + \cdots + [(n^2 - 1)^{1/2}] \qquad (n = 2, 3, 4, \ldots),$$

$$B_n = [1^{1/3}] + [2^{1/3}] + [3^{1/3}] + \cdots + [(n^3 - 1)^{1/3}] \qquad (n = 2, 3, 4, \ldots),$$

where, as usual, $[x]$ stands for the largest integer $\leq x$.

(442) Let a be the positive solution of the quadratic equation $x^2 - x - 1 = 0$. Show that for each $n \in \mathbb{N}$, we have

$$[a^2 n] = [a[an] + 1].$$

(443) Show that for each $n \in \mathbb{N}$, the positive solution a of the equation $x^2 - x - 1 = 0$ verifies the equation

$$2[a^3 n] = [a^2[2an] + 1].$$

(444) Show that the number N of positive integer solutions x, y of the inequality $xy \leq n$, where n is a fixed positive integer, is given by

$$N = \left[\frac{n}{1}\right] + \left[\frac{n}{2}\right] + \cdots + \left[\frac{n}{n}\right] = 2 \sum_{k=1}^{[\sqrt{n}]} \left[\frac{n}{k}\right] - [\sqrt{n}]^2.$$

(445) Let $n \in \mathbb{N}$. For each integer $k \geq 0$, find the number of integers i ($1 \leq i \leq n$) which are divisible by 2^k but not by 2^{k+1}. Establish also that

$$\sum_{j=1}^{\infty} \left[\frac{n}{2^j} + \frac{1}{2}\right] = n.$$

Thus, by doing so, it will have been proved that one can evaluate the sum $\frac{n}{2} + \frac{n}{4} + \frac{n}{8} + \cdots$ by substituting each term by its closest integer (choosing the largest one if two such numbers exist).

(446) Show that for each $\alpha \in \mathbb{R}$, we have $\lim\limits_{n \to \infty} \dfrac{[n\alpha]}{n} = \alpha$.

(447) Show that for each nonnegative real number α and for each positive integer k, we have $[\alpha^{1/k}] = [[\alpha]^{1/k}]$.

8. Arithmetical Functions

(448) One of the nice properties of Euler's ϕ function is given by the formula $\sum_{d|n} \phi(d) = n$. Using computer software, write a program which allows one to compute the value of $\sum_{d|n} f(d)$, where f is a given arithmetical function, for different values of n, for example, for $n = 1, \ldots, 100$.

(449) Use the program called for in Problem 448 to show that, for $n = 1, \ldots, 1000$,

$$\sum_{d|n} \tau(d) \equiv 0 \pmod{3}$$

except for $n = k^3$, $k \in \mathbb{N}$. To do so, write a program that confirms that indeed this is true for $n = 1, \ldots, 1000$.

(450) Is it true that, for each integer $n \geq 0$, there exists a perfect number located between $10^n + 1$ and $10^{n+1} + 1$? Is it true that the last digit of the perfect numbers alternates between 6 and 8? Here, the use of a computer may prove useful.

(451) Show that if f and g are multiplicative functions, then their product fg is also a multiplicative function. If f is a multiplicative function, can one say that kf, for $k \in \mathbb{R}$, is also a multiplicative function? What about $f + g$?

(452) Does there exist a multiplicative function f such that

$$f(30) = 0, \quad f(105) = 1 \quad \text{and} \quad f(70) = 1?$$

(453) Let $t_1 = 1, t_2 = 3, t_3 = 6, \ldots, t_k = k(k+1)/2, \ldots$ be the sequence of triangular numbers. Let f be the arithmetical function defined by $f(n) = 1/k$, where k is the only integer satisfying $t_{k-1} < n \leq t_k$, so that

$$\sum_{n \leq t_k} f(n) = 1 + \frac{1}{2} + \frac{1}{2} + \frac{1}{3} + \frac{1}{3} + \frac{1}{3} + \cdots + \underbrace{\frac{1}{k} + \cdots + \frac{1}{k}}_{k} = k,$$

for each positive integer k. Prove that

$$f(n) = \left\| \frac{1}{2} \sqrt{8n - 7} \right\|^{-1},$$

where $\|x\|$ stands for the closest integer to x.

(454) Let f be the arithmetical function defined by

$$f(n) = [\sqrt{n} - 1] + [\sqrt[3]{n} - 1] + [\sqrt[4]{n} - 1] + \cdots \qquad (n = 1, 2, \ldots).$$

Show that $\limsup_{n \to \infty} (f(n) - f(n-1)) = +\infty$.

(455) Indicate which amongst the following functions are totally multiplicative:

$$\gamma(n) = \prod_{p|n} p, \qquad \sigma_2(n) = \sum_{d|n} d^2,$$

$$g(n) = 2^{\omega(n)}, \qquad h(n) = \sum_{d|n} \mu^2(d) d, \qquad \rho(n) = 2^{\Omega(n)}.$$

(456) Show that the function $f(n) = [\sqrt{n}] - [\sqrt{n-1}]$ is multiplicative. Is this function totally multiplicative?

(457) A function f is said to be *strongly multiplicative* if it is multiplicative and if also $f(p^\alpha) = f(p)$ for each prime number p and each $\alpha \in \mathbb{N}$. Identify those functions, amongst the ones given below, that are strongly multiplicative:

$$\gamma(n) = \prod_{p|n} p, \quad \sigma_2(n) = \sum_{d|n} d^2, \quad g(n) = 2^{\omega(n)}, \quad h(n) = \sum_{d|n} \mu^2(d) d.$$

(458) Let f be a multiplicative function. Is the function g defined by $g(n) = \sum_{d|n} \mu^2(d) f(d)$ necessarily strongly multiplicative?

(459) Let f be an arithmetical function which is both strongly multiplicative and totally multiplicative. Is it true that $\{f(n) : n = 1, 2, 3, \ldots\}$ contains at most two elements? Explain.

(460) Let g be an arithmetical function defined, for each $n \geq 1$, by

$$g(n) = \begin{cases} 1 & \text{if } n \equiv 0 \pmod{3}, \\ 2 & \text{if } n \equiv 1 \pmod{3}, \\ 3 & \text{if } n \equiv 2 \pmod{3}. \end{cases}$$

Is it true that g is a multiplicative function? Explain.

(461) Prove that an arithmetical function f such that $f(1) = 1$ is multiplicative if and only if, for each $m, n \in \mathbb{N}$,

$$f((m, n)) f([m, n]) = f(m) f(n).$$

(462) Let $\gamma(n)$ be the arithmetical function which represents the kernel of n, that is $\gamma(n) = \prod_{p|n} p$. Show that:
   (a) γ is a multiplicative function;
   (b) $\gamma(n) = \sum_{d|n} |\mu(d)| \phi(d)$ for each integer $n \geq 1$.

(463) Show that if the *abc* conjecture (see page 12) is true, then for all $\varepsilon > 0$, there exists a constant $M = M(\varepsilon)$ such that for each integer $n \geq 2$, we have

$$n < M \cdot \gamma(n^2 - 1)^{1+\varepsilon},$$

where $\gamma(m)$ is the product of the prime factors of m.

(464) Let f be a multiplicative function and let k be a positive integer such that $f(k) \neq 0$. Show that the arithmetical function g defined by

$$g(n) := \frac{f(kn)}{f(k)} \quad \text{is also multiplicative.}$$

(465) Let $f : \mathbb{N} \to \mathbb{N}$ be a strictly increasing function such that $f(2) = 2$ and $f(mn) = f(m) f(n)$ if m and n are relatively prime. Show that f is the identity function, that is that $f(n) = n$ for each $n \geq 1$.

(466) Let f be an additive function. Assume that, for each positive integer n, $\lim_{k \to \infty} \dfrac{f(n^k)}{k}$ exists. Show that the function g defined by $g(n) = \lim_{k \to \infty} \dfrac{f(n^k)}{k}$ is totally additive.

(467) Let f and g be two multiplicative functions. Show that the function h defined by

$$h(n) = \sum_{\substack{dr=n \\ (d,r)=1}} f(d) g(r) \quad (n = 1, 2, \ldots)$$

is also a multiplicative function.

(468) Let f and g be two multiplicative functions. Show that the function h defined by

$$h(n) = \sum_{[d,r]=n} f(d)g(r) \qquad (n = 1, 2, \ldots),$$

where the sum runs over all ordered pairs (d, r) such that $[d, r] = n$ is a multiplicative function.

(469) Let f be a multiplicative function such that, for each prime number p, $\lim_{k\to\infty} f(p^k) = 0$. Do we necessarily have that $\lim_{n\to\infty} f(n) = 0$? Explain.

(470) Let f be a multiplicative function such that, for each positive integer k, $\lim_{p\to\infty} f(p^k) = 0$. Do we necessarily have that $\lim_{n\to\infty} f(n) = 0$? Explain.

(471) Let f be a totally additive function which is monotonically increasing; prove that there exists a constant $c \geq 0$ such that $f(n) = c\log n$ for each integer $n \geq 1$.

(472) Consider the arithmetical function f defined by $f(1) = 1$ and for $n > 1$ by

$$f(n) = \begin{cases} n/2 & \text{if } n \text{ is even,} \\ 3n + 1 & \text{if } n \text{ is odd.} \end{cases}$$

Define the functions f^0, f^1, $f^2, \ldots$ as follows: $f^0(n) = n$, $f^1(n) = f(n)$, $f^2(n) = f(f(n))$, $f^3(n) = f(f^2(n)), \ldots, f^k(n) = f(f^{k-1}(n)), \ldots$ The *Collatz Problem* (also called the *Syracuse Problem*) consists of attempting to establish that for each positive integer n, there exists $k \in \mathbb{N}$ such that $f^k(n) = 1$. This result is most likely true, but no one has ever been able to prove it. However, partial results have been obtained.

(a) Let α and j be two positive integers. What is the value of $f^j(2^\alpha)$?

(b) Let $\alpha \in \mathbb{N}$. For which values of $j \in \mathbb{N}$ is it true that $f^j(2^\alpha) = 1$?

(c) What is the smallest value of $n \in \mathbb{N}$ such that $f^k(n) = 11$ for a certain positive integer k?

(d) Show that, if n is a positive odd integer, then $f^3(n) < \frac{3}{4}n+1$ if $n \equiv 1$ (mod 4) while $f^3(n) > 4n$ if $n \equiv 3$ (mod 4).

(e) Find an integer n such that $f^{2k+1}(n)$ is odd for $k = 0, 1, 2, 3, 4$ and such that $f^{2k}(n)$ is even for $k = 0, 1, 2, 3, 4, 5$.

(f) Is it true that if $f^3(n) > n$, then $f^3(n+2) < n+2$?

(g) Given an odd positive odd n, what is the probability that $f^3(n)$ is larger than n?

(h) Consider the arithmetical function g defined by $g(1) = 1$ and for $n > 1$ by

$$g(n) = \begin{cases} n/2 & \text{if } n \text{ is even,} \\ 5n + 1 & \text{if } n \text{ is odd,} \end{cases}$$

and then define the functions g^0, g^1, $g^2, \ldots$ as we did for the function f. Show that the conjecture to the effect that "for each $n \in \mathbb{N}$ there exists $k \in \mathbb{N}$ such that $g^k(n) = 1$" is false.

(i) Let $n \geq 3$. We introduce the function $\mathsf{Syr}(n)$ which stands for the smallest positive integer α such that $f^\alpha(n) = 1$. For instance,

$\mathsf{Syr}(8) = 3$. Show that $\mathsf{Syr}(n) \geq \log_2 n$, where $\log_2(n)$ stands for the logarithm of n in basis 2.

(j) Let $n \geq 3$ and let $\mathsf{Syr}(n)$ be the function introduced above. Prove that if n is odd, then $\mathsf{Syr}(n) \geq \log_2 n + 3$.

(k) Let $\alpha \in \mathbb{N}$ and consider the number $n = 2^{\alpha+1} - 1$. Show that

$$f^{2k}(n) = 3^k \cdot 2^{\alpha-k+1} - 1 \qquad \text{for each integer } k, \ 1 \leq k \leq \alpha,$$

and therefore that the sequence of iterations

$$f^2(n), f^4(n), f^6(n), \ldots, f^{2\alpha}(n)$$

is strictly increasing.

(l) Let α and n be as in (k). Show that

$$f^{2\alpha}(n) = 2 \cdot 3^\alpha - 1.$$

(m) Let $\alpha \in \mathbb{N}$ and consider the sequence of integers $n_0, n_1, n_2, \ldots$ defined by $n_j = j2^\alpha + (2^\alpha - 1)$. Show that

$$f^{2\alpha}(n_j) = (j+1)3^\alpha - 1, \qquad \text{for each } j.$$

(n) Given an arbitrary large real number $C > 0$, show that there exist two positive integers n and k such that $f^k(n) > Cn$.

(o) Consider the arithmetical function f_* defined by $f_*(1) = 1$ and for $n > 1$ by

$$f_*(n) = \begin{cases} n/2^\beta & \text{if } n = 2^\beta r, \text{ with } \beta \geq 1 \text{ and } r \text{ odd,} \\ 3n+1 & \text{if } n \text{ is odd,} \end{cases}$$

and then define the iteration functions $f_*^0, f_*^1, f_*^2, \ldots$ as we did for the function f, and establish a table of the values $f_*^0(n), f_*^1(n), f_*^2(n), \ldots,$ $f_*^{10}(n)$ for $n = 1, 2, 3, \ldots, 50$. Now, given an arbitrary positive integer n, can we conclude that there necessarily exists a positive integer k such that $f_*^k(n) = 1$? Is there here an analogy with the "standard" Collatz Problem?

(473) Let a, b, c be positive integers such that $(a, b, c) = 1$. Is it true that

$$\tau(abc) = \tau(a)\tau(b)\tau(c) ?$$

(474) Find the smallest positive integer n such that
(a) $\tau(n) = 9$;   (b) $\tau(n) = 10$;   (c) $\tau(n) = 15$.

(475) Identify all natural numbers having exactly 14 divisors.

(476) Find the largest prime number p such that
(a) $p|\tau(20!)$;   (b) $p|\sigma(20!)$;   (c) $p^2|\tau(35!)$;   (d) $p^2|\sigma(35!)$.

(477) How many positive integers n are there dividing at least one of the two numbers 10^{40} and 20^{30}?

(478) Prove that

$$\sum_{n=1}^{\infty} \frac{\tau(n)}{2^n} = \sum_{n=1}^{\infty} \frac{1}{2^n - 1}.$$

(479) Consider the sequence (b_k) defined by

$$b_1 = 2, \qquad b_{k+1} = 1 + b_1 b_2 \cdots b_k \quad (k = 1, 2, \ldots).$$

Show that
(i) for each integer $k \geq 1$, $b_{k+1} = b_k^2 - b_k + 1$,

(ii) $\displaystyle\sum_{j=1}^{\infty} \frac{1}{b_j} = 1.$

Use this to show that the arithmetical function g defined by

$$g(n) = n - 1 - \sum_{j=1}^{\infty} \left[\frac{n-1}{b_j}\right]$$

has the representation

$$g(n) = \sum_{j=1}^{\infty} \left\{\frac{n-1}{b_j}\right\},$$

where $\{x\}$ stands for the fractional part of x.

(480) Let $\tau_1(n)$ be the number of odd divisors of n. Prove that τ_1 is a multiplicative function.

(481) Given a positive integer n, show that the number of ordered pairs of positive integers a, b such that $ab = n$ and $(a, b) = 1$ is $2^{\omega(n)}$.

(482) Let n be a positive integer. Show that the number of ordered pairs of positive integers a, b such that $[a, b] = n$ is $\tau(n^2)$.

(483) Let d and n be positive integers such that $d^2 | n$. Show that the number of ordered pairs of positive integers a, b such that $(a, b) = d$ and $ab = n$ is $2^{\omega(n/d^2)}$. Use this to show that

$$\tau(n) = \sum_{d^2|n} \omega(n/d^2).$$

(484) Let n be a positive integer and let 2^α be the largest power of 2 that divides n. To which of the following five values is the quotient $\tau(2n)/\tau(n)$ equal:

$$2, \quad \frac{\alpha+3}{\alpha+2}, \quad \frac{\alpha+2}{\alpha+1}, \quad \frac{\alpha+1}{\alpha}, \quad \frac{\alpha}{\alpha-1} \ ?$$

Explain.

(485) Show that $\tau(n)$ is odd if and only if n is a perfect square.

(486) Show that if $\sigma(n)$ is a prime number for a certain positive integer n, then $\tau(n)$ must also be a prime number.

(487) Show that $\sigma(n)$ is odd if and only if n is a square or two times a square.

(488) Show that, for each positive integer n,

$$\prod_{d|n} d = n^{\tau(n)/2}.$$

What happens if $\tau(n)$ is odd?

(489) Prove that for each integer $n \geq 1$, we have $\sigma_2(n) \geq n\tau(n)$, where $\sigma_2(n) = \sum_{d|n} d^2$.

(490) Find the minimal value of $\tau(n(n+1))$ as n runs through the positive integers greater or equal to 3.

(491) For each integer $n \geq 1$, consider the functions $f_1(n)$ and $f_2(n)$ which stand respectively for the product of the odd divisors of n and for the product of the even divisors of n. Establish the following formulas:

$$\begin{aligned} f_1(n) &= m^{\tau(m)/2}, \\ f_2(n) &= \left(2^{\alpha(\alpha+1)} m^\alpha\right)^{\tau(m)/2} = (2n)^{\alpha\tau(m)/2}, \end{aligned}$$

where m and α are defined implicitly by $n = 2^\alpha m$, m odd.

(492) Show that $\displaystyle\prod_{m=1}^{n} m^{2[n/m]-\tau(m)} = 1$, where $[y]$ stands for the largest integer $\leq y$.

(493) Given a positive integer n, consider the corresponding sequence

$$n,\ \tau(n),\ \tau(\tau(n)),\ \tau(\tau(\tau(n))),\ldots .$$

Identify those positive integers n for which the above sequence contains no perfect squares.

(494) For each real number a, define the function σ_a by $\sigma_a(n) = \sum_{d|n} d^a$. It is clear that $\tau(n)$ and $\sigma(n)$ are particular cases of $\sigma_a(n)$. Prove that

$$\sigma_a(n) = \begin{cases} \displaystyle\prod_{p^\alpha \| n} \frac{p^{a(\alpha+1)}-1}{p^a-1} & \text{if } a \neq 0, \\[2em] \displaystyle\prod_{p^\alpha \| n} (\alpha+1) & \text{if } a = 0. \end{cases}$$

(495) Assume that p and q are odd prime numbers, and a and b are positive integers such that $p^a > q^b$. Show that, if p^a divides $\sigma(p^a)\sigma(q^b)$, then $p^a = \sigma(q^b)$.

(496) Let $\sigma^*(n)$ be the sum of the odd divisors of n. Show that σ^* is a multiplicative function.

(497) Show that $3|\sigma(3n-1)$ and $4|\sigma(4n-1)$ for each positive integer n. Is it true that $12|\sigma(12n-1)$ for each $n \geq 1$? Explain.

(498) Let p be a prime number and let a and b be nonnegative integers. Show that $\sigma(p^a)|\sigma(p^b)$ if and only if $(a+1)|(b+1)$.

(499) Show that $\sigma_{-a}(n) = n^{-a}\sigma_a(n)$ for each real number a and each positive integer n. In particular, show that the sum of the reciprocals of the divisors of a positive integer n is equal to $\sigma(n)/n$.

(500) Show that n is an even perfect number if and only if there exists a positive integer k such that $n = 2^{k-1}(2^k - 1)$, where $2^k - 1$ is a prime number.

(501) In 1958, Perisatri proved that if n is an odd perfect number, then

$$\frac{1}{2} < \sum_{p|n} \frac{1}{p} < 2\log\frac{\pi}{2}.$$

Is it true that these inequalities still hold for each even perfect number n? Explain.

(502) Show that if n is an even perfect number, then $8n+1$ is a perfect square.

(503) Let a be a positive integer and p a prime number. Can p^a be a perfect number?

(504) Show that every even perfect number ends with the digit 6 or 8.

(505) Show that every even perfect number larger than 6 can be written as the sum of consecutive odd cubes.

(506) Show that each odd perfect number must have at least three distinct prime factors.

(507) Find all natural numbers n having the property that n and $\sigma(\sigma(n))$ are perfect numbers, or otherwise show that no such number n exists.

(508) A natural number n is said to be *tri-perfect* if $\sigma(n) = 3n$. Show that each odd tri-perfect number must be a perfect square.

(509) Show that the only tri-perfect numbers of the form $2^\alpha m$ with $1 \le \alpha \le 10$, m odd and $\mu^2(m) = 1$ are the numbers 120, 672, 523 776 and 459 818 240.

(510) A positive integer n is called respectively *deficient* or *abundant* if the sum of its divisors is $<$ or $> 2n$. Show that if the greatest common divisor of two positive integers a and b is deficient, then there exist
  (a) infinitely many deficient numbers n such that $n \equiv a \pmod{b}$;
  (b) infinitely many abundant numbers n such that $n \equiv a \pmod{b}$.

(511) Let n be an even perfect number. Show that

$$\tau(n) = [\log_2 n] + 2,$$

where $\log_2 n$ stands for the logarithm of n in basis 2.

(512) In 1997, Gordon Spence discovered the 36-th Mersenne prime, namely $2^{2976221} - 1$. Establish first a general formula allowing one to quickly compute the number of digits of a given large integer, and then use this formula to determine the number of digits contained in the prime number discovered by Spence.

(513) Let k be an arbitrarily large natural number. Prove that there exists an integer n such that

$$\frac{\sigma(n)}{n} > k.$$

(514) Show that an even perfect number is a triangular number, that is a number of the form $n(n+1)/2$.

(515) Show that a perfect number having k distinct prime factors has at least one prime factor which does not exceed k.

(516) Let $q_1, \ldots, q_k$ be distinct prime numbers. Show that

$$\frac{(q_1+1)(q_2+1)\cdots(q_k+1)}{q_1 q_2 \cdots q_k} \le 2 < \frac{q_1 q_2 \cdots q_k}{(q_1-1)(q_2-1)\cdots(q_k-1)}$$

is a necessary condition for $n = \prod_{i=1}^{k} q_i^{\alpha_i}$ to be a perfect number.

(517) Show that if n is an even perfect number, then $\phi(n) = 2^{k-1}(2^{k-1}-1)$ for a certain positive integer k.

(518) Is it true that $\phi(n)$ is a multiple of 10 for infinitely many positive integers n?

(519) Calculate the number of positive integers ≤ 600 which have a factor > 1 in common with 600 and then count the number of positive integers ≤ 1200 which are relatively prime with 600.

(520) Count the number of positive integers ≤ 4200 which are relatively prime with 600 by observing that $4200 = 7 \cdot 600$.

(521) If m and k are positive integers, show that the number of positive integers $\le mk$ which are relatively prime with m is equal to $k\phi(m)$.

(522) Let $m, n \in \mathbb{N}$. Show that

$$\phi(mn) = \phi(m)\phi(n)\frac{d}{\phi(d)}, \qquad \text{where } d = (m, n).$$

(523) Show that for $n > 2$, $\phi(n)$ is an even number.

(524) Show that the number of fractions a/b, $(a, b) = 1$, such that $0 < a/b \le 1$, b being a fixed positive integer, is $\phi(b)$.

(525) Show that if $m|n$, then $\phi(m)|\phi(n)$.

(526) Identify all positive integers n such that $\phi(n)|n$.

(527) If $d|n$ and $k \in \mathbb{N}$, show that $\phi(nd^k) = d^k\phi(n)$.

(528) Identify all positive integers n such that $5\phi(n)|2n$.

(529) Characterize the set of positive integers n such that
$$\text{(a) } \phi(2n) > \phi(n); \qquad \text{(b) } \phi(2n) = \phi(n); \qquad \text{(c) } \phi(2n) = \phi(3n).$$

(530) Let p be an odd prime number such that $2p + 1$ is also a prime number. Show that if $n = 4p$, then $\phi(n + 2) = \phi(n) + 2$.

(531) Show that for each integer $n \geq 2$, the sum of positive integers $\leq n$ and relatively prime with n is equal to $n\phi(n)/2$.

(532) Let $n > 1$ be an integer. Show that $2^{\omega(n)-1}|\phi(n)$.

(533) Show that $\phi(n)$ is a power of 2 if and only if $n = 2^\alpha F_1 \cdots F_r$, where $\alpha \geq 0$ and $F_i = 2^{2^i} + 1$, $i = 1, 2, \ldots, r$, are Fermat primes.

(534) Find the largest prime number p such that
$$\text{(a) } p|\phi(95!); \quad \text{(b) } p^2|\phi(95!); \quad \text{(c) } p^3|\phi(95!); \quad \text{(d) } p^4|\phi(95!).$$

(535) Find the largest positive integer n such that $\phi(n) \leq 500$.

(536) Is it true that $\phi(8m + 4) = 2\phi(4m + 2)$ for each integer $m \geq 0$?

(537) Let $a, b \in \mathbb{N}$ such that $a|b$. Show that for each integer $n \geq 0$,
$$\frac{\phi(a^2n + ab)}{\phi(abn + a^2)} = \frac{\phi(an + b)}{\phi(bn + a)}.$$

(538) Show that $\phi(n) > n/7$ for all natural numbers n such that $\omega(n) \leq 9$.

(539) Given an odd integer $n > 3$, show that there exists a prime number p which divides $(2^{\phi(n)} - 1)$ but not n.

(540) Show that an integer $n \geq 2$ is prime if and only if $\phi(n)|(n - 1)$ and $(n + 1)|\sigma(n)$.

(541) Show that if e runs through the even divisors of n and d runs through the odd divisors of n, then
$$\frac{\displaystyle\sum_{e|n} \phi(n/e)}{\displaystyle\sum_{d|n} \phi(n/d)} = \begin{cases} 0 & \text{if } n \text{ is odd,} \\ \\ 1 & \text{if } n \text{ is even.} \end{cases}$$

(542) Show that for $m > 2$,
$$\sum_{\substack{k=1 \\ (k,m)=1}}^{\phi(m)} \frac{1}{k}$$
is never an integer.

(543) Let $f(n)$ be the product of all the positive divisors of n. Does $f(m) = f(n)$ automatically implies that $m = n$?

(544) Let (i, n) be the greatest common divisor of the positive integers i and n. Express
$$\sum_{i=1}^{n} 2^{\omega((i,n))}$$
in terms of the prime factors appearing in the factorization of n.

(545) Let f be the arithmetical function defined by

$$f(n) = \begin{cases} 1 & \text{if } n \text{ is even,} \\ 2 & \text{if } n \text{ is odd.} \end{cases}$$

Let $S(x) = \sum_{n \leq x} f(n)$. Find the value of $\lim_{x \to \infty} \dfrac{S(x)}{x}$.

(546) Show that the expression $\sum_{p|n} \dfrac{1}{p}$ can become arbitrarily large if n is chosen appropriately.

(547) Show that

$$(*) \qquad \sum_{ab=n} f(a)f(b) = \tau(n)f(n) \qquad (n = 1, 2, \ldots)$$

if and only if $f(n)$ is totally multiplicative.

(548) Prove that for each positive integer n, we have

$$2^{\omega(n)} \leq \tau(n) \leq 2^{\Omega(n)}.$$

(549) Let

$$H(n) = \frac{\tau(n)}{\sum_{d|n} 1/d}$$

be the *harmonic mean* of the divisors of n. Show that n is an even perfect number if and only if $n = 2^{H(n)-1}(2^{H(n)} - 1)$.

(550) Show that

$$\tau(n)^2 = \sum_{c|n} \sum_{b|c} \sum_{a|b} \mu^2(a) \qquad (n = 1, 2, \ldots).$$

(551) Let $f : \mathbb{N} \to \mathbb{Z}$ be a function satisfying $f(n + m) \equiv f(n) \pmod{m}$ for all integers $m, n \geq 1$; any polynomial with integer coefficients is such a function. Let $g(n)$ be the number of values (counting repetitions) amongst $f(1), f(2), \ldots, f(n)$ which are divisible by n and let $h(n)$ be the number of these values which are relatively prime with n. Show that g and h are multiplicative functions and that

$$h(n) = \sum_{d|n} \mu(d)g(d) \frac{n}{d} = n \prod_{p|n} \left(1 - \frac{g(p)}{p}\right) \qquad (n = 1, 2, \ldots).$$

(552) Show that

$$\sum_{d|n} \lambda(d) = \begin{cases} 1 & \text{if } n = m^2 \text{ for a certain integer } m \geq 1, \\ 0 & \text{otherwise,} \end{cases}$$

where λ stands for the Liouville function.

(553) Let f be a multiplicative function such that $f(2) = 1$. Prove that if n is an even integer, then

$$\sum_{d|n} \mu(d)f(d) = 0.$$

(554) Let $f\colon [0,1]\cap\mathbb{Q}\to\mathbb{R}$ and let, for each integer $n\geq 1$,

$$F(n)=\sum_{k=1}^{n} f\left(\frac{k}{n}\right) \quad\text{and}\quad F^{*}(n)=\sum_{d\mid n}\ \sum_{\substack{1\leq k\leq d\\(k,d)=1}} f\left(\frac{k}{d}\right).$$

Show that

$$F(n)=\sum_{d\mid n} F^{*}(d) \qquad (n=1,2,\ldots).$$

(555) Given an integer m, let

$$\phi_{m}(n)=\sum_{\substack{k=1\\(k,n)=1}}^{n} k^{m} \qquad (n=1,2,\ldots).$$

Show, using the result of Problem 554, that

$$\sum_{d\mid n}\frac{\phi_{m}(d)}{d^{m}}=\frac{1^{m}+2^{m}+\cdots+n^{m}}{n^{m}} \qquad (n=1,2\ldots).$$

Use this equation to show that

$$\sum_{\substack{k=1\\(k,n)=1}}^{n} k^{m}=\sum_{d\mid n} d^{m}\mu(d)\left(1^{m}+2^{m}+\cdots+\left(\frac{n}{d}\right)^{m}\right) \qquad (n=1,2\ldots).$$

(556) Prove that for each positive integer n,

$$\sum_{\substack{k=1\\(k,n)=1}}^{n} k=\frac{n}{2}\phi(n)+\frac{n}{2}\sum_{d\mid n}\mu(d).$$

(557) Prove that for each positive integer n,

$$\sum_{\substack{k=1\\(k,n)=1}}^{n} k^{2}=\frac{n^{2}}{3}\phi(n)+\frac{n^{2}}{2}\sum_{d\mid n}\mu(d)+\frac{n}{6}\prod_{p\mid n}(1-p).$$

(558) Prove that for each positive integer n,

$$\sum_{\substack{k=1\\(k,n)=1}}^{n} k^{3}=\frac{n^{3}}{4}\phi(n)+\frac{n^{3}}{2}\sum_{d\mid n}\mu(d)+\frac{n^{2}}{4}\prod_{p\mid n}(1-p).$$

(559) For each positive integer n, set $f(n)=\displaystyle\sum_{d\mid n}\frac{\mu^{2}(d)}{\tau(d)}$. Establish a formula for $f(n)$ in terms of the canonical representation of n.

(560) Is it true that $n=\displaystyle\sum_{d\mid n}\mu(d)\sigma(n/d)$ for each positive integer n?

(561) Show that, for each positive integer n,

$$\sum_{d\mid n}\frac{1}{d^{2}}=\frac{\sigma_{2}(n)}{n^{2}}.$$

(562) Show that $\displaystyle\sum_{d|n} \tau^3(d) = \left(\sum_{d|n} \tau(d)\right)^2$ for each positive integer n.

(563) Show that each of the following relations holds for each integer $n \geq 1$:

$$\sum_{d|n} |\mu(d)| = 2^{\omega(n)}; \quad \sum_{d|n} \mu(d)\tau(d) = (-1)^{\omega(n)}; \quad \sum_{d|n} \mu(d)\lambda(d) = 2^{\omega(n)};$$

$$\sum_{d|n} \mu(d)\sigma(d) = (-1)^{\omega(n)} \prod_{p|n} p \text{ and } \sum_{d|n} \mu(d)\phi(d) = (-1)^{\omega(n)} \prod_{p|n}(p-2).$$

(564) Show that, for each positive integer n,

$$\frac{n}{\phi(n)} = \sum_{d|n} \frac{\mu^2(d)}{\phi(d)}.$$

(565) Let g be a multiplicative function. For each positive integer n, set

$$F(n) = \sum_{d|n} \mu(d)g(n/d)$$

and show that

$$F(n) = \prod_{p^\alpha \| n} \left(g(p^\alpha) - g(p^{\alpha-1})\right).$$

(566) Let f be the function defined by

$$f(n) = \sum_{[d,r]=n} \phi(d)\phi(r) \qquad (n = 1, 2, \ldots),$$

where the sum runs over all ordered pairs (d, r) such that $[d, r] = n$ (see Problem 468). Show that

$$f(n) = n^2 \prod_{p|n} \left(1 - \frac{1}{p^2}\right).$$

(567) Let Λ be the von Mangoldt function. Show that

$$\sum_{d|n} \Lambda(d) = \log n \qquad (n = 1, 2, \ldots).$$

(568) Prove that for each positive integer n,

$$\Lambda(n) = \sum_{d|n} \mu(d) \log(n/d).$$

(569) Let f be a multiplicative function. Show that

$$\sum_{d|n}(-1)^{n/d} f(d) =$$

$$\begin{cases} -\displaystyle\sum_{d|n} f(d), & \text{if } n \text{ is odd}, \\ \displaystyle\sum_{d|n} f(d) - 2f(2^k) \sum_{d|m} f(d), & \text{if } n = 2^k m,\ (m, 2) = 1,\ k \geq 1. \end{cases}$$

(570) Show that if n is an even integer, then

$$\sum_{d|n}(-1)^{n/d}\phi(d) = 0.$$

What if n is an odd integer?

(571) Let f be an arithmetical function verifying the equation $\sum_{d|n} f(d) = n$ for each positive integer n. Show that $f(n) = \phi(n)$ for each $n \geq 1$.

(572) For each function g defined implicitly above, find a formula for $g(n)$ in terms of the canonical representation of n:

(a) $n^2 = \sum_{d|n} g(d)$;   (b) $\mu(n) = \sum_{d|n} g(d)$.

(573) Show that the function $2^{\omega(n)}n/\phi(n)$ is multiplicative and find a formula for $g(n)$ in terms of the canonical representation of n, knowing that

$$2^{\omega(n)}\frac{n}{\phi(n)} = \sum_{d|n} g(d), \quad \text{for each integer } n \geq 1.$$

(574) For each positive integer n, show that

$$\sum_{d|n}(-1)^{\Omega(d)}\mu(d) = 2^{\omega(n)}, \quad \sum_{d|n}(-1)^{\Omega(d)}\mu\left(\frac{n}{d}\right) = (-1)^{\Omega(n)}2^{\omega(n)}$$

and that

$$\sum_{d|n}(-1)^{\Omega(d)}2^{\omega(n/d)} = 1.$$

(575) Show that, for each positive integer n,

$$\sum_{d|n}\mu(d)\lambda\left(\frac{n}{d}\right) = (-1)^{\Omega(n)}\,2^{\omega(n)}.$$

(576) Let g be an arithmetical function such that $g(n) > 0$ for each positive integer n and let

$$f(n) = \prod_{d|n} g(d) \qquad (n = 1, 2, \ldots).$$

Show that

$$g(n) = \prod_{d|n}(f(n/d))^{\mu(d)} \qquad (n = 1, 2, \ldots).$$

(577) Let k be a real number. Show that, for each positive integer n,

$$\prod_{d|n} d^{(k/2)\tau(d)\mu(n/d)} = n^k.$$

(578) Let f be a totally multiplicative function and let F be defined by

$$F(n) = \sum_{d|n} f(d) \qquad (n = 1, 2, \ldots).$$

Do we necessarily have that F is also totally multiplicative?

(579) Prove that for each integer $n \geq 1$,

$$\sum_{\substack{d|n \\ \mu^2(d)=1}} \mu(n/d) = \begin{cases} \mu(\sqrt{n}) & \text{if } n = m^2, \\ 0 & \text{otherwise.} \end{cases}$$

(580) Let f be an arithmetical function. Show that for each integer $n \geq 1$,

$$\prod_{d|n} d^{f(d)+f(n/d)} = n^{\sum_{d|n} f(d)}.$$

Use this to show that

$$(*) \qquad \prod_{d|n} d^{\phi(d)+\phi(n/d)} = n^n \qquad (n = 1, 2, \ldots).$$

(581) Letting as usual $f * g$ stand for the Dirichlet product of the arithmetical functions f and g, show that if A stands for the set of arithmetical functions $f : \mathbb{N} \to \mathbb{R}$ such that $f(1) \neq 0$, then A is a commutative group with respect to the operation $*$; thus, prove successively that:
   (a) the Dirichlet product is commutative; that is if f and g are arithmetical functions, then $f * g = g * f$;
   (b) the Dirichlet product is associative; that is if f, g and h are arithmetical functions, then $(f * g) * h = f * (g * h)$;
   (c) the arithmetical function E defined by $E(1) = 1$ and $E(n) = 0$ for $n > 1$ is such that $f * E = E * f = f$ for each arithmetical function f;
   (d) for each arithmetical function f such that $f(1) \neq 0$, there exists a function f^{-1} called the *inverse function* of f (with respect to the Dirichlet product $*$) such that $f^{-1} * f = f * f^{-1} = E$ and that f^{-1} is given by the recurrence formula

$$f^{-1}(1) = \frac{1}{f(1)}, \quad f^{-1}(n) = -\frac{1}{f(1)} \sum_{\substack{d|n \\ d<n}} f\left(\frac{n}{d}\right) f^{-1}(d) \quad \text{for } n > 1.$$

(582) Let f and g be two arithmetical functions. Show that if g and $f * g$ are multiplicative, then f is also multiplicative.

(583) Show that if f is a multiplicative function such that $f(1) \neq 0$, then its inverse f^{-1}, with respect to the Dirichlet product $*$, is also multiplicative.

(584) Let r be a real number and let ι_r be the arithmetical function defined by $\iota_r(n) = n^r$ for each positive integer n. Show that
   (a) $\mu * \iota_0 = E$, where μ is the Moebius function.
   (b) $\sigma_r = \iota_r * \iota_0$, where $\sigma_r(n) = \sum_{d|n} d^r$.
   (c) $\phi = \iota_1 * \mu$, where ϕ is Euler's function.
   (d) $\mu * \sigma = \iota_1$, where $\sigma = \sigma_1$.
   (e) $\phi * \sigma_r = \iota_1 * \iota_r$.
   (f) $\iota_1 * \iota_1 = \iota_1 \tau$, where $\tau = \sigma_0$.
   (g) $f = \iota_0 * \iota_0 * \iota_0$, where $f(n) = \sum_{d|n} \tau(d)$.

(585) Show that $\iota_0^{-1} = \mu$ and more generally that $\iota_r^{-1} = \mu \iota_r$ for each real number r.

(586) For each of the arithmetical functions f given below, determine its inverse f^{-1} (with respect to the Dirichlet product $*$):

(i) $f(n) = \iota_0(n);$ (ii) $f(n) = E(n);$ (iii) $f(n) = |\mu(n)|.$

(587) Let f and g be two arithmetical functions such that $f(1) \neq 0$ and $g(1) \neq 0$. Show that
$$(f * g)^{-1} = f^{-1} * g^{-1}.$$

(588) Let ϕ^{-1} be the inverse (with respect to the Dirichlet product $*$) of the Euler function ϕ. Show that
$$\phi^{-1}(n) = \prod_{p|n}(1 - p)   (n = 1, 2, \ldots).$$

(589) Let f be a multiplicative function. Show that f is totally multiplicative if and only if $f^{-1}(n) = \mu(n)f(n)$ for each integer $n \geq 1$.

(590) Show that the inverse, with respect to the Dirichlet product $*$, of the Liouville function λ is
$$\lambda^{-1}(n) = \lambda(n)\mu(n) = \begin{cases} 1 & \text{if } n \text{ is squarefree,} \\ 0 & \text{otherwise.} \end{cases}$$

(591) Show that the inverse, with respect to the Dirichlet product $*$, of σ_a is given by
$$\sigma_a^{-1}(n) = \sum_{d|n} d^a \mu(d)\mu(n/d)   (n = 1, 2, \ldots).$$

(592) Let m and n be positive integers. Show that
$$\sigma(n)\sigma(m) = \sum_{d|(m,n)} d\sigma(mn/d^2).$$

(593) Let f be a multiplicative function. Show that the following three statements are equivalent:

(a) There exists a multiplicative function F such that for all positive m and n,

(1) $$f(mn) = \sum_{d|(m,n)} f(m/d)f(n/d)F(d).$$

(b) There exists a totally multiplicative function g such that for all integers m and n,

(2) $$f(m)f(n) = \sum_{d|(m,n)} f(mn/d^2)\,g(d).$$

(c) For each prime number p and each integer $a \geq 1$,

(3) $$f(p^{a+1}) = f(p)f(p^a) + f(p^{a-1})\Big(f(p^2) - f^2(p)\Big).$$

(594) Show that, for each positive integer n,
$$\sum_{d|n} \sigma(d) = n \sum_{d|n} \frac{\tau(d)}{d}.$$

(595) Show that, for each positive integer n,
$$\sum_{d|n} \phi(d)\tau\left(\frac{n}{d}\right) = \sigma(n).$$

(596) Show that, for each positive integer n,
$$\sum_{d|n} \phi(d)\sigma\left(\frac{n}{d}\right) = n\tau(n).$$

(597) Show that, for each positive integer n,
$$\sum_{d|n} \frac{n}{d}\sigma(d) = \sum_{d|n} d\tau(d).$$

(598) Show that, for each positive integer n,
$$\sum_{d|n} d\sigma(d) = \sum_{d|n} \left(\frac{n}{d}\right)^2 \sigma(d).$$

(599) Let k and r be real numbers. Show that
$$\sum_{d|n} d^r \sigma_{k-r}(d) = \sum_{d|n} \left(\frac{n}{d}\right)^k \sigma_r(d).$$

(600) Show that, for each positive integer n,
$$\sum_{d|n} \sigma(d)\sigma\left(\frac{n}{d}\right) = \sum_{d|n} d\,\tau(d)\tau\left(\frac{n}{d}\right).$$

(601) Let r be a real number. Show that
$$\sum_{d|n} \sigma_r(d)\sigma_r\left(\frac{n}{d}\right) = \sum_{d|n} d^r\tau(d)\tau\left(\frac{n}{d}\right).$$

(602) Show that, for each positive integer n,
$$\sum_{d|n} \mu(d)\tau(n/d) = 1.$$

(603) Show that $\Lambda = \mu * \log$, where Λ is the von Mangoldt function.

(604) Show that $\sum_{d|n} \mu^2(d)\Lambda(d) = \log\gamma(n)$ for each positive integer n, where $\gamma(n) = \prod_{p|n} p$ and $\gamma(1) = 1$ and Λ stands for the von Mangoldt function.

(605) Let f be a totally arithmetic function which only takes the values $+1$ and -1. Let $I = [N, N+M]$, where $M \geq 3\sqrt{N}$. Assume that there exists an integer $n_0 < \sqrt{N}$ such that $f(n_0) = -1$. Prove that this function f cannot be constant on the interval I.

(606) Show that the Liouville function λ does necessarily take the two values $+1$ and -1 on any interval of the form $[N, N+3\sqrt{N}]$, $N \geq 2$.

(607) Given an arbitrary real number $x \geq 1$, show that
$$\sum_{1\leq d\leq x} \mu(d)\left[\frac{x}{d}\right] = 1.$$

(608) Given an arbitrary real number $x \geq 1$, show that

$$\left| \sum_{1 \leq n \leq x} \frac{\mu(n)}{n} \right| \leq 1.$$

(609) Let $\delta(n)$ be the largest odd divisor of the positive integer n. Show that for each integer $m \geq 1$,

$$\left| \sum_{n=1}^{m} \frac{\delta(n)}{n} - \frac{2m}{3} \right| < 1.$$

(610) It is clear that any positive integer n can be written uniquely in the form $n = mr^2$ where m is squarefree. In light of this, justify the chain of equations

$$\mu^2(n) = \mu^2(mr^2) = E(r) = \sum_{d|r} \mu(d) = \sum_{d^2|n} \mu(d).$$

(611) Let f be a strongly multiplicative function such that $0 \leq f(p) \leq 1$ for each prime number p and such that $f(2) = f(3) = 0$. Show that, for each positive integer N,

$$\sum_{n \leq N} f(n) \leq \frac{N}{3} + 2.$$

(612) Is it possible to construct a multiplicative function f such that $f(2) = 0$ and such that, as $N \to \infty$,

$$\sum_{n \leq N} f(n) \sim \frac{3N}{4} \ ?$$

(613) Establish that the number $A(N)$ of squarefree integers $\leq N$ satisfies the relation

$$A(N) = \sum_{d \leq \sqrt{N}} \mu(d) \left[\frac{N}{d^2} \right].$$

(614) Let $\phi(n)$ be Euler's function and let $\tau(n)$ be the function which counts the number of positive divisors of n. Show that

$$\liminf_{n \to \infty} \frac{\phi(n)}{n\tau(n)} = 0 \quad \text{and} \quad \limsup_{n \to \infty} \frac{\phi(n)}{n\tau(n)} = \frac{1}{2}.$$

(615) Consider the sequence $u_n = 2^n$, $n \geq 1$. For each positive integer n, choose the smallest prime number q_n satisfying $u_n \leq q_n < u_{n+1}$; according to Bertrand's Postulate, such a prime number exists. What can be said about the convergence or the divergence of the series $\sum_{n=1}^{\infty} \frac{1}{q_n}$? Explain.

(616) Prove that for each positive integer n, $\tau(2^n + 1) > \tau_1(n)$, where $\tau_1(n)$ stands for the number of odd divisors of n.

(617) Let $n > 1$ be a composite number. Show that $\sigma(n) > n + \sqrt{n}$. Use this to prove that $\lim_{n \to \infty} (\sigma(p_n + 1) - \sigma(p_n)) = +\infty$.

(618) Show that, for each positive integer n, we have

$$\frac{\sigma(n)}{n} < \begin{cases} \left(\frac{3}{2}\right)^{\omega(n)} & \text{if } n \text{ is odd,} \\ 2\left(\frac{3}{2}\right)^{\omega(n)-1} & \text{if } n \text{ is even.} \end{cases}$$

(619) Show that $\sigma(n) < n\tau(n)$ for each integer $n \geq 2$.

(620) Find infinitely many integers n such that $\sigma(n) \leq \sigma(n-1)$.

(621) Show that for each integer $n \geq 1$, $n \leq \sigma(n) \leq n^2$.

(622) Let $\sigma_p(n)$ stand for the sum of even divisors of the integer $n \geq 1$. Show that

$$\sigma_p(n) \geq \alpha\tau(m)\sqrt{2n},$$

where α and m are defined implicitly by $n = 2^\alpha m$, m odd.

(623) Prove that for each integer $n \geq 2$, $\sigma(n) \geq \phi(n) + \tau(n)$, with equality if and only if n is prime.

(624) Let f and g be two multiplicative functions taking only positive values. Show that for each integer $n \geq 2$,

$$\sum_{d|n} f(d)g(n/d) \geq f(n) + g(n),$$

with equality if and only if n is prime.

(625) Show that for each integer $n \geq 2$, we have $\sigma(n) + \phi(n) \leq n\tau(n)$, with equality if and only if n is prime.

(626) Let f, g and h be three multiplicative functions. If for each integer $n \geq 1$, $f(n) + g(n) \geq 0$ and $h(n) \geq 0$, show that,

$$\sum_{d|n} f(d)h(n/d) + \sum_{d|n} g(d)h(n/d) \geq 2h(n) \quad (n = 1, 2, \ldots),$$

with equality if and only if $n = 1$ or else for each $d|n$, $d > 1$, $f(d)+g(d) = 0$.

(627) Show that for each integer $n \geq 1$, $\sigma(n) + \phi(n) \geq 2n$, with equality if and only if $n = 1$ or n is prime.

(628) Prove that for each integer $n \geq 2$, we have $\sigma(n) > n + (\omega(n) - 1)\sqrt{n}$.

(629) Show that for each integer $n > 2$, we have

$$\phi(n^2) + \phi((n+1)^2) < 2n^2.$$

More generally, show that for all integers $n > 2$ and $k \geq 2$,

$$\phi(n^k) + \phi((n+1)^k) < 2n^2(n+1)^{k-2}.$$

(630) Show that

$$\limsup_{n\to\infty} \frac{\phi(n+1)}{\phi(n)} = \infty \quad \text{and} \quad \liminf_{n\to\infty} \frac{\phi(n+1)}{\phi(n)} = 0.$$

(631) Can one find arbitrarily large integers N such that $\phi(n) \geq \phi(N)$ for each integer $n \geq N$, while $\phi(n) \leq \phi(N)$ for each integer $n \leq N$?

(632) Show that

$$\phi(n)\tau^2(n) \leq n^2$$

for all the positive integers $n \neq 4$. For which values of n does equality hold?

(633) Show that, for each integer $n \geq 2$, $\phi(n) \leq n - n^{1-\frac{1}{\omega(n)}}$, with equality if and only if n is prime.

(634) If n is a composite integer, show that $\phi(n) \leq n - \sqrt{n}$.

(635) Show that if $\omega(n) = r$ for positive integers n and r, then

$$\phi(n) \geq \frac{n}{2^r}.$$

(636) For each $n \in \mathbb{N}$, let $\sigma(n) = \sum_{d|n} d$ and $\sigma_2(n) = \sum_{d|n} d^2$. Show that

$$\frac{\sigma^2(n)}{\tau(n)} \leq \sigma_2(n) \leq \sigma^2(n) \qquad (n = 1, 2, \ldots).$$

(637) Show that the mean value of the divisors of the positive integer n is larger or equal to $\prod_{d|n} d^{1/\tau(n)}$.

(638) Show that

$$\frac{\sigma(n)}{\tau(n)} \geq \sqrt{n} \qquad (n = 1, 2, \ldots).$$

(639) Show that $\prod_{d|n} d = n^3$ if and only if $n = p^5$ or $n = p^2 q$, with p and q distinct prime numbers.

(640) For each integer $n \geq 1$, show that $\tau(n) \leq 2\sqrt{n}$.

(641) Prove that for each integer $n \geq 3$, we have $\sigma(n) < n\sqrt{n}$.

(642) Prove that for each integer $n \geq 2$,

$$\sum_{\substack{d|n \\ d \geq 2}} \frac{d-1}{\log d} > \frac{2\tau(n)(\sqrt{n}-1)}{\log n} - 1.$$

(643) Prove that for each integer $n \geq 2$,

$$\sum_{\substack{d|n \\ d \geq 2}} \frac{d^2-1}{\log d} > \frac{2\tau(n)(n-1)}{\log n} - 2.$$

(644) Prove that for each integer $n \geq 2$,

$$\prod_{p|n} \left(1 - \frac{1}{p}\right) \leq 1 - \frac{2^{\omega(n)} - 1}{n},$$

with equality if and only if n is prime.

(645) Prove that for each integer $n \geq 2$,

$$\sum_{p|n} \frac{1}{p} \geq \frac{\omega(n)}{n^{1/\omega(n)}}.$$

(646) Let h be the arithmetical function defined by $h(1) = 0$ and $h(p^a) = 0$ if p is prime and a a positive integer, and otherwise, that is if $n = q_1^{a_1} \cdots q_r^{a_r}$, with $r \geq 2$, q_i prime, by

$$h(n) = \sum_{i=2}^{\omega(n)} \frac{1}{q_i - q_{i-1}}.$$

For each positive integer n such that $\omega(n) \geq 2$, show that

$$h(n) \geq \frac{(\omega(n) - 1)^2}{P(n) - p(n)},$$

where $p(n)$ and $P(n)$ stand respectively for the smallest and largest prime factors of n.

(647) Let H be the arithmetical function defined by $H(1) = 0$, and for $n > 1$ by

$$H(n) = \sum_{i=2}^{\tau(n)} \frac{1}{d_i - d_{i-1}},$$

where $1 = d_1 < d_2 < \ldots < d_{\tau(n)} = n$ represent the divisors of n. Show that, for each positive integer n,

$$H(n) \geq \frac{(\tau(n) - 1)^2}{n - 1}.$$

(648) Show that $\tau(2^n - 1) \geq \tau(n)$ for each integer $n \geq 1$.

(649) Let m and n be positive integers; show that $\phi(mn) \leq m\phi(n)$. On the other hand, if each prime number dividing m also divides n, then show that $\phi(mn) = m\phi(n)$.

(650) For each integer $n \geq 1$, show that

$$\phi\left(n\left[\frac{\sigma(n)\tau(n)}{n^{3/2}}\right]\right) \leq 2n,$$

where $[y]$ stands for the largest integer $\leq y$.

(651) Show that $\phi(n)\tau(n) \geq n$ for each positive integer n.

(652) Find all the solutions of the equation $\phi(n)\tau(n) = n$, where $n \in \mathbb{N}$.

(653) Let m and n be integers larger than 2; show that $\phi(mn) + \phi((m+1)(n+1)) < 2mn$.

(654) Consider the arithmetical function $\Psi(n)$ defined by

$$\Psi(n) = n \prod_{p|n} \left(1 + \frac{1}{p}\right) \qquad (n = 1, 2 \ldots).$$

It is clear that the function Ψ is multiplicative.

  (a) Show that $\Psi(n) \leq \sigma(n)$, where $\sigma(n)$ represents the sum of the divisors of n.

  (b) Show that $\Psi(n) = \sigma(n)$ if and only if n is squarefree.

  (c) We say that a natural number n is Ψ-perfect if $\Psi(n) = 2n$. Prove that a number n is Ψ-perfect if and only if it is of the form $2^a \cdot 3^b$, where a and b are positive integers.

(655) Let f be a polynomial with integer coefficients and let

$$\phi^*(n) = \#\{k \mid 1 \leq k \leq n, (f(k), n) = 1\}.$$

Observe that in the case $f(n) = n$, we find that $\phi^*(n) = \phi(n)$, that is Euler's function.

  (a) Show that ϕ^* is a multiplicative function.

  (b) Show that, for each positive integer n,

$$\phi^*(n) = n \prod_{p|n} \left(1 - \frac{b_p}{p}\right),$$

where $b_p = p - \phi^*(p) = \#\{k \mid 1 \leq k \leq p, \ p|f(k)\}$.

(656) Let n be a positive integer. Find the number of terms of the sequence
$$1 \cdot 2, \ 2 \cdot 3, \ 3 \cdot 4, \ldots, n(n+1)$$
which are relatively prime with n.

(657) Let n be a positive integer. Find a formula which gives the number of positive integers $k \le n$ such that $(k, n) = (k+1, n) = 1$.

(658) Let n be a positive integer. Find an expression for the number of terms of the sequence
$$1 \cdot 2 \cdot 3, \ 2 \cdot 3 \cdot 4, \ldots, n(n+1)(n+2)$$
which are relatively prime with n.

(659) Let $f(n)$ be the n-th positive integer which is not a perfect square. Hence $f(1) = 2$, $f(2) = 3$, $f(3) = 5$ and $f(4) = 6$. Show that, for each integer $n \ge 1$, $f(n) = n + \|\sqrt{n}\|$, where $\|x\|$ stands for the closest integer to x.

(660) Show that
$$\sum_{d^k | n} \mu(d) = \begin{cases} 1 & \text{if } n = 1, \\ 1 & \text{if } p^\alpha | n \Rightarrow \alpha < k, \\ 0 & \text{otherwise,} \end{cases}$$
thus generalizing the result of Problem 610.

(661) Prove that the Liouville function λ takes infinitely many times each of the values $+1, -1$ when applied to the sequence of integers $2, 5, 10, 17, \ldots, n^2 + 1, \ldots$.

(662) Let n be a positive integer. If $\{a_1, a_2, \ldots, a_k\}$ is the set of positive integers $i \le n$ with $(i, n) = 1$, show that
$$\sum_{i=1}^{k} \frac{a_i}{n - a_i} \ge \phi(n) \qquad (n = 1, 2, \ldots).$$

(663) Let $f \colon \mathbb{N} \to \mathbb{R}^+$ and let $F(n) = \sum_{d|n} f(d)$. Show that
$$\prod_{d|n} f(d) \le \left(\frac{F(n)}{\tau(n)} \right)^{\tau(n)} \qquad (n = 1, 2, \ldots).$$
Use this result to show that
$$\prod_{d|n} \phi(d) \le \left(\frac{n}{\tau(n)} \right)^{\tau(n)} \qquad (n = 1, 2, \ldots).$$

(664) Show that there exists a positive constant C such that, for each integer $n \ge 2$,
$$C < \frac{\sigma(n)\phi(n)}{n^2} < 1.$$

(665) Define the *derivative* f' of an arithmetical function f by
$$f'(n) = f(n) \log n \qquad (n = 1, 2 \ldots).$$
Show that, given any arithmetical functions f and g, we have
(a) $(f + g)' = f' + g'$,
(b) $(f * g)' = f' * g + f * g'$,
(c) $(f^{-1})' = -f' * (f * f)^{-1}$ provided that $f(1) \ne 0$.

(666) Given an arithmetical function f, define

$$(*) \qquad \overline{f}(n) = \frac{1}{\tau(n)} \sum_{d|n} f(d) \qquad (n = 1, 2, \ldots),$$

where $\tau(n)$ stands for the number of divisors of n. Show that if f is multiplicative, then $\overline{f}$ is also multiplicative.

(667) Let $\overline{f}$ be the function introduced in Problem 666. Show that if f is additive, then $\overline{f}$ is also additive.

(668) Let $1(n) = 1$ for each positive integer n and let μ be the Moebius function. What represents the functions $\overline{1}$ and $\overline{\mu}$, where $\overline{f}$ is defined by the relation $(*)$ of Problem 666?

(669) Let $\omega(n) = \sum_{p|n} 1$. Determine the values of the function $\overline{\omega}$, where $\overline{f}$ is defined by the relation $(*)$ of Problem 666.

(670) Let $f(n) = 2^{\omega(n)}$, where $\omega(n) = \sum_{p|n} 1$. Prove that

$$\overline{f}(n) = \frac{\tau(n^2)}{\tau(n)} \qquad (n = 1, 2, \ldots),$$

where $\overline{f}$ is defined by the relation $(*)$ of Problem 666.

(671) Let $f(n) = 2^{\Omega(n)}$, where $\Omega(n) = \sum_{p^\alpha \| n} \alpha$. Show that

$$\overline{f}(n) = \prod_{p^\alpha \| n} \frac{2^{\alpha+1} - 1}{\alpha + 1} \qquad (n = 1, 2, \ldots),$$

where $\overline{f}$ is defined by the relation $(*)$ of Problem 666.

(672) Let λ be the Liouville function. Show that $\overline{\lambda}(n)\tau(n) = \chi(n)$, where $\chi(n)$ is the characteristic function of the set of perfect squares, that is

$$\chi(n) = \begin{cases} 1 & \text{if } n = m^2, \\ 0 & \text{otherwise,} \end{cases}$$

and where $\overline{f}$ is defined by the relation $(*)$ of Problem 666.

(673) Given a multiplicative function g, show that there exists a multiplicative function f such that $g = \overline{f}$, where $\overline{f}$ is defined by the relation $(*)$ of Problem 666.

(674) Let $g(n) = 2^{\omega(n)}$. Find the function f such that $g = \overline{f}$, where $\overline{f}$ is defined by the relation $(*)$ of Problem 666.

(675) Given an arithmetical function f, define

$$(**) \qquad \widehat{f}(n) = \frac{1}{2^{\omega(n)}} \sum_{d|n} \mu^2(d) f(d) \qquad (n = 1, 2, \ldots),$$

where $\omega(n) = \sum_{p|n} 1$ and μ stands for the Moebius function. Show that if f is multiplicative, then $\widehat{f}$ is multiplicative.

(676) Let $1(n) = 1$ for each positive integer n and λ stand for the Liouville function. Determine the functions $\widehat{1}$ and $\widehat{\lambda}$, where $\widehat{f}$ is defined by the relation $(**)$ of Problem 675.

(677) We know that $\tau(n)$ represents the number of ways of writing a positive integer n as a product of two positive integers, taking into account the

order of the factors. In other words,

$$\tau(n) = \sum_{d_1 d_2 = n} 1.$$

More generally, given an integer $k \geq 2$, let $\tau_k(n)$ be the number of ways of writing a positive integer n as a product of k positive integers, taking into account the order of the factors. In other terms,

$$\tau_k(n) = \sum_{d_1 d_2 \cdots d_k = n} 1.$$

Show that

$$\tau_k = \underbrace{1 * 1 * \ldots * 1}_{k}.$$

(678) Show that if $F(k) = \sum_{d|k} f(d)$ for $k = 1, 2, \ldots$, then $\sum_{k=1}^{n} F(k) = \sum_{k=1}^{n} \left[\frac{n}{k}\right] f(k)$ for each positive integer n.

(679) Show that

$$\sum_{k=1}^{2n} \tau(k) - \sum_{k=1}^{n} \left[\frac{2n}{k}\right] = n \qquad (n = 1, 2, \ldots).$$

(680) Show that $\sum_{d=1}^{n} \phi(d) \left[\frac{n}{d}\right] = \frac{1}{2}n(n+1)$ for each positive integer n.

(681) Show that $\sum_{k=1}^{n} \Lambda(k) \left[\frac{n}{k}\right] = \log n!$ for each positive integer n.

(682) Show that $\sum_{k=1}^{n} \lambda(k) \left[\frac{n}{k}\right] = [\sqrt{n}]$ for each positive integer n.

(683) Given an integer $n \geq 2$ and p a prime divisor of n, let p^{a_p} be the largest power of p not exceeding n, meaning that a_p is the only positive integer satisfying $p^{a_p} \leq n < p^{a_p+1}$. Finally, let

$$S(n) = \sum_{p|n} p^{a_p} \qquad (n = 2, 3, \ldots).$$

Show that there exist infinitely many integers n such that $S(n) > n$.

(684) Let n be a positive integer. Show that

$$\sum_{k=1}^{n} \tau(k) = \sum_{k=1}^{n} \left[\frac{n}{k}\right].$$

(685) Let n be a positive integer. Show that $\sum_{k=1}^{n} \sigma(k) = \sum_{k=1}^{n} k \left[\frac{n}{k}\right]$.

(686) Let n be a positive integer. Show that

$$\sum_{k=1}^{n} \phi(k) = \frac{1}{2} \sum_{k=1}^{n} \mu(k) \left[\frac{n}{k}\right] \left(\left[\frac{n}{k}\right] + 1\right).$$

(687) Let n be a positive integer. Show that

$$\sum_{k=1}^{n} k \sum_{d|k} \lambda(d) = \frac{[\sqrt{n}]([\sqrt{n}]+1)(2[\sqrt{n}]+1)}{6}.$$

(688) For each positive integer n, let $S(n)$ be the set of all positive integers k such that the fractional part of n/k is $\geq 1/2$. Let f be an arbitrary arithmetical function and let

$$g(n) = \sum_{k=1}^{n} f(k) \left[\frac{n}{k}\right] \qquad (n=1,2,\ldots).$$

Show that

$$\sum_{k \in S(n)} f(k) = g(2n) - 2g(n) \qquad (n=1,2,\ldots).$$

Use this result to show that for each integer $n \geq 1$,

$$\sum_{k \in S(n)} \phi(k) = n^2, \qquad \sum_{k \in S(n)} \mu(k) = -1,$$

$$\sum_{k \in S(n)} \Lambda(k) = \log\left(\binom{2n}{n}\right), \qquad \sum_{k \in S(n)} \lambda(k) = [\sqrt{2n}] - 2[\sqrt{n}],$$

where ϕ, μ, Λ and λ are respectively the Euler function, the Moebius function, the von Mangoldt function and the Liouville function.

(689) Let x be a real number such that $|x| < 1$. Show that

$$\sum_{n=1}^{\infty} \frac{\phi(n) \, x^n}{1 - x^n} = \frac{x}{(1-x)^2} \qquad (n=1,2,\ldots),$$

where ϕ is the Euler function.

(690) Let f and g be two arithmetical functions tied by the relation $f(n) = \sum_{d|n} g(d)$, $n \geq 1$, and x a real number such that $|x| < 1$. Show that

$$\sum_{n=1}^{\infty} g(n) \frac{x^n}{1 - x^n} = \sum_{n=1}^{\infty} f(n) \, x^n \qquad (n=1,2,\ldots).$$

(691) Let n be a positive integer. Consider the square matrix $M_{n \times n} = (b_{ij})_{n \times n}$, where the element $b_{ij} = (i,j)$, that is the GCD of i and j. Use the fact that $\sum_{d|n} \mu(d)\frac{n}{d} = \phi(n)$ in order to prove that

$$\det M = \phi(1)\phi(2)\cdots\phi(n),$$

where ϕ stands for Euler's function.

(692) Let n be a positive integer and let $M = (a_{ij})_{n \times n}$ be the matrix whose a_{ij} element is defined by $a_{ij} = \tau((i,j))$, where $\tau(m)$ represents the number of divisors of m. Show that $\det M = 1$.

(693) Let n be a positive integer and let $M = (a_{ij})_{n \times n}$ be the matrix whose a_{ij} element is defined by $a_{ij} = \sigma((i,j))$, where $\sigma(m)$ represents the sum of the divisors of m. Show that $\det M = n!$.

(694) Let n be a positive integer and let $M = (a_{ij})_{n \times n}$ be the matrix whose a_{ij} element is defined by $a_{ij} = \mu((i,j))$, where μ stands for the Moebius function. Show that $\det M \neq 0$ for $1 \leq n \leq 7$ and then that $\det M = 0$ for $n \geq 8$.

(695) Let n be a positive integer and let $M = (a_{ij})_{n \times n}$ be the matrix whose a_{ij} element is defined by $a_{ij} = [i,j]$, that is the LCM of i and j. Show that

$$\det M = \prod_{k=1}^{n} (-1)^{\omega(k)} \phi(k) \gamma(k),$$

where $\omega(k) = \sum_{p|k} 1$, $\gamma(k) = \prod_{p|k} p$ and ϕ stands for Euler's function.

(696) Let k be a positive integer and let f be an arithmetical function. Show that if

$$g(x) = \sum_{\substack{n \leq x \\ (n,k)=1}} f(x/n),$$

then

$$f(x) = \sum_{\substack{n \leq x \\ (n,k)=1}} \mu(n) g(x/n).$$

(697) Let f be an arithmetical function. Show that

$$\sum_{\substack{n \leq N \\ (n,k)=1}} f(n) = \sum_{d|k} \sum_{m \leq N/d} \mu(d) f(md).$$

(698) Let $M(x) := \sum_{n \leq x} \mu(n)$. Show that

$$\sum_{n \leq x} M(x/n) = 1.$$

(699) Let $p(n)$ be the smallest prime factor of n, $p(1) = 1$. Show that

$$\sum_{n \leq x} p(n(n+1)) = 2[x].$$

(700) Recently, when Canada celebrated its 125th anniversary, mathematicians at the University of Manitoba introduced the notion of "Canada perfect number". A composite integer n is called a *Canada perfect number* if the sum of the square of its digits is equal to the sum of its proper divisors > 1. In other words, n is "Canada perfect" if and only if

(∗) $$\sum_{1 \leq i \leq c(n)} \ell_i^2 = \sum_{\substack{d|n \\ 1<d<n}} d,$$

where $\ell_1, \ell_2, \ldots, \ell_{c(n)}$ are the digits appearing in the decimal representation of n and where $c(n)$ is the number of digits of n. One easily checks that 125 is "Canada perfect", since

$$1^2 + 2^2 + 5^2 = 30 = 5 + 25.$$

Show that the only Canada perfect numbers are 125, 581, 8549 and 16999:
  (a) by using a computer to identify all Canada perfect numbers $\leq 10^6$,
  (b) by proving that no Canada perfect number larger that 10^6 exists.

9. Solving Equations Involving Arithmetical Functions

(701) Using a computer,
    (a) find all values of $n < 10000$ for which $4\tau(n+2) = \phi(n)$.
    (b) write a program that gives the positive integers $1 \le n \le 2000$, for which $\sigma(n) = 2n - 1$.

(702) Without using a computer, find at least six solutions of $\phi(\sigma(n)) = n$.

(703) Show how one can obtain from each solution of $\phi(\sigma(n)) = n$ a corresponding solution of the equation $\sigma(\phi(n)) = n$, and then use this argument to find six solutions of $\sigma(\phi(n)) = n$ with the help of Problem 702.

(704) Show that if p and $(3^p - 1)/2$ are two prime numbers, then $n = 3^{p-1}$ is a solution of $\sigma(\phi(n)) = \phi(\sigma(n))$. Use a computer to obtain explicitly three of these solutions. Are there any other solutions besides those of this particular type?

(705) Show that the equation $\phi(\tau(n)) = \tau(\phi(n))$ has infinitely many solutions.

(706) Find all the solutions of $\tau(\gamma(n)) = \gamma(\tau(n))$, where $\gamma(n) := \prod_{p|n} p$.

(707) Consider the arithmetical function δ defined by $\delta(1) = 1$ and, for $n \ge 2$, by $\delta(n) = \prod_{p\|n} p$. In particular, if n is squarefree, we have $\delta(n) = n$. Use a computer to obtain the three smallest nonsquarefree solutions n of

$$(*) \qquad\qquad \delta(n+1) - \delta(n) = 1,$$

and then prove that the equation $(*)$ has infinitely many nonsquarefree solutions.

(708) Equation $\gamma(\sigma(n)) = n$ has only two solutions. What are they?

(709) Let k be a positive integer which is not a multiple of 8. Show that the only possible values of the smallest positive integer n which divides $\sigma_k(n)$ are 6, 10 and 34.

(710) Show that all the solutions of the equation $\dfrac{\phi(n)}{n} = \dfrac{2}{3}$ are of the form $n = 3^k$, $k = 1, 2, \ldots$.

(711) Prove that a positive integer n is a solution of the equation

$$\frac{\phi(n)}{n} = \frac{4}{7} \Longleftrightarrow n = 3^\alpha 7^\beta \text{ with } \alpha = 1, 2, \ldots, \ \beta = 1, 2, \ldots.$$

(712) Let $S(n) = \sum_{d|n} \tau(d)$. Determine all values of n such that $n = S(n)$.

(713) Find all positive integers n such that
    (a) $\sigma(n) = 24$;   (b) $\sigma(n) = 57$.

(714) What is the smallest positive integer n such that $\sigma(x) = n$ has exactly one solution?

(715) What is the smallest positive integer n such that $\sigma(x) = n$ has exactly two solutions?

(716) What is the smallest positive integer n such that $\sigma(x) = n$ has exactly three solutions?

(717) Let n be a fixed positive integer. Is the number of solutions of the equation $\sigma(x) = n$ finite or infinite? What about the equation $\tau(x) = n$?

(718) Is it true that n is prime if and only if $\sigma(n) = n + 1$?

(719) Let a be a rational number $\geq \frac{35}{16}$ and let n be an odd solution of the equation $\sigma(n) = an$. Show that n has at least four distinct prime factors.

(720) Let a be a rational number $\geq \frac{15}{4}$ and let n be an arbitrary solution of the equation $\sigma(n) = an$. Show that n has at least four distinct prime factors.

(721) Find two integers n for which

$$\frac{\sigma(n)}{n} = \frac{9}{4}.$$

(722) Show that there exist infinitely many positive integers m such that equation $\sigma(n) = m$ has at least three solutions.

(723) Let $\hat{\sigma}(n)$ be the total number of subgroups of the *dihedral* group D_n of the symmetries of the regular polygon with n sides. It is possible to show that $\hat{\sigma}(n) = \tau(n) + \sigma(n)$ (see S. Cavior [**5**]). A number n is said to be *dihedral perfect* if $\hat{\sigma}(n) = 2n$. Characterize all such numbers which are also of the form $n = 2^k p$, where p is prime and k is a positive integer. Use a computer to find the five smallest dihedral perfect numbers of this form.

(724) Find all the solutions x of the equation $\phi(x) = 24$.

(725) Show that if $\phi(x) = 2^r N$, where $(2, N) = 1$, then x has at most r distinct odd prime factors.

(726) Find all positive integers n such that $4 \nmid \phi(n)$.

(727) Show that if $m = 2 \cdot 3^{6k+1}$ with $k \geq 1$, then

$$\phi(n) = m \iff n = 3^{6k+2} \text{ or } n = 2 \cdot 3^{6k+2}.$$

Use this to show that there exist infinitely many positive integers m such that $\#\{n : \phi(n) = m\} = 2$.

(728) Show that there does not exist any positive integer n such that $\phi(n) = 2 \cdot 7^m$, where $m \geq 1$.

(729) Let $n \geq 2$. Show that $\phi(n) = n - 1$ if and only if n is prime.

(730) Let p be a prime number such that $2p + 1$ is composite. Show that $\phi(x) = 2p$ has no solutions.

(731) Show that $\phi(n) = n/2$ if and only if $n = 2^k$, for a certain integer $k \geq 1$.

(732) Show that $\phi(n) = 2n/5$ if and only if $n = 2^r 5^s$, $r, s \in \mathbb{N}$.

(733) Show that there exist infinitely many positive integers n such that $\phi(n) = n/3$.

(734) Are there any positive integers n such that $\phi(n) = n/4$?

(735) Let $n, a \in \mathbb{N}$ and let p be a prime number. Show that $\phi(p^a) = 2(6n + 1)$ if and only if $p > 6$, $p \equiv 11 \pmod{12}$ and a is even.

(736) Let $n \in \mathbb{N}$.
   (a) Show that $\frac{1}{2}\sqrt{n} \leq \phi(n) \leq n$.
   (b) Show that the equation $\phi(x) = n$ has only a finite number of integer solutions x.

(737) Find the smallest positive integer n such that $\phi(x) = n$ has no solutions. Find the smallest positive integer n such that $\phi(x) = n$ has exactly one solution, and finally find the smallest positive integer n such that $\phi(x) = n$ has exactly two solutions.

(738) Show that if a certain arithmetical function f satisfies

$$\frac{1}{\tau(n)} \sum_{d|n} f(d) = f(n)$$

for each positive integer n, then necessarily there exists a constant c such that $f(n) = c$ for each $n \in \mathbb{N}$.

(739) Show that if a certain multiplicative function f satisfies

$$\frac{1}{\tau(n)} \sum_{d|n} f(d) = f(n)$$

for each positive integer n, then necessarily $f(n) = 1$ for each $n \geq 1$.

(740) Show that the equation $(*)$ $\Omega(n)^{\Omega(n)} = n$ has infinitely many solutions.

(741) Consider the equation $(*)$ $\sum_{d|n} \gamma(d) = n$, where $\gamma(1) = 1$ and, for $n \geq 2$, $\gamma(n) = \prod_{p|n} p$. Show that the only solution $n > 1$ of $(*)$ is $n = 56$.

(742) Show that the equation $\sigma(n) - \phi(n) = (-1)^n \tau(n)$ has only one solution.

(743) Find all pairs of positive integers m and n such that

$$\phi(mn) = \phi(m) + \phi(n).$$

(744) Show that the only solutions of $\phi(n) = \gamma(n)$ are $n = 1, 4, 18$.

(745) Show that the only solutions of $\phi(n) = \gamma(n)^2$ are $n = 1, 8, 108, 250, 6174$ and 41154.

(746) Consider the equation $(*)$ $\sigma(n) = \gamma(n)^2$. Show that each solution $n > 1$ of $(*)$ must satisfy the following properties:
  (a) n is an even number.
  (b) n cannot be squarefree.
  Then, use a computer to find the only solution $n > 1$ of $(*)$ which is smaller than 10^8.

(747) Let k be an arbitrary positive integer. Prove that there exists infinitely many positive integers n such that $\gamma(n)^k$ divides $\sigma(n)$.

(748) Show that the equation $\phi(n) + \gamma(n) = \sigma(n)$ has only one solution.

(749) Show that $\dfrac{\phi(n) + \sigma(n)}{\gamma(n)^2}$ is an integer for infinitely many positive integers n.

(750) Find all positive integers n such that $2^{\phi(n)} \leq 2n$.

10. Special Numbers

(751) Squaring 12 gives 144. By reversing the digits of 144, we notice that 441 is also a perfect square. Using computer software, write a program to find all those integers n, $1 \le n \le N$, verifying this property.

(752) A positive integer which is divisible by the sum of its digits is called a *Niven number*. For example, 81 is a Niven number since it divisible by $8+1 = 9$; but 71 is not a Niven number since it is not divisible by $7+1 = 8$. Using computer software, write a program which finds all Niven numbers $n \in [12476, 12645]$.

(753) A positive integer is said to be a *palindrome* if by reversing the order of its digits, we obtain the same number, such as is the case with the number 12321. Use a computer to show that 26 is the smallest positive integer which is not a palindrome, but such that its square is a palindrome. Find other integers having this property.

(754) A positive integer N is called a *Cullen number* if it is of the form $n \cdot 2^n + 1$, $n > 1$. Find the Cullen prime numbers smaller than 1000.

(755) Write a program which allows one to find the positive integers $\le N$ which can be written as the sum of two squares. Use this program to determine all the positive integers ≤ 300 with this property.

(756) Carmichael's conjecture states that for each positive integer n, there exists an integer $m \ne n$ such that $\phi(m) = \phi(n)$, where ϕ stands for Euler's function (see Schlafly and Wagon [**36**]). Write a program which verifies this conjecture for a given integer n.

(757) A positive integer N is called a *Silverbach number* if it can be written as the sum of two prime numbers in three different ways. Using computer software, write a program which allows one to write any integer n as the sum of two prime numbers in one way, in two distinct ways, in three distinct ways, and so on.

(758) A prime number p is called a *Wilson prime* if $(p-1)! \equiv -1 \pmod{p^2}$. Using a computer, find the three smallest Wilson primes.

(759) Let $k \ge 1$ be an integer. A positive integer n is said to be k–*hyperperfect* if

$$n = 1 + k \sum_{\substack{d|n \\ 1<d<n}} d.$$

A 1-hyperperfect number is simply a perfect number.
  (a) Show that a positive integer n is k-hyperperfect if and only if $k\sigma(n) = (k+1)n + k - 1$.
  (b) Show that a positive integer n is k-hyperperfect if and only if $\sigma(n) = n + 1 + \frac{n-1}{k}$.
  (c) Show that if n is k-hyperperfect, then $n \equiv 1 \pmod{k}$.
  (d) Show that if n is k-hyperperfect, then the smallest prime factor of n is larger than k.
  (e) Prove that no prime power can be a k-hyperperfect number, for any integer $k \ge 1$.
  (f) Use a computer to find all 2-hyperperfect numbers smaller than 10^6.

(g) Construct an algorithm which allows one to identify all 2-hyperperfect numbers $< 10^9$ of the form $3^\alpha \cdot p$, where α is a positive integer and where $p > 3$ is a prime number.

(760) Show that if we add the digits of an even perfect number larger than 6 and we then add the digits of the number thus obtained, and so on until we obtain a one-digit number, then this digit must be 1.

(761) Show that if $\{t_k\}$ stands for the increasing sequence of triangular numbers, then, for each positive integer n,

$$\sum_{k=1}^{n} t_k = \frac{n(n+1)(n+2)}{6}.$$

(762) Show that the Catalan number $\dfrac{(2n)!}{n!(n+1)!}$ is an integer for each integer $n \geq 0$.

(763) Show that the following method, invented by Thabit ben Korrah (826–901), an Arabic mathematician of the ninth century, for finding amicable numbers does work: if $p = 3 \cdot 2^{k-1} - 1$, $q = 3 \cdot 2^k - 1$ and $r = 9 \cdot 2^{2k-1} - 1$ are primes for a certain positive integer k, then the numbers

$$M = 2^k pq \quad \text{and} \quad N = 2^k r$$

form an amicable pair.

(764) Show that the quotient of two triangular numbers can never be 4.

(765) A positive integer n is said to be *abundant* if $\sigma(n) > 2n$. Use a computer to find the smallest odd abundant number, and then prove that there exist infinitely many abundant numbers.

(766) Let n be a positive integer. Show that $\dfrac{\sigma(n)}{n} \geq \dfrac{\sigma(d)}{d}$ for each divisor d of n. Use this result to show that a positive integer n which is a multiple of 6 is a nondeficient number, that is such that $\sigma(n) \geq 2n$.

(767) Show that there exist infinitely many positive integers n such that $n|(2^n + 1)$.

(768) Show that if n is an integer larger than 1 such that $n|(2^n + 1)$, then n is a multiple of 3.

(769) Prove that a Fermat number $F_m = 2^{2^m} + 1$ cannot be equal to p^k, where p is prime and k is an integer ≥ 2.

(770) Does there exist a prime number p which is a factor of two Mersenne numbers (that is numbers of the form $2^q - 1$, where q is a prime number)?

(771) Use a computer to find the two smallest nondeficient consecutive numbers; that is find the smallest number n such that $\sigma(n-1)/(n-1) \geq 2$ and $\sigma(n)/n \geq 2$. Proceed in the same manner to find the three smallest nondeficient consecutive numbers. Finally, show that given an arbitrary integer $k \geq 2$, there exist k nondeficient consecutive numbers.

(772) Show that there exists a positive integer n such that $\sigma(n) \geq 3n$ and $\sigma(n+1) \geq 3(n+1)$.

(773) Show that each odd tri-perfect number must have at least eight distinct prime factors.

(774) Show that if a and b are two positive integers such that $ab+1$ is a perfect square, then the set

$$A = \{a, b, a + b + 2\sqrt{ab+1}, 4(a + \sqrt{ab+1})(b + \sqrt{ab+1})\sqrt{ab+1}\}$$

is such that if $x, y \in A$, $x \neq y$, then $xy + 1$ is also a perfect square. Then, find two sets A with this property.

(775) Show that for each positive integer n equal to twice a triangular number, the corresponding expression

$$\sqrt{n + \sqrt{n + \sqrt{n + \sqrt{n + \cdots}}}}$$

represents an integer.

(776) Prove Cassiny's identity

$$F_{n-1}F_{n+1} - F_n^2 = (-1)^n \qquad (n = 2, 3, \ldots),$$

where F_n stands for the n-th Fibonacci number.

(777) Show that the set $A = \{F_{2n}, F_{2n+2}, F_{2n+4}, 4F_{2n+1}F_{2n+2}F_{2n+3}\}$, where F_m stands for the m-th Fibonacci number, is such that if $x, y \in A$, $x \neq y$, then $xy + 1$ is a perfect square.

(778) Show that, for each integer $n \geq 1$, the number

$$\frac{n(n + 1)(n + 2)(n + 3)}{8}$$

is a triangular number.

(779) Show that there exist infinitely many prime numbers whose last four digits are 7777. Find five such primes.

(780) Use a computer to find the three smallest integers $n > 1$ which have the property of being divisible by the sum of the squares of their digits as well as by the product of the squares of their digits. Deduct the existence of a fourth one.

(781) We know that $\phi(p) = p - 1$ if p is prime. In 1932, Derrick Henry Lehmer (1905–1991) conjectured that there does not exist any composite number n such that $\phi(n)$ is a proper divisor of $n - 1$. Show that if such a number exists, it must be a Carmichael number.

(782) Let us write the integer $n > 9$ in the form $n = d_1 d_2 \cdots d_r$, where $d_1, d_2, \ldots, d_r$ are the r digits of n. Show that there exist only a finite number of integers n such that

$$n = d_1^1 + d_2^2 + d_3^3 + \cdots + d_r^r$$

and use a computer to find the eight smallest such numbers $n > 9$.

(783) Let us write the integer $n > 9$ in the form $n = d_1 d_2 \cdots d_r$, where $d_1, d_2, \ldots, d_r$ are the r digits of n. Show that there exists no number n such that

$$n = d_1^r + d_2^{r-1} + d_3^{r-2} + \cdots + d_r^1.$$

(784) Show that there exist infinitely many numbers n such that $\sigma(n) = 2n - 1$.

(785) Given a positive integer $n \equiv 2 \pmod{3}$, show that each odd prime divisor of $n^2 + n + 1$ is congruent to 1 modulo 3.

11. Diophantine Equations

(786) For which positive integer(s) x is it true that

$$(*) \qquad x^3 + (x+1)^3 + (x+2)^3 = (x+3)^3 \ ?$$

(787) Show that the Diophantine equation

$$x^3 + 5 = 117y^3$$

has no solutions.

(788) One day, as the English mathematician Godfrey Harold Hardy (1877–1947) was visiting Srinivasa Ramanujan (1885–1920) at the hospital, the patient commented to his visitor that the number on the license plate of the taxi that had brought him, namely 1729, was a very special number: it is the smallest positive integer which can be written as the sum of two cubes in two different ways, namely

$$1729 = 1^3 + 12^3 = 9^3 + 10^3.$$

Using the identity

$$(*) \quad (3a^2 + 5ab - 5b^2)^3 + (4a^2 - 4ab + 6b^2)^3 = (-5a^2 + 5ab + 3b^2)^3$$
$$+ (6a^2 - 4ab + 4b^2)^3$$

due to Ramanujan, show that there exist infinitely many positive integers which can be written as a sum of two cubes in two different ways. Does this identity allow one to find the "double" representation of 1729?

(789) Let $a, b, c \in \mathbb{Z}$. Show that $ax + by = b + c$ is solvable in integers x and y if and only if $ax + by = c$ is also solvable.

(790) Let $a, b, c \in \mathbb{Z}$. Show that $ax + by = c$ is solvable in integers x and y if and only if $(a, b) = (a, b, c)$.

(791) Let a and b be positive integers such that $(a, b) = 1$. Show that $ax + by = n$ has positive integer solutions if $n > ab$, while it has no positive integer solution if $n = ab$.

(792) Find the positive integer solution(s) of the system of equations

$$x + y + z = 100,$$
$$2x + 5y + \frac{z}{10} = 100.$$

(793) The triangle whose sides have lengths 5, 12 and 13 respectively has the property that its perimeter is equal to its area. There exist exactly five such triangles with integer sides. Which are they?

(794) Identify all the integer solutions, if any, to the equation

$$x^2 + 3y = 5.$$

(795) Show that if the positive integers x, y, z are the respective lengths of the sides of a rectangular triangle, then at least one of these three numbers is a multiple of 5.

(796) Let a, b and c be three real nonnegative numbers. Show that the system of equations

$$\begin{aligned} ax + by + cxy &= a + b + c, \\ by + cz + ayz &= a + b + c, \\ cz + ax + bzx &= a + b + c \end{aligned}$$

has one and only one solution in nonnegative integers x, y, z. What is this solution? Why is it so?

(797) Let a and b be positive integers such that $(a, b) = 1$. Show that $ax + by = ab - a - b$ has no solutions in integers $x \geq 0$ and $y \geq 0$.

(798) Let a and b be positive integers such that $(a, b) = 1$. Show that the number of nonnegative solutions of $ax + by = n$ is equal to

$$\left[\frac{n}{ab} \right] \quad \text{or} \quad \left[\frac{n}{ab} \right] + 1.$$

(799) At the fruit counter in a store, apples are sold 5 cents each and oranges are sold 7 cents each. Say Peter purchases four apples and twelve oranges. Peter notices that Paul also bought apples and oranges and that he pays the same total amount as you did, but with a different number of apples and oranges. Knowing that Paul purchased at least three oranges, does Peter have enough information to determine the exact number of apples and oranges purchased by Paul?

(800) Determine the set of solutions of the Diophantine equation $3x + 7y = 11$ located in the third quadrant of the cartesian plane.

(801) Determine the set of solutions of the Diophantine equation $5x + 7y = 11$ located above the line $y = x$.

(802) Assume that the set E of solutions of the Diophantine equation

$$(*) \qquad\qquad ax + by = 11$$

is given by

$$E = \{(x, y) : x = 5 - 4t \text{ and } y = 1 - 3t, \text{ where } t \in \mathbb{Z}\}.$$

Determine the values of a and b.

(803) Find the primitive solutions of $x^2 + 3y^2 = z^2$, that is those solutions x, y, z which have no common factor other than 1.

(804) Show that the only nonzero integer solutions (x, y, z) to the system of equations

$$x + y + z = x^3 + y^3 + z^3 = 3$$

are $(1, 1, 1)$, $(-5, 4, 4)$, $(4, -5, 4)$ and $(4, 4, -5)$.

(805) Show that the equation

$$x^2 = y^3 + z^5$$

has infinitely many solutions in positive integers x, y z.

(806) Find the four different ways of writing 136 as a sum of two positive integers, one of which is divisible by 5 and the other by 7.

(807) Any solution in positive integers x, y, z of $x^2 + y^2 = z^2$ is called a Pythagorean triple, since in such a case there exists a rectangular triangle whose sides have x, y, z for their respective lengths. Find all Pythagorean triples whose terms form an arithmetic progression.

(808) Find the dimensions of the Pythagorean triangle whose hypotenuse is of length 281.

(809) Show that 60 divides the product of the lengths of the sides of a Pythagorean triangle.

(810) Find every Pythagorean triangle whose area is equal to three times its perimeter.

(811) Find every Pythagorean triangle whose perimeter is equal to twice its area.

(812) Show that $\{x, y, z\} = \{3, 4, 5\}$ is the only solution of $x^2 + y^2 = z^2$ with consecutive integers x, y, z.

(813) Show that $n^2 + (n + 1)^2 = 2m^2$ is impossible for $n, m \in \mathbb{N}$.

(814) Show that the equation $x^2 + y^2 = 4z + 7$ has no integer solution.

(815) Find all integer solutions of $x^2 + y^2 = z^4$ such that $(x, y, z) = 1$.

(816) Find all integer points on the line $x + y = 1$ which are located inside the circle centered at the origin and of radius 3.

(817) Find all primitive solutions of the Diophantine equation

$$x^2 + 3136 = z^2.$$

(818) Find all integer solutions of the equation

$$x^2 + y^2 = xy.$$

(819) Find the solutions of the Diophantine equation

$$(*) \qquad\qquad x^2 + 2y^2 = 4z^2.$$

(820) Find a triangle such that each of its sides is of integer length and for which an interior angle is equal to twice another interior angle.

(821) Find all positive integer solutions to the equation $x^2 + y^2 = 10$. Do the same for $x^2 + y^2 = 47$.

(822) Find all positive rational solutions of $x^2 + y^2 = 1$.

(823) Find all primitive Pythagorean triangles such that the length of one of their sides is equal to 24.

(824) Show that the radius of any circle inscribed in a Pythagorean triangle is an integer.

(825) Show that the equation $x^2 + y^2 + z^2 = 2239$ has no solutions in positive integers x, y, z.

(826) Show that

$$t^2 = x^2 + y^2 + z^2$$

has no nontrivial integer solution with t even and with x, y, z having no common factor.

(827) Find all primitive solutions of the Diophantine equation $x^2 + 2y^2 = z^2$.

(828) Find all positive integer solutions to the system of equations

$$\begin{cases} a^3 - b^3 - c^3 = 3abc, \\ a^2 = 2(b + c). \end{cases}$$

(829) Find all integer solutions of $y^2 + y = x^4 + x^3 + x^2 + x$.

(830) Find the smallest prime number which can be written in each of the following forms: $x^2 + y^2, x^2 + 2y^2, \ldots, x^2 + 10y^2$.

(831) Determine the set of quadruples (x, y, x, w) verifying $x^3 + y^3 + z^3 = w^3$ and such that x, y, z and w are positive integers in arithmetical progression.

(832) Consider the sequence $8, 26, 56, 98, 152, \ldots$, that is the sequence $\{x_n\}$ defined by $x_1 = 8$ and $x_{n+1} = x_n + 6(2n + 1)$, $n \geq 1$, and show that for $n > 1$, x_n cannot be the cube of an integer.

(833) Show that $x^n + 1 = y^{n+1}$ has no solutions in positive integers x, y, n $(n \geq 2)$ with $(x, n + 1) = 1$.

(834) Show that neither of the equations

$$3^a + 1 = 5^b + 7^c \quad \text{and} \quad 5^a + 1 = 3^b + 7^c$$

has a solution in integers a, b, c other than $a = b = c = 0$.

(835) Find all integer triples (x, y, z) such that $4^x + 4^y + 4^z$ is a perfect square.

(836) Show that there exist solutions in positive integers a, b, c, x, y to the system of equations

$$\begin{cases} a + b + c = x + y, \\ a^3 + b^3 + c^3 = x^3 + y^3. \end{cases}$$

Show, in particular, that there exist infinitely many solutions such that a, b, c are in arithmetic progression.

(837) Solve each of the following Diophantine equations: (here m is a nonnegative integer)

$$x^m(x^2 + y) = y^{m+1},$$
$$x^m(x^2 + y^2) = y^{m+1}.$$

(838) Can the following equations be verified for an appropriate choice of integers x, a, b, c, d?

$$(x + 1)^2 + a^2 = (x + 2)^2 + b^2 = (x + 3)^2 + c^2 = (x + 4)^2 + d^2.$$

(839) Does the equation

$$x^2 + y^2 + z^2 = xyz - 1$$

have integer solutions?

(840) Find all pairs of real numbers (x, y) which satisfy the two equations:

$$(*) \qquad\qquad 2x^3 - x^2 + y^2 = 1,$$
$$(**) \qquad\qquad 2y^3 - y^2 + x^2 = 1.$$

(841) Find all positive integer solutions x, y of $x^y = y^{x-y}$.

(842) Find all integers solutions x, y, z to the system of equations

$$\begin{cases} 2x(1 + y + y^2) = 3(1 + y^4), \\ 2y(1 + z + z^2) = 3(1 + z^4), \\ 2z(1 + x + x^2) = 3(1 + x^4). \end{cases}$$

(843) Prove that there exist infinitely many integers a, b, c, d such that $a > b > c > d > 1$ and $a!\, d! = b!\, c!$.

(844) Show that the equation $x^3 + y^3 + z^3 = 4$ has no solutions in integers. What about the equation $x^3 + y^3 + z^3 = 5$?

(845) Does the Diophantine equation $x^4 = 4y^2 + 4y - 80$ have any solutions? If so, what are they? If no, explain why.

(846) Does the Diophantine equation $x^4 + y^4 + z^4 = 363932239$ have any solutions? If so, what are they? If no, explain why.

(847) Let a be an arbitrary integer. Does the Diophantine equation

$$303x + 57y = a^2 + 1$$

have any solution?

(848) Does the Diophantine equation
$$x^4 = 4y^2 + 4y - 15$$
have any solution?

(849) Do integers x, y, z exist such that
$$x^4 + (2y + 1)^4 = z^2?$$

(850) Determine the set of positive solutions of the Diophantine equation
$$x^2 = y^4 + 8.$$

(851) Let p be an odd prime number. Assume that $q = p + 8$ is also a prime number. Analyze the set of solutions of the Diophantine equation
$$x^2 = y^4 + pq$$
and give one such solution explicitly.

(852) Does the Diophantine equation
$$x^2 + y^2 + 2x + 4y + 4z + 2 = 0$$
have any solution?

(853) Does the Diophantine equation
$$x^4 + y^4 + z^4 + u^4 = 3xyzu$$
have any nonzero solution?

(854) Does the Diophantine equation
$$x^3 + 2y^3 = 4z^3$$
have any nonzero solution?

(855) Find all integer solutions of $x^2 + y^2 = 8z + 7$.

(856) Show that $x^4 + y^4 = 7z^2$ has no solutions in $\mathbb{N}$. What about the equation $x^4 + y^4 = 5z^2$?

(857) Does the equation $x^4 + x^2 = y^4 + 5$ have any solution in integers x and y?

(858) Let $0 < x < y < z$ be integers such that $x^2 + y^2 = z^2$. Show that for each integer $n > 2$, $x^n + y^n = z^n$ is impossible.

(859) Prove that the equation
$$x^3 + 3y^3 = 9z^3$$
has no nontrivial integer solution.

(860) Let p be a prime number. Does the Diophantine equation
$$x^4 + py^4 + p^2 z^4 = p^3 w^4$$
have any trivial solution?

(861) Show that
$$x^2 + y^2 + z^2 = 2xyz$$
has no nontrivial integer solution.

(862) Determine all rational solutions of the equation
$$x^3 + y^3 = x^2 + y^2.$$

(863) Show that the Diophantine equation
$$\frac{1}{x_1} + \frac{1}{x_2} + \cdots + \frac{1}{x_n} + \frac{1}{x_1 x_2 \cdots x_n} = 1$$
has at least one solution for each integer $n \geq 1$.

(864) Show that the Diophantine equation

$$x^2 + y^2 + z^2 = x^2 y^2$$

has no nontrivial solution.

(865) Show that the equation

$$x^8 + x^7 + x^6 + x^5 + x^4 + x^3 + x^2 + x + 1 = 1\,234\,567\,891\,314$$

has no integer solution.

(866) Prove that the Diophantine equation

$$(x + y)^2 + (x + z)^2 = (y + z)^2$$

has no solutions in odd integers x, y, z.

(867) Let p be a fixed prime number. Find all positive integer solutions of $x^2 + py^2 = z^2$.

(868) Show that there exist infinitely many solutions to the Diophantine equation $x^2 + 4y^2 = z^3$.

(869) Find all solutions, for x, y integers and n positive integers, to the Diophantine equation $x^n + y^n = xy$.

(870) Show that the equation $n^x + n^y = n^z$ has positive integer solutions only if $n = 2$.

(871) Show that the equation $n^x + n^y + n^w = n^z$ has positive integer solutions only if $n = 2$ or 3.

(872) Show that the abc conjecture implies the following result: The equation $x^p + y^q = z^r$ has no solutions in positive integers p, q, r, x, y, z with $z \geq z_0$ and

$$(*) \qquad \frac{1}{p} + \frac{1}{q} + \frac{1}{r} < 1,$$

so that in particular the Fermat equation $x^n + y^n = z^n$ has no nontrivial solution for $n \geq 4$ and z sufficiently large.

(873) Show that if the abc conjecture is true, then there can exist only a finite number of triples of consecutive powerful numbers.

(874) Show that if the abc conjecture is true, then there exist only a finite number of positive integers n such that $n^3 + 1$ is a powerful number. Moreover, find two numbers n with this property.

(875) Erdős conjectured that the equation $x + y = z$ has only a finite number of solutions in 4-powerful integers x, y, z pairwise coprime. Show that the abc conjecture implies this conjecture.

(876) Show that if the abc conjecture is true, then there exist only a finite number of 4-powerful numbers which can be written as the sum of two 3-powerful numbers pairwise coprime.

(877) Given an integer $n \geq 2$, let $P(n)$ stand for the largest prime factor of n. Prove that it follows from the abc conjecture that, for each real number $y > 0$, the set $A_y := \{p \text{ prime} : P(p^2 - 1) \leq y\}$ is a finite set and therefore has a largest element $p = p(y)$.

(878) In 1877, Edouard Lucas (1842–1891) observed that although 2701 is a composite number, we have that $2^{2700} \equiv 1 \pmod{2701}$, thus providing a counter-example to the reverse of Fermat's Little Theorem. More generally, show that one can construct a large family of such counter-examples

by considering the numbers $n = pq$, where p and q are prime numbers such that $p \equiv 1 \pmod 4$ and $q = 2p - 1$.

(879) Show that if the *abc* conjecture is true, then for any $\varepsilon > 0$, there exists a positive constant $M = M(\varepsilon)$ such that for all triples (x_1, x_2, x_3) of positive integers, pairwise coprime and verifying $x_1 + x_2 = x_3$, we have that

$$(*) \qquad \min(x_1, x_2, x_3) \le M \left(\gamma(x_i) \right)^{3+\varepsilon} \qquad (i = 1, 2, 3).$$

(880) In 1979, Enrico Bombieri naively claimed that: "the equation

$$\binom{x}{n} + \binom{y}{n} = \binom{z}{n} \qquad (n \ge 3)$$

had no solutions in positive integers x, y, z." Was Bombieri right? If so, prove it; if no, provide a counter-example.

(881) Let p be an odd prime number and let $\alpha_1, \alpha_2, \ldots, \alpha_r$ be positive integers not exceeding $p - 1$. Show that the Diophantine equation

$$n^p = x_1^{\alpha_1} + x_2^{\alpha_2} + \cdots + x_r^{\alpha_r}$$

has solutions in positive integers $n, x_1, x_2, \ldots, x_r$.

(882) Even though, according to Fermat's Last Theorem, for each prime number $p \ge 3$, the equation $x^p + y^p = z^p$ has no solutions in positive integers x, y, z, show that the equation $x^{p-1} + y^{p-1} = z^p$ always has solutions (besides the trivial one $x = y = z = 2$).

12. Quadratic Reciprocity

(883) Characterize all prime numbers $p > 11$ for which

$$x^2 \equiv 11 \pmod{p}$$

has a solution.

(884) Which of the following congruences have solutions?
  (a) $x^2 \equiv 1 \pmod 3$;
  (b) $x^2 \equiv -1 \pmod 3$;
  (c) $x^2 + 4x + 8 \equiv 0 \pmod 3$;
  (d) $x^2 + 8x + 16 \equiv -1 \pmod{17}$.

(885) Find the solutions of the congruence $2x^2 + 3x + 1 \equiv 0 \pmod 7$.

(886) Show that $(1!)^2 + (2!)^2 + \cdots + (n!)^2$ is never a perfect square, whatever the integer $n > 1$.

(887) Let $n \in \mathbb{N}$. Show that the odd prime divisors of $n^2 + 1$ are of the form $12k + 1$ or of the form $12k + 5$.

(888) Let $p > 3$ be a prime number. Show that p divides the sum

$$\sum_{\substack{j=1 \\ \left(\frac{j}{p}\right)=1}}^{p-1} j.$$

(889) Assuming that m is a positive integer such that $p = 4m + 3$ and $q = 2p + 1$ are two prime numbers, show that $q | M_p = 2^p - 1$. Use this result to show that the Mersenne number $M_{1\,122\,659}$ is composite.

(890) Show that 9239 divides $2^{4619} - 1$.

(891) Show that 5 is a nonquadratic residue of all the prime numbers of the form $6^n + 1$.

(892) Does there exist a perfect square of the form $1997k - 1$?

(893) Show that there exist infinitely many prime numbers of the form $3k + 1$.

(894) Does there exist a perfect square of the form $1! + 2! + \cdots + k!$ with $k > 3$?

(895) Show that for each integer $n > 1$, $(2^n - 1) \nmid (3^n - 1)$.

(896) Let p and q be two odd prime numbers, and a an integer. If $p = q + 4a$, is it true that $\left(\dfrac{p}{q}\right) = \left(\dfrac{a}{q}\right)$?

(897) If p is a prime number of the form $24k + 1$, is it true that $\left(\dfrac{3}{p}\right) = 1$?

(898) Does the congruence $x^2 \equiv 52 \pmod{159}$ have any solutions?

(899) If p is a prime number of the form $8k + 3$ and if $q = \frac{p-1}{2}$ is a prime number, can one conclude that q is a quadratic residue modulo p?

(900) Show that 3 is a nonquadratic residue of all Mersenne primes larger than 3.

(901) If p is a prime number of the form $p = 8k + 7$, show that

$$p | 2^{\frac{p-1}{2}} - 1.$$

(902) Does the congruence $x^2 \equiv 2 \pmod{231}$ have any solution? If so, what are they? If not, explain why.

(903) Does there exist a positive integer n and a prime number p of the form $p = 100k + 3$ such that $p | n^2 + 1$? Explain.

(904) Is it true that there exist infinitely many positive integers n such that $23 | n^2 + 14n + 47$? Explain.

(905) Does there exist an integer x such that the prime number 541 divides $x^2 - 3x - 1$? Explain.

(906) If p is a prime number, $p \equiv 1 \pmod{24}$, does the congruence $x^2 \equiv 6 \pmod{p}$ have any solution? Explain.

(907) Let n be a positive integer such that $p = 4^n + 1$ is a prime number. Does the congruence $x^2 \equiv 3 \pmod{p}$ have any solution? Explain.

(908) Let A be the set of integers a, $1 \leq a \leq 43$, for which there exists a prime number $p \equiv a \pmod{44}$ such that the corresponding congruence

$$x^2 \equiv 11 \pmod{p}$$

has solutions. Determine A.

(909) Find all prime numbers p for which $\left(\dfrac{5}{p}\right) = -1$.

(910) Let p and q be odd prime numbers such that $p = q + 4a$, $a \in \mathbb{N}$. Show that

$$\left(\frac{a}{p}\right) = \left(\frac{a}{q}\right).$$

(911) Of which prime numbers is the number -2 a quadratic residue?

(912) Let p be an odd prime number. Show that $x^2 \equiv 2 \pmod{p}$ has solutions if and only if $p \equiv 1$ or $7 \pmod{8}$. Using this result, prove that $2^{4n+3} \equiv 1 \pmod{8n+7}$ for each integer $n \geq 0$. In particular, find a proper divisor of the Mersenne number $2^{131} - 1$.

(913) Observing that $2717 = 11 \cdot 13 \cdot 19$, determine if the quadratic congruence $x^2 \equiv 1237 \pmod{2717}$ has solutions.

(914) Let a be an integer such that $(a, p) = 1$. Determine all prime numbers p such that $\left(\dfrac{a}{p}\right) = \left(\dfrac{p-a}{p}\right)$.

(915) Does the congruence $x^2 \equiv 131313 \pmod{1987}$ have any solutions?

(916) Show that the equation $x^2 - y^3 = 7$ has no integer solution.

(917) Determine all prime numbers p for which 15 is a quadratic residue modulo p.

(918) Show that the statement of the law of quadratic reciprocity can be written (as Gauss did) as

$$\left(\frac{p}{q}\right) = \left(\frac{(-1)^{(q-1)/2}q}{p}\right).$$

(919) Does the congruence $x^2 \equiv 34561 \pmod{1234577}$ have any solution?

(920) Show that if r is a quadratic residue modulo $m > 2$, then

$$r^{\phi(m)/2} \equiv 1 \pmod{m}.$$

(921) Let a be an integer > 1 and let n be a positive integer. Show that $n | \phi(a^n - 1)$.

(922) Show that if p is an odd prime number, then

$$\sum_{j=1}^{p-1} \left(\frac{j}{p}\right) = 0.$$

(923) Let p be an odd prime number. Show that

$$\sum_{k=1}^{p-2} \left(\frac{k(k+1)}{p} \right) = -1.$$

(924) Let $p > 5$ be a prime number. Using the Problem 923, show that one can always find two consecutive integers which are quadratic residues modulo p, as well as two consecutive integers which are quadratic nonresidues modulo p.

(925) Find all prime numbers p such that the corresponding numbers $5p+1$ are perfect squares. Is it possible to find prime numbers p for which $5p + 2$ are perfect squares?

(926) Let $f \colon \mathbb{Z} \to \mathbb{Z}$ be a polynomial function and let a, b be integers. Set $\left(\dfrac{m}{p} \right) = 0$ if $p|m$. If $(a, p) = 1$, show that

$$\sum_{k=0}^{p-1} \left(\frac{f(ak+b)}{p} \right) = \sum_{k=0}^{p-1} \left(\frac{f(k)}{p} \right).$$

Use this to prove that if $(a, p) = 1$, then

$$\sum_{k=0}^{p-1} \left(\frac{ak+b}{p} \right) = 0.$$

(927) Let $a, b \in \{-1, 1\}$, p be an odd prime number and

$$N(a, b) = \# \left\{ k \mid 1 \leq k \leq p - 2, \quad \left(\frac{k}{p} \right) = a, \quad \left(\frac{k+1}{p} \right) = b \right\}.$$

Show that

$$N(a, b) = \frac{1}{4} \left(p - 2 - b - ab - a(-1)^{(p-1)/2} \right).$$

Use this to prove that the number of pairs of consecutive quadratic residues modulo p is given by

$$N(1, 1) = \frac{p - 4 - (-1)^{(p-1)/2}}{4}.$$

(928) Let p be a prime number satisfying $p \equiv 1 \pmod 4$. Show that

$$\sum_{j=1}^{(p-1)/2} \left(\frac{j}{p} \right) = 0.$$

(929) Let p be a prime number such that $p \equiv 1 \pmod 4$. Show that

$$\sum_{k=1}^{p-1} k \left(\frac{k}{p} \right) = 0.$$

(930) Let p be a prime number such that $p \equiv 1 \pmod 4$. Show that

$$\sum_{\substack{k=1 \\ \left(\frac{k}{p} \right) = 1}}^{p-1} k = \frac{p(p-1)}{4}.$$

(931) Let p be a prime number such that $p \equiv 3 \pmod 4$. Show that

$$\sum_{k=1}^{p-1} k^2 \left(\frac{k}{p} \right) = p \sum_{k=1}^{p-1} k \left(\frac{k}{p} \right).$$

(932) Show that the equation $x^2 - 33y^2 = 5$ has no integer solutions.

13. Continued Fractions

(933) Express each of the numbers $\sqrt{2}$ and $\sqrt{2}/2$ as a simple infinite continued fraction.

(934) Using continued fractions, find a solution of the equation $12x + 5y = 13$; do the same for $13x - 19y = 1$.

(935) Find the irrational number represented by the infinite continued fraction $[3, \overline{1,4}]$.

(936) Let α be an irrational number > 1 whose representation as a simple infinite continued fraction is $[a_1, a_2, \ldots]$. Express $1/\alpha$ as a simple infinite continued fraction.

(937) Find a rational number which provides a good approximation of $\sqrt{5}$; that is, find a rational number a/b such that
$$|\sqrt{5} - a/b| < 10^{-4}.$$

(938) Let $\{p_n\}$ and $\{q_n\}$ be the sequences defined in Definition 24. Show that
$$\frac{p_n}{p_{n-1}} = [a_n, a_{n-1}, \ldots, a_1], n \geq 1, \quad \frac{q_n}{q_{n-1}} = [a_n, a_{n-1}, \ldots, a_2], n \geq 2.$$

(939) Show that if $ax^2 - bx - c = 0$ where $abc \neq 0$ and $b^2 + 4ac$ is not a perfect square, then the continued fraction $\left[\dfrac{b}{a}, \dfrac{b}{c}\right]$ is a real root of the quadratic equation.

(940) Find an approximation for the real roots of $2x^2 - 5x - 4 = 0$ which is accurate up to the first decimal.

(941) For each $n \in \mathbb{N}$, show that $\sqrt{n^2 + 1} = [n, \overline{2n}]$.

(942) Given an integer $n > 1$, show that the continued fraction which represents $\sqrt{n^2 - 1}$ is $[n - 1, \overline{1, 2n - 2}]$.

(943) For each $n \in \mathbb{N}$, show that $\sqrt{n^2 + 2} = [n, \overline{n, 2n}]$.

(944) Given an integer $n > 1$, show that the continued fraction that represents $\sqrt{n^2 - 2}$ is $[n - 1, \overline{1, n - 2, 1, 2n - 2}]$.

(945) Find the continued fraction of $\sqrt{38}$, that of $\sqrt{47}$ and that of $\sqrt{120}$.

(946) If n is a positive integer, show that the continued fraction which represents $\sqrt{n^2 + n}$ is $[n, \overline{2, 2n}]$.

(947) Given an integer $n > 1$, show that the continued fraction which represents $\sqrt{n^2 - n}$ is $[n - 1, \overline{2, 2n - 2}]$.

(948) Given an integer $n > 1$, show that the continued fraction which represents $\sqrt{9n^2 + 3}$ is $[3n, \overline{2n, 6n}]$.

(949) Find the real number r whose expansion in a continued fraction is $q = [1, \overline{1, 2}]$ by multiplying the quantities $q + 1$ and $q - 1$.

(950) Find the best rational approximation a/b of π when $b < 1000$. Do the same for e and then for $\sqrt{5}$.

(951) Find an approximation of the irrational number $[1, 2, 3, 4, 5, 6, 7, \ldots]$ correct up to the sixth decimal.

(952) Knowing that
$$e = [2, 1, 2, 1, 1, 4, 1, 1, 6, 1, 1, 8, 1, \ldots],$$
find a rational number which is a correct approximation of the number e up to the fourth decimal.

(953) Knowing that
$$\pi = [3, 7, 15, 1, 292, 1, 1, 1, 2, 1, 3, 1, 14, 2, \ldots],$$
find a rational number which is a correct approximation of the number π up to the sixth decimal.

(954) Find an approximation correct up to 10^{-4} of the number
$$[4, \overline{2, 1, 3, 1, 2, 8}].$$

(955) Let $k \geq 1$ and let $C_k = p_k/q_k$ be the k–th convergent of the irrational number α. Assume that a and b are integers, with b positive. Show that if $|\alpha - a/b| < |\alpha - p_k/q_k|$, then $b > q_{k+1}/2$.

(956) Assume that $|\sqrt{3} - a/b| < |\sqrt{3} - 26/15|$, where $b > 0$. Show that $b \geq 21$.

(957) Let $k \geq 1$ and let $C_k = p_k/q_k$ be the k–th convergent of the irrational number α. If a and b are integers with $a \geq 1$, $b \geq 1$ and
$$\alpha < \frac{a}{b} < \frac{p_k}{q_k},$$
show that $b > q_{k+1}$.

(958) Let a and b be positive integers such that $\sqrt{3} < a/b < 26/15$. Show that $b > 41$.

(959) Let a/b be a rational number such that $a/b \neq 333/106$. If $0 \leq b \leq 56$, show that $|\pi - 333/106| < |\pi - a/b|$, and thus that $333/106$ is a better approximation of π then any other rational number whose denominator is smaller or equal to 56.

(960) Let $[a_1, a_2, \ldots]$ be a simple continued fraction. Show that
$$q_n \geq 2^{(n-1)/2}, \quad \text{for } n \geq 3.$$

(961) Show that for $n \geq 4$,
$$p_n q_{n-3} - q_n p_{n-3} = (a_n a_{n-1} + 1)(p_{n-4} q_{n-3} - q_{n-4} p_{n-3})$$
$$= (-1)^n (a_n a_{n-1} + 1).$$

(962) Given an integer $n \geq 2$, let $a_1, a_2, \ldots$ and $b_1, b_2, \ldots$ be integers such that the a_j's and b_j's are positive for each $2 \leq j \leq n$. If $a_i = b_i$ for $1 \leq i < n$ and $a_n < b_n$, show that
$$[a_1, a_2, \ldots] < [b_1, b_2, \ldots] \qquad \text{if } n \text{ is odd,}$$
$$[a_1, a_2, \ldots] > [b_1, b_2, \ldots] \qquad \text{if } n \text{ is even.}$$

(963) Let $a_1, a_2, \ldots, b_1, b_2, \ldots$ and $c_1, c_2, \ldots$ be integers such that a_j, b_j and c_j are positive for each $j \geq 2$. If $c_i \leq a_i \leq b_i$ for $i \geq 1$, then show that
$$[b_1, c_2, b_3, c_4, b_5, \ldots] \leq [a_1, a_2, a_3, \ldots] \leq [c_1, b_2, c_3, b_4, c_5, \ldots].$$

(964) Let a_i be integers taking the value 1 or 2. Show that if $\alpha = [a_1, a_2, \ldots]$, then
$$\frac{1 + \sqrt{3}}{2} \leq \alpha \leq 1 + \sqrt{3}.$$

(965) A complete orbit of the Earth around the Sun takes approximately 365 days, 5 hours, 48 minutes and 46 seconds. Thus, the actual length of a year exceeds 365 days by $\frac{20926}{86400}$ of a day. In 45 B.C., Julius Ceasar used the correction $1/4$, namely by adding the 29-th of February every four

years. This approximation created an error of 10 days every 1500 years. This necessitated a modification which was done by Pope Gregory XIII in 1582. Our present calendar, known as the *Gregorian calendar*, gives an additional day on each year divisible by 4, except on the years divisible by 100 but not by 400. This correction corresponds to adding 97 days (one day per bissextile year) for a period of 400 years, which is a fairly good approximation of $20926/86400$. Find an even better approximation.

(966) Let $k \geq 1$ and $C_k = p_k/q_k$ be the k-th convergent of the irrational number $\alpha = [a_1, a_2, a_3, \ldots]$. Show that

$$
p_k = \det \begin{pmatrix}
a_1 & -1 & 0 & \ldots & 0 & 0 \\
1 & a_2 & -1 & \ldots & 0 & 0 \\
0 & 1 & a_3 & \ldots & 0 & 0 \\
. & . & . & \ldots & . & . \\
. & . & . & \ldots & -1 & . \\
. & . & . & \ldots & a_{k-1} & -1 \\
0 & 0 & 0 & \ldots & 1 & a_k
\end{pmatrix}.
$$

Find a similar expression for q_k.

(967) A simple infinite continued fraction is said to be *periodic* if it is of the form $[a_1, a_2, \ldots, a_n, \overline{b_1, b_2, \ldots, b_m}]$. If it is of the form $[\overline{b_1, b_2, \ldots, b_m}]$, we say that it is *purely periodic*. The smallest positive integer m satisfying the above relation is called the *period* of the simple infinite continued fraction. Show that any simple continued fraction which is purely periodic must be a *quadratic irrational number* (that is an irrational number which is a root of a quadratic equation whose coefficients are integers).

(968) Show that every periodic simple continued fraction is a quadratic irrational number.

(969) Let α be an irrational root of $f(x) := ax^2 + bx + c = 0$, where a, b, c are integers. If

$$\alpha = [a_1, a_2, \ldots], \qquad \alpha_n = [a_{n+1}, a_{n+2}, \ldots] \text{ for each } n \geq 1,$$

show that α_n $(n \geq 1)$ is a root of the polynomial $A_n x^2 + B_n x + C_n = 0$, where

$$
A_n = q_n^2 f\left(\frac{p_n}{q_n}\right) = ap_n^2 + bp_n q_n + cq_n^2,
$$
$$
B_n = 2ap_n p_{n-1} + bp_n q_{n-1} + bp_{n-1} q_n + 2cq_n q_{n-1},
$$
$$
C_n = q_{n-1}^2 f\left(\frac{p_{n-1}}{q_{n-1}}\right) = ap_{n-1}^2 + bp_{n-1} q_{n-1} + cq_{n-1}^2,
$$

and where $B_n - 4A_n C_n = b^2 - 4ac$. Use this to prove that $A_n C_n < 0$.

(970) Let α be a quadratic irrational number and write

$$\alpha = [a_1, a_2, \ldots, a_n, \alpha_n] \qquad (n = 1, 2, \ldots),$$

where

$$\alpha_n = [a_{n+1}, a_{n+2}, \ldots] \text{ for each } n \geq 1.$$

Show that there exists a finite number of quadratic polynomials with integer coefficients, say

$$A_1 x^2 + B_1 x + C_1,$$
$$A_2 x^2 + B_2 x + C_2,$$

$$\ldots$$

$$A_N x^2 + B_N x + C_N,$$

of which α_n is a root.

(971) Show that each quadratic irrational number has a periodic expansion as a simple continued fraction.

(972) Given any integer $D > 1$ which is not a perfect square, the following result is known: *Let $\alpha = a + b\sqrt{D}$ and $\overline{\alpha} = a - b\sqrt{D}$, where α is a quadratic irrational number. If $\alpha > 1$ and $-1 < \overline{\alpha} < 0$, then the continued fraction which represents α is a simple continued fraction which is purely periodic.* Show that this is the case for the quadratic irrational numbers $(3+\sqrt{23})/7$, $2 + \sqrt{7}$ and $(5 + \sqrt{37})/3$.

(973) If D is a positive integer which is not a perfect square, show that the continued fraction which represents $\sqrt{D}$ is a periodic continued fraction whose period begins after the first term. In particular, show that

$$\sqrt{D} = [a_1, \overline{a_2, a_3, \ldots, a_n, 2a_1}].$$

14. Classification of Real Numbers

(974) Show that the sequence $[\sqrt{2}]$, $[2\sqrt{2}]$, $[3\sqrt{2}]$, $[4\sqrt{2}]$, ... contains infinitely many powers of 2.

(975) Assume that $0 < r \in \mathbb{Q}$ is given as an approximation of $\sqrt{2}$. Show that the number $\frac{r+2}{r+1}$ represents an even better approximation.

(976) Consider each of the following numbers and indicate if it is rational or irrational:

(a) $\sqrt{676}$; (b) $\sqrt{75} + \sqrt{2}$.

(977) Let $a = 12$, $b = 245$, $c = 363$, $d = 375$. Consider each of the following numbers and indicate if it is rational or irrational:

(a) $\sqrt{ab}$; (b) $\sqrt{ac}$; (c) $(6ad)^{1/3}$; (d) $\log a$.

(978) Let $f : \mathbb{R} \to \{0, 1\}$ be the *Dirichlet function* defined by

$$f(x) = \begin{cases} 1 & \text{if } x \text{ is rational,} \\ 0 & \text{if } x \text{ is irrational.} \end{cases}$$

Show that this function has the representation

$$f(x) = \lim_{m,n \to \infty} (\cos(m!\pi x))^n.$$

(979) Show that the positive root of the equation $x^5 + x = 10$ is irrational.

(980) Let r and s be two positive integers. If the equation $x^2 + rx + s = 0$ has a root $x_0 \in \mathbb{Q}$, show that $x_0 \in \mathbb{Z}$.

(981) Let p and q be two prime numbers. For which integers m and n is the number $m\sqrt{p} + n\sqrt{q}$ an integer?

(982) Show that the number $\alpha = 0.0110101000101\ldots$, where the j-th decimal after the dot is 1 if j is prime and 0 otherwise, is an irrational number.

(983) Let p and q be two prime numbers. Show that $\sqrt{p} + \sqrt{q}$ is necessarily an irrational number.

(984) Consider the three numbers

$$\alpha = 1 + \cfrac{1}{1 + \cfrac{1}{1 + \cfrac{1}{1 + \ldots}}}, \quad \beta = \sqrt{1 + \sqrt{1 + \sqrt{1 + \ldots}}}, \quad \delta = \frac{1 - \sqrt{5}}{2}.$$

Find the only number t satisfying the equation

$$\alpha + \beta + t\delta = 2.$$

(985) Is the number $2^{1/3} + 3^{1/3}$ an irrational number?

(986) Is the number $\log_{10} 2$ irrational?

(987) Let $0 < \xi < 1$ be a rational number. Prove that there exists only a finite number of solutions p/q to the inequality

$$\left| \frac{p}{q} - \xi \right| < \frac{1}{q^2}.$$

(988) In 1934, Gelfond and Schneider established that if α and β are algebraic, $\alpha \neq 0, 1$, and β is irrational, then α^β is transcendental. Use this result to prove that $\dfrac{\log 3}{\log 2}$ is transcendental.

(989) Let $m \in \mathbb{Q}$, $m > 0$. Prove that $m + \dfrac{1}{m}$ is an integer if and only if $m = 1$.

(990) Find the polynomial of minimal degree of which the real number $\sqrt{2} + \sqrt{7}$ is a root.

(991) Determine the roots of the polynomial $p(x) = x^3 + 2x^2 - 1$ and indicate those which are rational numbers as well as those which are irrational numbers.

(992) Let e stand for the Euler number. Is it possible to find integers a and b such that
$$e = \frac{4}{\sqrt{ae + b}}?$$
If so, find them. If not, explain why.

(993) Is it true that the interval $[\frac{7}{2}, \frac{9}{2}]$ contains at least one transcendental number? Explain.

(994) Using the fact that
$$\left(\sqrt{2}^{\sqrt{2}} \right)^{\sqrt{2}} = 2,$$
show that there exist irrational numbers α and β such that α^β is rational.

(995) Show that
  (a) if y is a real nonnegative number such that e^y is rational, then y is irrational;
  (b) π is an irrational number.

(996) Show that $\log 2$ (the neperian logarithm of 2) is an irrational number.

(997) Show that $\dfrac{1 + 7^{1/3}}{2}$ is an algebraic number of degree three by finding its minimal polynomial.

(998) Show that $1 + \sqrt{2} + \sqrt{3}$ is an algebraic number of degree four by finding its minimal polynomial.

(999) Is the number $2^{1/2} + 3^{1/3}$ irrational, algebraic, transcendental? Explain.

(1000) Without using a computer, find all rational roots of the polynomial $x^5 + 39x^4 + 83x^3 + 325x^2 - 348x - 1924$.

(1001) Does there exist a rational number x such that
$$\pi^5 x + 2\pi^4 x^2 + 3\pi^3 x^3 + 4\pi^2 x^4 + 5\pi x^5 + 6 = 0 \,?$$
Explain.

Part 3

Solutions

Solutions

(1) (a) If $n = 1$, the result is true. Assume that the result is true for n and let us prove it for $n+1$. Since

$$1 + 2 + \cdots + n + (n+1) = \frac{n(n+1)}{2} + (n+1) = \frac{(n+1)(n+2)}{2},$$

the result follows.

(b) If $n = 1$, the result is true. Assume that the result is true for n and let us prove it for $n+1$. Since

$$1^2 + 2^2 + \cdots + n^2 + (n+1)^2 = \frac{n(n+1)(2n+1)}{6} + (n+1)^2$$
$$= \frac{(n+1)(n+2)(2n+3)}{6},$$

the result follows.

(c) If $n = 1$, the result is true. Assume that the result is true for n and let us prove it for $n+1$. Since

$$1^3 + 2^3 + \cdots + n^3 + (n+1)^3 = \frac{n^2(n+1)^2}{4} + (n+1)^3$$
$$= \frac{(n+1)^2(n+2)^2}{4},$$

the result follows.

(2) We only need to observe that

$$n^3 = (1^3 + 2^3 + \cdots + n^3) - (1^3 + 2^3 + \cdots + (n-1)^3)$$
$$= \left(\frac{n(n+1)}{2}\right)^2 - \left(\frac{(n-1)n}{2}\right)^2,$$

where we used the identity of Problem 1 (c).

(3) The given expression can be written as $\dfrac{1}{2}\left(\dfrac{3}{2} - \dfrac{2n+1}{n(n+1)}\right)$ for each positive integer n.

(4) By using the formula of Problem 1 (a), we obtain

$$2 + 4 + \cdots + 2n = 2(1 + 2 + \cdots + n) = 2 \cdot \frac{n(n+1)}{2} = n(n+1).$$

(5) Since $\sum_{j=1}^{n} j = \frac{n(n+1)}{2}$, it is enough to check that

$$(*) \qquad \sum_{j=1}^{n} (-1)^j j^2 = (-1)^n \frac{n(n+1)}{2}.$$

We use induction. First of all, $(*)$ is true for $n = 1$. Assume that $(*)$ is true for $n = k$; we will show that this implies that it is true for $n = k+1$.

Indeed, this implies successively

$$\sum_{j=1}^{k+1}(-1)^j j^2 = \sum_{j=1}^{k}(-1)^j j^2 + (-1)^{k+1}(k+1)^2$$

$$= (-1)^k \frac{k(k+1)}{2} + (-1)^{k+1}(k+1)^2$$

$$= (-1)^{k+1}\left((k+1)^2 - \frac{k(k+1)}{2}\right)$$

$$= (-1)^{k+1}\frac{(k+1)(k+2)}{2},$$

as required.

(6) This result can be proved by induction by observing that

$$a^{2n+1} + b^{2n+1} = a^{2n+1} + a^2 b^{2n-1} - a^2 b^{2n-1} + b^{2n+1}$$
$$= a^2(a^{2n-1} + b^{2n-1}) - b^{2n-1}(a^2 - b^2).$$

(7) This result can be proved by induction by observing that

$$a^{4n+4} - b^{4n+4} = a^{4n+4} - a^4 b^{4n} + a^4 b^{4n} - b^{4n+4}$$
$$= a^4(a^{4n} - b^{4n}) + b^{4n}(a^4 - b^4).$$

(8) This result can be proved using induction by observing that

$$a^{n+1} - b^{n+1} = a^{n+1} - a^n b + a^n b - b^{n+1} = a^n(a - b) + b(a^n - b^n).$$

(9) We use induction as well as the relation

$$\sum_{j=1}^{n+1} j \cdot j! = \sum_{j=1}^{n} j \cdot j! + (n+1)(n+1)!.$$

(10) Multiplying the given inequality $(2n)! < 2^{2n}(n!)^2$ by the trivial inequality $(2n+1) < 2^2(n+1)^2$, then using induction, one easily proves the inequality.

(11) Multiplying the relation $(n+1)^3/n^3 < (n+1)$ (valid for $n \geq 3$) by the given inequality $n^3 < n!$ allows one to use induction and thereby obtain the result.

(12) This follows from

$$(1+\theta)^{n+1} = (1+\theta)^n(1+\theta) \geq (1+n\theta)(1+\theta) = 1 + \theta + n\theta + n\theta^2$$
$$> 1 + (n+1)\theta.$$

REMARK: This inequality is often called the *Bernoulli inequality*, being attributed to Jacques Bernoulli (1654–1705).

(13) By using the induction hypothesis and by observing that

$$(1+\theta)^{n+1} = (1+\theta)(1+\theta)^n \geq (1+\theta)(1+n\theta + \frac{n(n-1)}{2}\theta^2)$$
$$\geq 1 + (n+1)\theta + \frac{(n+1)n}{2}\theta^2,$$

the result follows.

(14) We prove this result by induction. For $n = 1$, the result is true. Assume that it is true for n and let us prove it for $n + 1$. Since

$$\frac{1}{3}\big((n+1)^3 + 2(n+1)\big) = \frac{1}{3}(n^3 + 3n^2 + 3n + 1 + 2n + 2)$$

$$= \frac{1}{3}(n^3 + 2n) + n^2 + n + 1$$

is an integer because of the induction hypothesis, the result follows.

(15) Let $f(n) = 10^n + 3 \cdot 4^{n+2} + 5$. Since $f(0) = 54$ is divisible by 9 and since for each integer $n \geq 0$, $\dfrac{f(n+1) - f(n)}{9} = 10^n + 4^{n+2}$ is an integer, the result follows.

(16) The first equation is equivalent (after simplification) to

$$\frac{1}{n-k} = \frac{1}{k+1},$$

which in turn is equivalent to $n = 2k + 1$, as was to be shown.

(17) (a) By adding the relations

$$2^{2n} = (1+1)^{2n} = \sum_{k=0}^{2n} \binom{2n}{k} \text{ and } 0 = (1-1)^{2n} = \sum_{k=0}^{2n} (-1)^k \binom{2n}{k},$$

we obtain

$$2^{2n} = \sum_{k=0}^{2n} (1 + (-1)^k) \binom{2n}{k} = 2 \sum_{k=0}^{n} \binom{2n}{2k},$$

which yields the result.

(b) This follows essentially from part (a) and the fact that $\sum_{k=0}^{2n} \binom{2n}{k} = 2^{2n}$.

(18) Comparing the geometric mean with the arithmetic mean, we obtain

$$(n!)^{1/n} \leq \frac{1 + 2 + \cdots + n}{n} = \frac{n(n+1)}{2n} = \frac{n+1}{2},$$

so that the result follows by raising each side to the power n.

(19) (Gelfand [**13**]) For $n = 8$ the result is true. Assume that k can be written as a sum of 3's and 5's. Then, this sum contains one 5 (possibly many) or none at all. In the first case, we replace a 5 by two 3's. The new number $k + 1$ then contains 3's or 5's. In the second case, there is at least three 3's, and we can replace them by two 5's. The new number $k + 1$ then contains 3's or 5's. This proves the result.

(20) Let P_n be the following proposition: the number of lines thus created by n points for which no combination of three of these points are on a straight line is $n(n-1)/2$. Since two points determine a straight line and since $2(2-1)/2 = 1$, P_2 is true. Assume now that P_n is true for an integer $n \geq 2$. If a new point is added to the collection of n points in such a way that it cannot be on a straight line created by two of the points, then n additional lines will thus be added and the new collection of $n + 1$ points will determine $n(n-1)/2 + n = \frac{n(n+1)}{2}$ lines. The result then follows by induction.

(21) The proof is done by induction. If

$$1 + \frac{1}{\sqrt{2}} + \frac{1}{\sqrt{3}} + \cdots + \frac{1}{\sqrt{k}} > \sqrt{k}$$

then, since $\sqrt{k+1} - \sqrt{k} < 1/\sqrt{k+1}$, the sum of these two inequalities gives the result.

(22) Let

$$R_{2k-1} := 1^3 + 3^3 + 5^3 + \cdots + (2k-1)^3.$$

In light of Problem 1 (c), we know that

$$S_n := 1^3 + 2^3 + 3^3 + \cdots + n^3 = \left(\frac{n(n+1)}{2}\right)^2.$$

For $n = 2k - 1$, this last sum can be written as

$$
\begin{aligned}
S_{2k-1} &= \left(1^3 + 3^3 + \cdots + (2k-1)^3\right) + \left(2^3 + 4^3 + \cdots + (2k-2)^3\right) \\
&= \left(1^3 + 3^3 + \cdots + (2k-1)^3\right) + 2^3 \left(1^3 + 2^3 + 3^3 + \cdots + (k-1)^3\right) \\
&= R_{2k-1} + 2^3 S_{k-1}.
\end{aligned}
$$

It follows that

$$
\begin{aligned}
R_{2k-1} &= S_{2k-1} - 2^3 S_{k-1} = \left(\frac{(2k-1)(2k)}{2}\right)^2 - 8\left(\frac{(k-1)k}{2}\right)^2 \\
&= k^2(2k-1)^2 - 2k^2(k-1)^2 = k^2(2k^2 - 1),
\end{aligned}
$$

as was to be shown.

(23) Using the Binomial Theorem, we have

$$
\begin{aligned}
(n+1)^{k+1} - 1 &= \left((n+1)^{k+1} - n^{k+1}\right) + \left(n^{k+1} - (n-1)^{k+1}\right) + \cdots \\
&\qquad\qquad + \left(2^{k+1} - 1^{k+1}\right) \\
&= \sum_{j=1}^{n} \left\{(j+1)^{k+1} - j^{k+1}\right\} \\
&= \sum_{j=1}^{n} \left\{ j^{k+1} + \binom{k+1}{1} j^k + \binom{k+1}{2} j^{k-1} + \cdots \right. \\
&\qquad\qquad\qquad \left. + \binom{k+1}{k+1} j^0 - j^{k+1} \right\} \\
&= \sum_{j=1}^{n} \left\{ \binom{k+1}{1} j^k + \binom{k+1}{2} j^{k-1} + \cdots + \binom{k+1}{k+1} j^0 \right\} \\
&= \sum_{j=1}^{n} \left\{ \binom{k+1}{k} j^k + \binom{k+1}{k-1} j^{k-1} + \cdots + \binom{k+1}{0} j^0 \right\} \\
&= \sum_{j=1}^{n} \sum_{r=0}^{k} \binom{k+1}{r} j^r \\
&= \sum_{r=0}^{k} \binom{k+1}{r} \sum_{j=1}^{n} j^r = \sum_{r=0}^{k} \binom{k+1}{r} S_r(n).
\end{aligned}
$$

Therefore, if $S_r(n)$ is a polynomial of degree $r+1$ for each positive integer $r \leq k-1$, we may conclude, using induction, that $S_k(n)$ is a polynomial of degree $k+1$.

(24) (a) The required formula is $\displaystyle\prod_{i=2}^{n}\left(1-\frac{1}{i}\right) = \frac{1}{n}$, since

$$\prod_{i=2}^{n}\left(1-\frac{1}{i}\right) = \frac{1}{2}\cdot\frac{2}{3}\cdot\frac{3}{4}\cdots\frac{n-2}{n-1}\cdot\frac{n-1}{n} = \frac{1}{n}.$$

(b) The required formula is $\displaystyle\prod_{i=2}^{n}\left(1-\frac{1}{i^2}\right) = \frac{n+1}{2n}$, since

$$\prod_{i=2}^{n}\left(1-\frac{1}{i^2}\right) = \frac{3}{2^2}\cdot\frac{(3-1)(3+1)}{3^2}\cdot\frac{(4-1)(4+1)}{4^2}\cdots$$

$$\cdot\frac{(5-1)(5+1)}{5^2}\cdots\frac{(n-2)n}{(n-1)^2}\cdot\frac{(n-1)(n+1)}{n^2} = \frac{n+1}{2n}.$$

(25) Let S_n be the given sum. Since

$$i^4 + i^2 + 1 = (i^2+1)^2 - i^2 = (i^2+i+1)(i^2-i+1)$$

and since

$$\frac{i}{(i^2+i+1)(i^2-i+1)} = \frac{1}{2}\left(\frac{1}{i^2-i+1} - \frac{1}{i^2+i+1}\right),$$

we have

$$\begin{aligned}
S_n &= \frac{1}{2}\left(\sum_{i=1}^{n}\frac{1}{i^2-i+1} - \sum_{i=1}^{n}\frac{1}{i^2+i+1}\right)\\
&= \frac{1}{2}\left(\sum_{i=1}^{n}\frac{1}{(i-1)i+1} - \sum_{i=1}^{n}\frac{1}{i(i+1)+1}\right)\\
&= \frac{1}{2}\left(1 - \frac{1}{n(n+1)+1}\right),
\end{aligned}$$

and the result follows.

(26) We will show that the choice $(m,n,r) = (2,3,7)$, for which $S = 41/42$, maximizes the sum S in the sense that for any other choice (m,n,r), with $S < 1$, we must have $S < \frac{41}{42}$. So let us consider such a triple (m,n,r). Without any loss in generality, we may assume that $2 \leq m \leq n \leq r$. We shall first show that $m = 2$. Indeed, if $m \geq 3$, then

$$S \leq \frac{1}{3} + \frac{1}{3} + \frac{1}{4} = \frac{11}{12} < \frac{41}{42}.$$

Hence, $m = 2$. Let us now show that $n = 3$. If $n \geq 4$, we have

$$S \leq \frac{1}{2} + \frac{1}{4} + \frac{1}{5} = \frac{19}{20} < \frac{41}{42}.$$

Hence, $n = 3$. It remains to show that $r = 7$. Two situations need to be considered: $3 \leq r \leq 6$ and $r \geq 8$. In the first case, we have

$$S = \frac{1}{2} + \frac{1}{3} + \frac{1}{r} \geq \frac{1}{2} + \frac{1}{3} + \frac{1}{6} = 1,$$

a contradiction. In the second case, we have

$$S = \frac{1}{2} + \frac{1}{3} + \frac{1}{r} \leq \frac{1}{2} + \frac{1}{3} + \frac{1}{8} = \frac{23}{24} < \frac{41}{42}.$$

We may therefore conclude that $r = 7$, thus completing the proof.

(27) The problem is equivalent to the combinatorial problem which consists in distributing k balls in ℓ urns with the restriction that there must be at least one ball in the first urn. We shall call upon the combinatorial theorem according to which there are $\binom{k-1}{\ell-1}$ distinct vectors with (positive) integer components satisfying the relation

$$x_1 + x_2 + \cdots + x_\ell = k.$$

We then place the first ball in the first urn and distribute the $k - 1$ remaining balls in the ℓ urns. The above result then yields

$$A_k(\ell) = \binom{k + \ell - 2}{\ell - 1},$$

as required.

(28) We prove the first formula; the proof of the second formula is similar. First of all, it is true for $n = 1$. Assume that it is true for each odd number $n \leq k$, k odd. Since

$$1^2 + 3^2 + 5^2 + \cdots + k^2 + (k+2)^2 = \binom{k+2}{3} + (k+2)^2,$$

the result will follow if we manage to show that

$$\binom{k+2}{3} + (k+2)^2 = \binom{k+4}{3}$$

or similarly that

$$\binom{k+4}{3} - \binom{k+2}{3} = (k+2)^2.$$

But this relation is true, since

$$
\begin{aligned}
\binom{k+4}{3} - \binom{k+2}{3} &= \frac{(k+4)!}{3!(k+1)!} - \frac{(k+2)!}{3!(k-1)!} \\
&= \frac{(k+2)!}{3!(k-1)!}\left(\frac{(k+3)(k+4)}{k(k+1)} - 1\right) \\
&= \frac{(k+2)!}{3!(k-1)!}\frac{(6k+12)}{k(k+1)} \\
&= \frac{k+2}{6}(6k+12) = (k+2)^2,
\end{aligned}
$$

which completes the proof.

(29) (*CRUX, 1975*). Since the sum of the elements of any subset cannot exceed $90 + 91 + \cdots + 99 = 945$, the sum of the elements of the subsets of S can be found amongst the numbers $1, 2, \ldots, 945$. Since the set S contains 10 elements, we can form $2^{10} - 1 = 1023$ nonempty different subsets. The Pigeonhole Principle then allows us to conclude that there exist (at least) two subsets having the same sum. By removing the elements which are common to these two subsets, we obtain two disjoint subsets with the same sum.

(30) There are exactly 50 possible remainders when we divide the numbers by 50, and these remainders are the numbers: 0,1,2,..., 49. Since we have 51 integers and only 50 possible remainders, it follows that using the Pigeonhole Principle, there are at least two numbers amongst these 51 integers having the same remainder. Then, the difference of these two integers has 0 as a remainder and is therefore divisible by 50.

(31) For each n-th day of the year, let a_n be the total number of solved problems between the first day and the n-th day inclusively. Then $a_1, a_2, \ldots$ is a strictly increasing sequence of positive integers. Consider another sequence $b_1, b_2, \ldots$ obtained by adding 20 to each element of the preceding sequence, that is $b_n = a_n + 20$, $n = 1, 2, \ldots$. The b_n's are strictly increasing and are also all distinct. But for a period of eight consecutive weeks (one needs to consider at least seven consecutive weeks) during the year, the student cannot solve more than $11 \cdot 8 = 88$ problems. Then, the numbers a_n are located between 1 and 88 inclusively, while the b_n's are between 21 and 108 inclusively. Since there are 56 days in eight weeks, the concatenation of the two sequences gives

$$a_1, a_2, \ldots, a_{56}, a_1 + 20, a_2 + 20, \ldots, a_{56} + 20,$$

which yields a total of 112 distinct integers all located between 1 and 108 inclusively. By the Pigeonhole Principle, at least two elements of the concatenated sequence must be equal. One of the two must be in the first half of the sequence and the other in the second part. Let a_j and $a_k + 20$ be these two integers. We then have $a_k - a_j = 20$, which implies that the student must solve exactly 20 problems between the $(j + 1)$-th day and the k-th day of the year.

(32) Divide the surface of the table into squares of 3 inches by 3 inches. We then have a total of 2000 squares. The diagonal of each of these squares is $\sqrt{18}$ inches long, that is approximately 4.25 inches. Therefore, a cylindrical glass of diameter 5 inches will cover entirely any given square. Hence, if we place seven marbles in each square, there will be a total of 14000 marbles on the table. Hence, by the Pigeonhole Principle, since we have a total of 14001 marbles, one of these will contain at least eight marbles.

(33) We easily see that the number N_1 of secants thus drawn is given by $N_1 = \binom{n}{2}$. Let N_2 be the number of points of intersection of these secants. For any group of points taken four by four, there is exactly two secants joining the points that intersect inside the circle, so that $N_2 = \binom{n}{4}$.

We are then ready to count the number of regions in terms of n. At each drawn secant, the circle is divided into an additional region. At each intersection point, the circle is divided into an additional region. The solution is therefore given by

$$1 + N_1 + N_2 = \binom{n}{2} + \binom{n}{4} + 1,$$

which can also be written in the polynomial form $n^4/24 - n^3/4 + 23n^2/24 - 3n/4 + 1$, provided that $n > 3$.

(34) Let $f(n)$ be the number of required moves, that is the number of moves that are necessary to succeed in transferring a tower of n disks. It is easy to see that we must

(a) first move the $n-1$ disks from the top of the first post to the second post (using in the process the third post);

(b) then move the largest disk to the third post;

(c) and finally move the $n-1$ disks from the second post to the third (using if need be the first post).

We then obtain, by setting $f(0) = 0$ and $f(1) = 1$,

$$f(n) = 2f(n-1) + 1 \qquad (n \geq 1).$$

We observe that $f(2) = 3$, $f(3) = 7, \ldots$, and we then conjecture that

$$f(n) = 2^n - 1,$$

a result which can easily be proved by induction.

(35) Let N be an arbitrary positive integer. Let F_{i_1} be such that $F_{i_1} \leq N < F_{i_1} + 1$. Set $\Delta_1 = N - F_{i_1}$. If $\Delta_1 = 0$, we are done, since $N = F_{i_1}$. Otherwise, let F_{i_2} with $i_2 < i_1$ be such that $F_{i_2} \leq \Delta_1 < F_{i_2} + 1$. If $\Delta_2 := \Delta_1 - F_{i_2} = 0$, we are done, since in this case $N = \Delta_1 + F_{i_1} = \Delta_2 + F_{i_2} + F_{i_1} = F_{i_1} + F_{i_2}$. Otherwise, we choose F_{i_3} such that $F_{i_3} \leq \Delta_2 < F_{i_3} + 1$, and so on. The process will end since the sequence of positive integers Δ_i is decreasing, so that eventually we will obtain $\Delta_r = 0$ for a certain positive integer r, in which case we have

$$N = F_{i_1} + F_{i_2} + \cdots + F_{i_r}.$$

(36) (Problem #360 in Barbeau, Klamkin & Moser [3]) We first observe that

$$
\begin{aligned}
1^2 - 0^2 &= 1 \\
3^2 - 2^2 &= 5 \\
5^2 - 4^2 &= 9 \\
7^2 - 6^2 &= 13 \\
&\vdots
\end{aligned}
$$

while

$$
\begin{aligned}
2^2 - 1^2 &= 3 \\
4^2 - 3^2 &= 7 \\
6^2 - 5^2 &= 11 \\
8^2 - 7^2 &= 15 \\
&\vdots
\end{aligned}
$$

Hence, it easily follows that

$$((m+3)^2 - (m+2)^2) - ((m+1)^2 - m^2) = 4 \qquad (m = 0, 1, 2, 3, \ldots);$$

that is

$$4 = m^2 - (m+1)^2 - (m+2)^2 + (m+3)^2 \qquad (m = 0, 1, 2, 3, \ldots).$$

It follows from this that if n can be written as

$$n = e_1 1^2 + e_2 2^2 + e_3 3^2 + e_4 4^2 + \cdots + e_k k^2,$$

the same is true for $n + 4$, since we then have

$$n + 4 = e_1 1^2 + e_2 2^2 + \cdots + e_k k^2 + (k+1)^2 - (k+2)^2$$
$$- (k+3)^2 + (k+4)^2.$$

As we mentioned in the statement of the problem, the numbers 1, 2, 3 and 4 can be written in the stated form. We may therefore conclude that all the integers can be written in this form. Our argument therefore establishes the result without however providing the explicit form taken by any given integer. Curiously, it is nevertheless possible to obtain explicitly such a representation. Here it is. Each integer ≥ 5 is of the form $4r + 1$, $4r + 2$, $4r + 3$ or $4r + 4$, with $r \geq 1$, and we can establish that

$$4r + 1 = 1^2 + \sum_{i=1}^{2r} (-1)^{i+1} ((2i)^2 - (2i+1)^2),$$

$$4r + 2 = -1^2 - 2^2 - 3^2 + 4^2 + \sum_{i=2}^{2r+1} (-1)^i ((2i+1)^2 - (2i+2)^2),$$

$$4r + 3 = -1^2 + 2^2 + \sum_{i=1}^{2r} (-1)^{i+1} ((2i+1)^2 - (2i+2)^2),$$

$$4r + 4 = -1^2 - 2^2 + 3^2 + \sum_{i=2}^{2r+1} (-1)^i ((2i)^2 - (2i+1)^2).$$

(37) We construct a procedure, using MAPLE software (here, we have used version 5), which gives the positive integers $n \leq N$ such that $(!n, n!) \neq 2$.

```
> kurepa:=proc(N)
> local n;
> for n from 1 to N do
> if gcd(sum(k!,k=0..n-1),n!)<>2
> then print(' n'=n) else fi;od; end:
> kurepa(1000);
```

(38) Assume that the integers a and b are increased by n. Then we have

$$a + n = (b + n)652 + 8634 - 651n.$$

Since the remainder must be positive, it follows that $651n \leq 8634$, that is $n \leq 13.26$. Hence, $n = 13$.

REMARK: More generally, with $a = bq + r$ instead of $(*)$, the quantity n is given by $n = \left[\frac{r}{q-1} \right]$.

(39) Let n be the number of "1"s in N. We then have

$$N = 1 + 2^1 + 2^2 + 2^3 + \cdots + 2^{n-1} = 2^n - 1.$$

Therefore,

$$\begin{aligned}
N^2 &= 2^{2n} - 2^{n+1} + 1 = 2^{n+1} \left(2^{n-1} - 1 \right) + 1 \\
&= 2^{n+1} \left(2^{n-2} + 2^{n-3} + \cdots + 2 + 1 \right) + 1 \\
&= 2^{2n-1} + 2^{2n-2} + \cdots + 2^{n+2} + 2^{n+1} + 1,
\end{aligned}$$

an expression which can be written as follows in basis 2:

$$\underbrace{11\ldots11}_{n-1}\underbrace{00\ldots00}_{n}1.$$

(40) Let $N = 7^{37} + 13^{37} + 19^{37}$. We will show that $3|N$ and that $13|N$. Indeed,

$$N = (6+1)^{37} + (12+1)^{37} + (18+1)^{37} = 3A + 3$$

for a certain integer A, while

$$N = (13-6)^{37} + 13^{37} + (13+6)^{37} = 13^{37}B + (-6)^{37} + 6^{37} = 13^{37}B$$

for a certain integer B. Since $(3,13) = 1$, it follows that $39|N$.

REMARK: This result remains true when the integer 37 is replaced by an arbitrary odd positive integer. Moreover, note that by using congruences, the proof is almost immediate.

(41) We first observe that

$$49 = 48 + 1, \quad 2352 = 7^2 \cdot 48 \quad \text{and} \quad 2304 = 48^2.$$

We therefore need to show that

(1) $$48^2 | 49^n - 49 \cdot 48 \cdot n - 1.$$

In fact, we will show the more general result

(2) $$(a-1)^2 | a^n - a(a-1)n - 1.$$

First of all, we observe that

$$\begin{aligned} a^n - a(a-1)n - 1 &= (a^n - 1) - a(a-1)n \\ &= (a-1)(a^{n-1} + a^{n-2} + \cdots + a + 1) - a(a-1)n \\ &= (a-1)\left(a^{n-1} + a^{n-2} + \cdots + a + 1 - an\right). \end{aligned}$$

Since the expression $a^{n-1} + a^{n-2} + \cdots + a + 1 - an$ vanishes when $a = 1$ and is divisible by $a - 1$, we have

$$a^n - a(a-1)n - 1 = (a-1)^2 \cdot N$$

for a certain positive integer N, which establishes (2) and therefore (1).

(42) Let n be an arbitrary positive integer and let $N = n^4 + 2n^3 + 2n^2 + 2n + 1$. It is clear that

$$\begin{aligned} N &= (n+1)^4 - (2n^3 + 4n^2 + 2n) \\ &= (n+1)^4 - 2n(n+1)^2 = (n+1)^2(n^2+1). \end{aligned}$$

Therefore, if N is a perfect square, there exists a positive integer a such that $(n+1)^2(n^2+1) = a^2$, in which case there exists another integer b such that $n^2 + 1 = b^2$. Since two perfect squares cannot be consecutive, the result is proved.

(43) We write $N = 10a + b$ where $0 < a \leq 9$, $0 \leq b \leq 9$ and $M = 10b + a$. In this case, $M - N = 9(b-a)$ and the first result is proved. In order to find the integers N such that $|M - N| = 18$, it is enough to choose $|b - a| = 2$. Therefore, $N = 13, 20, 24, 31, 35, 42, 46, 53, 57, 64, 68, 75, 79, 86, 97$.

(44) The answer is YES. Since $n = 3m + r$, $0 \leq r \leq 2$, it is obvious that $(3, n^2 + 1) = 1$, and the statement is verified.

(45) The answer is YES. By simply writing $n = 5m + r$, where $0 \leq r \leq 4$, we easily obtain the result. The result is clearly the same when we replace 5 by 7.

(46) Let $n = kp + r$, $0 \leq r \leq p - 1$. Using the Binomial Theorem, we obtain

$$
\begin{aligned}
N(n) &= a_0 + a_1(kp + r) + \cdots + a_{s-1}(kp + r)^{s-1} + a_s(kp + r)^s \\
&= N(r) + pM,
\end{aligned}
$$

for a certain integer M. Hence, $p|N(n)$ if and only if $p|N(r)$.

Setting $n = 7k + r$, $0 \leq r \leq 6$, we find that the required integers are those of the form $n = 7k + 1$ as well as those of the form $n = 7k + 4$, where $k \in \mathbb{Z}$.

(47) (*1987 American Invitational Mathematics Examination*). Let N be the number to compute. Since $324 = 18^2$ and since

$$a^4 + 18^2 = (a^2 + 18)^2 - 36a^2 = (a^2 + 18 + 6a)(a^2 + 18 - 6a),$$

the number N can be written as

$$
\begin{aligned}
N &= \prod_{k=0}^{4} \frac{(10 + 12k)^4 + 18^2}{(4 + 12k)^4 + 18^2} \\
&= \prod_{k=0}^{4} \frac{[(10 + 12k)^2 + 18 + 6(10 + 12k)][(10 + 12k)^2 + 18 - 6(10 + 12k)]}{[(4 + 12k)^2 + 18 + 6(4 + 12k)][(4 + 2k)^2 + 18 - 6(4 + 12k)]} \\
&= \prod_{k=0}^{4} \frac{(144k^2 + 312k + 178)(144k^2 + 168k + 58)}{(144k^2 + 168k + 58)(144k^2 + 24k + 10)} \\
&= \prod_{k=0}^{4} \frac{144k^2 + 312k + 178}{144k^2 + 24k + 10}.
\end{aligned}
$$

But since $144(k + 1)^2 + 24(k + 1) + 10 = 144k^2 + 312k + 178$, the number N can be written as

$$N = \frac{144 \cdot 4^2 + 312 \cdot 4 + 178}{10} = 373.$$

(48) Consider the number 10101 in basis $b \geq 2$. Then

$$
\begin{aligned}
10101 = 1 \cdot b^4 + 0 \cdot b^3 + 1 \cdot b^2 + 0 \cdot b + 1 \cdot b^0 &= b^4 + b^2 + 1 \\
&= (b^2 + b + 1)(b^2 - b + 1),
\end{aligned}
$$

a product of two integers larger than 1.

(49) The product of four consecutive integers is

$$N := n(n + 1)(n + 2)(n + 3).$$

Since a member of the product is divisible by 4 and another is divisible by 2, this shows that $8|N$. On the other hand, if we write $n = 3k + r$, $0 \leq r \leq 2$, we easily see that $3|N$. Since $(3, 8) = 1$, we conclude that $24|N$.

(50) If n is an odd positive integer, we know that $a + b|a^n + b^n$. Therefore,

$$7 = 1 + 6|1^{47} + 6^{47}, \quad 7 = 2 + 5|2^{47} + 5^{47}, \quad 7 = 3 + 4|3^{47} + 4^{47},$$

and the result follows.

(51) (*CRUX, 1987; solution given by Aage Bondesen*). Let $N = n(n+1)(n+2)(n+3)(n+4)$ be the given product. It is clear that N must contain two or three multiples of 2, one or two multiples of 3, only one multiple of 5 and no more than one multiple of any other prime number. Thus, if the product is a perfect square, each of the integers $n+j$ ($0 \leq j \leq 4$) can be written as

$$2^r 3^s 5^{2a} 7^{2b} \cdots \qquad (r \geq 2, s \geq 1, a \geq 1, b \geq 0, \ldots).$$

In short, each of the integers $n+j$ ($0 \leq j \leq 4$) is of one of the following forms:

(i) r even, s even:
$$n + j = 2^{2k} 3^{2m} 5^{2a} 7^{2b} \cdots = \left(2^k 3^m 5^a 7^b \cdots\right)^2, \text{ a perfect square;}$$

(ii) r odd, s even:
$$n + j = 2^{2k+1} 3^{2m} 5^{2a} 7^{2b} \cdots = 2\left(2^k 3^m 5^a 7^b \cdots\right)^2, \text{ twice a perfect square;}$$

(iii) r even, s odd:
$$n + j = 2^{2k} 3^{2m+1} 5^{2a} 7^{2b} \cdots = 3\left(2^k 3^m 5^a 7^b \cdots\right)^2, \text{ that is three times a perfect square;}$$

(iv) r odd, s odd:
$$n + j = 2^{2k+1} 3^{2m+1} 5^{2a} 7^{2b} \cdots = 6\left(2^k 3^m 5^a 7^b \cdots\right)^2, \text{ that is six times a perfect square.}$$

But we have five factors in the product N, each being of one of the above four types. Using the Pigeonhole Principle, we may conclude that two of the factors $n+j$ must be of the same type. Let us first examine the possibility that it is one of the types (ii), (iii) or (iv). This is not possible, since if we take for example type (ii), we would have that two amongst five consecutive numbers belong to the sequence 2, 8, 18, 32, 50, $\ldots$, that is numbers separated by at least 6. Therefore, two of the factors $n+j$ must be of type (i), that is perfect squares. But the only chain of five consecutive numbers which contains two perfect squares is 1, 2, 3, 4, 5, whose product is equal to 120, which is not a perfect square. This completes the proof.

(52) We only need to observe that $n^5 - n = (n-2)(n-1)n(n+1)(n+2) + 5n(n^2-1)$.

(53) Since n and $(n+1)$ are two consecutive integers, $2|n(n+1)$. To show that $3|n(n+1)(2n+1)$, it is enough to consider the three cases: $n = 3k + r$, $0 \leq r \leq 2$. If $n = 3k$ or $n = 3k + 2$, the result is immediate. When $n = 3k + 1$, we have that $3|(2n+1)$. The result is therefore true for each $n \geq 1$.

(54) We only need to observe that using the Binomial Theorem, there exists a positive integer M such that

$$(a+1)^{n+1} = 1 + (n+1)a + a^2 M.$$

(55) Observe that

$$n^2 + 1 = (n+1-1)^2 + 1 = (n+1)^2 - 2(n+1) + 2.$$

Hence, for the relation to be true, we must have that $(n+1)|2$, that is $n = 1$.

(56) Since

$$n^6 + 206 = (n^2 + 2 - 2)^3 + 206 = (n^2 + 2)^3 - 6(n^2 + 2)^2$$
$$+ 12(n^2 + 2) + 198,$$

the relation will be true if $(n^2+2)|198$. The only possibilities are therefore: $n^2 + 2 = 1, 2, 3, 6, 9, 11, 18, 22, 33, 66, 99, 198$, in which case the positive values of the required n are: 1, 2, 3, 4, 8 and 14. For example, with $b = 5$ and $a = 2$, we have $5|2^2 + 1$, $5 \nmid 2 \cdot 2^2$ and $5 \nmid 2^4 + 1 = 17$.

(57) First observe that $n^6 + 216 = (n^3 + 2 - 2)^2 + 216$. Then we proceed as in the preceding problem, and we obtain that the only possible integer n satisfying the given property is $n = 2$.

(58) The answer is NO. Indeed, if $b|a^2+1$, then there exists a positive integer k such that $a^2+1 = kb$. It follows that $a^4+1 = (kb-1)^2+1 = k^2b^2-2kb+2$ and therefore that in order to have $b|a^4 + 1$, we must have that $b|2$. It is easy to choose integers a and b in such a way that $b|a^2 + 1$ and $b \nmid 2a^2$. For instance, with $b = 5$ and $a = 2$, we obtain that $5|2^2 + 1$, $5 \nmid 2 \cdot 2^2$ and $5 \nmid 2^4 + 1 = 17$.

(59) (a) Since

$$n\left(\binom{n}{k}, \binom{n-1}{k-1} \right) = \left(n\binom{n}{k}, n\binom{n-1}{k-1} \right) = \left(n\binom{n}{k}, k\binom{n}{k} \right)$$
$$= \binom{n}{k}(n, k),$$

we obtain the result.

(b) This follows from the fact that

$$(n + 1 - k)\left(\binom{n}{k}, \binom{n}{k-1} \right) = \left((n + 1 - k)\binom{n}{k}, k\binom{n}{k} \right)$$
$$= \binom{n}{k}(n + 1 - k, k)$$

and from the fact that $(n + 1 - k, k) = (n + 1, k)$.

(c) Let

$$A = \left\{ x \in \mathbb{Z} \,\Big|\, \frac{x}{n+2-k}\binom{n}{k-1} \text{ is an integer} \right\}.$$

Obviously, $x = n + 2 - k \in A$. Since

$$\frac{k-1}{n+2-k}\binom{n}{k-1} = \frac{(k-1)n!}{(n+2-k)(k-1)!\,(n+1-k)!} = \binom{n}{k-2},$$

then $x = k - 1 \in A$. Hence, any linear combination of $(k - 1)$ and $(n+2-k)$ also belongs to A. In particular, $(n+2-k, k-1)$ belongs also to A and since $(n + 2 - k, k - 1) = (n + 1, k - 1)$, we obtain the result.

REMARK: Parts (a) and (b) of this problem could have been solved as in part (c).

(60) (*Putnam, 1984*). It is clear that

$$f(n + 2) - f(n + 1) = (n + 2)! = (n + 2)(n + 1)!$$
$$= (n + 2)\Big(f(n + 1) - f(n) \Big).$$

Therefore, choosing $P(x) = x + 3$ and $Q(x) = -x - 2$, the result follows.

(61) We only need to observe that

$$(2^{3(n+1)+3} - 7(n+1) - 8) - (2^{3n+3} - 7n - 8) = 7(2^{3n+3} - 1)$$

and that $7 | 2^{3n+3} - 1$, and thereafter use induction.

REMARK: This result follows also from Problem 54. Indeed,

$$7^2 | \{(7+1)^{3n} - 7n - (7+1)\} = 2^{3n+3} - 7n - 8.$$

(62) We have $a = 10q + r$, $0 \le r < 10$. Therefore, we must have that $10 | r^{10} + 1$, and this is why we must have $r = 3$ or 7.

(63) The answer is YES. Indeed,

$$
\begin{aligned}
2^{2n} - 1 &= 4^n - 1 = (3+1)^n - 1 \\
&= 3^n + \binom{n}{1} 3^{n-1} + \binom{n}{2} 3^{n-2} + \cdots + \binom{n}{n-1} 3 + 3^0 - 1 \\
&= 3^n + \binom{n}{1} 3^{n-1} + \binom{n}{2} 3^{n-2} + \cdots + \binom{n}{n-1} 3,
\end{aligned}
$$

an expression which is divisible by 3.

(64) Let $6k + 5$ be an integer. To show that this integer can be written in the form $3m - 1$, we must find an integer m such that $6k + 5 = 3m - 1$. To do so, it is enough to choose $m = 2k + 2$, thus ending the proof of the first part. For the second part, let $3k - 1$ be an integer. Can one find, for each positive integer k, an integer m such that $3k - 1 = 6m + 5$, that is such that $6m = 3k - 6$? The answer is NO, because if k is odd, it is clear that it is impossible to find such an integer m.

(65) The answer is YES. Indeed, if $n = 8k + 7 = 6\ell + 5$ for certain integers k and ℓ, then $4k = 3\ell - 1$, which happens if and only if $\ell = 3, 7, 11, 15, \ldots$, that is when ℓ is of the form $4m + 3$. Hence, all numbers n of the form

$$n = 6\ell + 5 = 6(4m + 3) + 5 = 24m + 23$$

are automatically of the two required forms, and, of course, there are infinitely many of them.

(66) Let $k \in \mathbb{N}$; then $M_k = 2p_2 p_3 \cdots p_k + 1 = 2(2r + 1) + 1 = 4r + 3$. But we know that each perfect square is of the form $4r$ or $4r + 1$, and certainly not of the form $4r + 3$.

(67) Let $N = n^2 = m^3$ for certain positive integers n and m. We easily see that n^2 is of the form $7k$, $7k + 1$, $7k + 2$ or $7k + 4$, while m^3 is of the form $7k$, $7k + 1$ or $7k + 6$. Hence, N must be of the form $7k$ or $7k + 1$.

(68) Since x^2 and y^2 are of the form $4n + 1$, we see that $x^2 + y^2$ is of the form $4m + 2$. But each perfect square is of the form $4k$ or of the form $4k + 1$. Thus the result.

(69) The Binomial Theorem gives

$$(n+1)^n - 1 = \sum_{k=1}^{n} n^k \binom{n}{k} = n^2 + \sum_{k=2}^{n} \binom{n}{k} n^k,$$

and since $\binom{n}{k}$ is an integer, the result is immediate.

(70) By using the Binomial Theorem, we obtain

$$n^k - 1 = [(n-1)+1]^k - 1$$
$$= (n-1)^k + k(n-1)^{k-1} + \cdots + k(n-1),$$

and we observe that all the terms of this last expression are divisible by $(n-1)^2$ except perhaps the term $k(n-1)$, thus the result. The more general case can be treated in a similar manner, by considering the relation $n^k - a^k = ((n-a)+a)^k - a^k$.

(71) (a) By using the Binomial Theorem, we find

$$a^n = (a-b+b)^n = (a-b)^n + \binom{n}{1}(a-b)^{n-1}b + \cdots$$

$$+ \binom{n}{n-2}(a-b)^2 b^{n-2} + \binom{n}{n-1}(a-b)b^{n-1} + b^n.$$

Hence, there exists an integer K such that

$$\frac{a^n - b^n}{a-b} = (a-b)^{n-1} + \binom{n}{1}(a-b)^{n-2}b + \cdots$$

$$+ \binom{n}{n-2}(a-b)b^{n-2} + nb^{n-1} = K(a-b) + nb^{n-1}.$$

It follows that

(1) $$\left(\frac{a^n - b^n}{a-b}, a-b \right) = \left(nb^{n-1}, a-b \right).$$

Similarly,

$$b^n = (a - (a-b))^n = a^n - \binom{n}{1}a^{n-1}(a-b) + \cdots$$

$$+ (-1)^{n-1}\binom{n}{n-1}a(a-b)^{n-1} + (-1)^n(a-b)^n,$$

and therefore, we find

$$\frac{a^n - b^n}{a-b} = na^{n-1} + L(a-b)$$

for a certain integer L. Hence,

(2) $$\left(\frac{a^n - b^n}{a-b}, a-b \right) = \left(na^{n-1}, a-b \right).$$

Using the equations (1) and (2), we obtain

$$\left(\frac{a^n - b^n}{a-b}, a-b \right) = \left(n(a,b)^{n-1}, a-b \right).$$

Indeed, let $d = (na^{n-1}, a-b) = (nb^{n-1}, a-b)$ and $g = \left(n(a,b)^{n-1}, a-b \right)$. Since $d|na^{n-1}$ and $d|nb^{n-1}$, it follows that

$$d|(na^{n-1}, nb^{n-1}) = n(a^{n-1}, b^{n-1}) = n(a,b)^{n-1}.$$

By using this relation and the fact that $d|(a-b)$, we have $d|g$. Conversely, since $g|n(a,b)^{n-1}$, then $g|na^{n-1}$; and since $g|(a-b)$, it follows that $g|(na^{n-1}, a-b) = d$. Hence, $g = d$, which gives the result.

(b) Setting $b = -B$, we have

$$\frac{a^n + b^n}{a + b} = \frac{a^n - B^n}{a - B}.$$

Part (a) allows us to conclude that

$$\left(\frac{a^n + b^n}{a + b}, a + b\right) = \left(\frac{a^n - B^n}{a - B}, a - B\right)$$
$$= \left(n(a, B)^{n-1}, a - B\right) = \left(n(a, b)^{n-1}, a + b\right),$$

as was required to prove.

(c) Part (b) allows us to conclude that

$$\left(\frac{a^p + b^p}{a + b}, a + b\right) = (p, a + b) = 1 \text{ or } p$$

depending whether p divides $a + b$ or not.

(72) If $2|n$ and $n > 2$, then $(n-1)!$ is even while $n^k - 1$ is odd. If $2 \nmid n$ and $n > 5$, then $n - 1$ is even and

$$n - 1 = 2\frac{n-1}{2} \mid (n-2)!$$

so that $(n-1)^2|(n-1)!$. Since $(n-1)! = n^k - 1$, we must have $(n-1)^2|n^k - 1$, and using Problem 70, we must have $(n-1)|k$ and therefore $k \geq n - 1$. In this case, for $n > 5$,

$$n^k - 1 \geq n^{n-1} - 1 > (n-1)!.$$

Hence, the only possible cases yielding a solution are $n = 2$, $n = 3$ and $n = 5$, the corresponding values of k then being 1, 1 and 2.

(73) (a) Since $b = aq_1 + r_1$ and since $a > r_1$, $q \geq 1$, then $b = aq_1 + r_1 > r_1q_1 + r_1 \geq 2r_1$. Similarly, $a = r_1q_2 + r_2 > r_2q_2 + r_2 \geq 2r_2$. Finally, for each $k \geq 1$,

$$r_k = r_{k+1}q_{k+2} + r_{k+2}, \quad 0 < r_{k+2} < r_{k+1},$$

so that

$$r_k = r_{k+1}q_{k+2} + r_{k+2} > r_{k+2}q_{k+2} + r_{k+2} \geq 2r_{k+2}.$$

(b) Since

$$b > 2r_1 > 2^2r_3 > 2^3r_5 > \ldots > 2^{(j+1)/2}r_j \geq 2^{(j+1)/2} > 2^{j/2},$$

we conclude that $j < 2\log b/\log 2$, and the result follows.

(74) Consider the numbers of the form $n = 3^k$, $k = 0, 1, 2, 3, \ldots$, so that

$$(*) \quad 2^n + 1 = 2^{3^k} + 1 = 2^{3^{k-1}3} + 1 = \left(2^{3^{k-1}} + 1\right)\left(\left(2^{3^{k-1}}\right)^2 - 2^{3^{k-1}} + 1\right).$$

We will show that

(i) the first factor on the right-hand side is divisible by 3^{k-1}, while
(ii) the second factor is divisible by 3.

From this, it follows that the left-hand side of $(*)$ is divisible by $3^{k-1} \cdot 3 = 3^k = n$.

To show (i), we will show by induction that $2^{3^m} + 1$ is divisible by 3. It is clear that this result is true for $m = 1$, since $3|9$. Assume now that $2^{3^{m-1}} + 1$ is divisible by 3. Thus, if we set $x = 2^{3^{m-1}}$, we may write

$$2^{3^m} + 1 = \left(2^{3^{m-1}}\right)^3 + 1 = x^3 + 1 = (x+1)(x^2 - x + 1)$$
$$= (2^{3^{m-1}} + 1)(x^2 - x + 1),$$

an expression divisible by 3, because of our induction hypothesis.

In order to show (ii), we only need to observe that if a is an odd positive integer (namely, here $a = 3^{k-1}$), then $(2^a)^2 - 2^a + 1$ is divisible by 3, which is indeed the case since

$$(2^a)^2 - 2^a + 1 = 4^a - 2^a + 1 \equiv 1 - 2 + 1 = 0 \pmod{3}.$$

(75) Using Euclidean division, we can write $m = nq + r$, $0 \le r < n$. Hence,

$$a^m - 1 = a^{nq+r} - a^{nq} + a^{nq} - 1 = a^{nq}(a^r - 1) + (a^n)^q - 1.$$

In light of Problem 8, $a^n - 1 | a^{nq} - 1$, so that we have

$$a^n - 1 | a^m - 1 \iff a^n - 1 | a^{nq}(a^r - 1).$$

But $(a^n - 1, a^{nq}) = 1$, meaning that $a^n - 1$ must divide $a^r - 1$, and this is true provided $r \ne 0$. Finally, the result follows.

(76) We only need to observe that $N_n = 10^{n-1} + 10^{n-2} + \cdots + 10 + 1 = (10^n - 1)/9$ and to use the fact that $10^n - 1 | 10^m - 1$ if and only if $n|m$ (see the preceding problem).

(77) Any number in the sequence is of the form $100k + 11 = 4(25k + 2) + 3$, where k is a nonnegative integer (made up entirely of the digit "1"). Then, any integer in the sequence leaves 3 as a remainder when it is divided by 4 and therefore cannot be a perfect square, since it is well known that any perfect square is of the form $4r$ or $4r + 1$.

(78) The required number is $2^6 \cdot 3^6$. For the general case, the smallest number is $2^{[n,m]} \cdot 3^{[n,m]}$.

(79) Since $371 = 370 + 1$, and since 370 is already in the list, it is obvious that 371 is this fourth number.

(80) There are

$$1000 - \left[\frac{1000}{2}\right] - \left[\frac{1000}{3}\right] - \left[\frac{1000}{5}\right] + \left[\frac{1000}{6}\right] + \left[\frac{1000}{10}\right]$$
$$+ \left[\frac{1000}{15}\right] - \left[\frac{1000}{30}\right] = 266$$

such numbers.

(81) We are looking for $a, b \in \mathbb{N}$ such that

$$p = a^2 - b^2 = (a+b)(a-b).$$

But since p is a prime number, we must have $a - b = 1$ and $a + b = p$. Hence,

$$a = \frac{p+1}{2} \quad \text{and} \quad b = \frac{p-1}{2}.$$

We have thus obtained that

$$p = \left(\frac{p+1}{2}\right)^2 - \left(\frac{p-1}{2}\right)^2.$$

(82) The result follows immediately from the fact that the system of equations $a-b=1$ and $a+b=p$ has a solution, namely $a=(p+1)/2$, $b=(p-1)/2$.

(83) The answer is YES. However, uniqueness does not hold because for instance $21=11^2-10^2=5^2-2^2$.

(84) This follows from the fact that $7=10-3$ is a divisor of 10^9-3^9.

(85) Let m be this integer. Then,

$$m=2k+1=a^2+b^2, \qquad a \text{ odd}, \ b \text{ even}.$$

We then have that there exist nonnegative integers M and N such that $a=2M+1$ and $b=2N$. Hence, $a^2=4M^2+4M+1=4K+1$ for a certain integer K, and $b^2=4N^2$. We have therefore established that

$$m=a^2+b^2=4K+1+4N^2=4n+1,$$

for a certain integer n, as required.

(86) If a and b are odd, that is $a=2m+1$ and $b=2n+1$, say, then we have $a^2+b^2=4k+2=c^2$, which is impossible since any perfect square is of the form $4k$ or $4k+1$.

(87) (*TYCM, March 85*). For $k=0$, we have $10^k-1=10^0-1=0=0^3$, a cube. If $k<0$, then 10^k-1 is not an integer. Hence, assume that $k \geq 1$ and that $10^k-1=n^3$. Setting $N_k=\frac{1}{9}(10^k-1)$, we then have

$$N_k=\underbrace{11\ldots1}_{k}=10^{k-1}+10^{k-2}+\cdots+10+1.$$

But, for $j \geq 1$, there exists a constant $A \geq 1$ such that $10^j=(3^2+1)^j=3A+1$, which allows us to conclude that there exists a constant $M \geq 1$ such that $N_k=3M+k$. Since $9|(10^k-1)=n^3$, it follows that $27|10^k-1$. Therefore $3|N_k$ so that $k=3r$ for a certain positive integer r. We have thus established that $10^{3r}-1$ and 10^{3r} are two consecutive cubes (with $r \geq 1$), a contradiction. Hence, the only integer k such that 10^k-1 is a cube is $k=0$.

(88) Since $a|42n+37-6(7n+4)=13$, the result follows.

(89) By hypothesis, we have that $(a+b)/ab$ is an integer and therefore that $ab|(a+b)$. Since $a|a+b$ and $a|a$, it follows that $a|b$. Moreover, $b|a+b$ and $b|b$ imply $b|a$. Now clearly, $a|b$ and $b|a$, with a,b positive, implies that $a=b$. It then follows that, for $1/a+1/a=2/a$ to be an integer, we must have $a|2$, which means that $a=1$ or 2.

(90) We write $a=4A$, $b=4B$ where $(A,B)=1$. Then, $(a^2,b^3)=16(A^2,4B^3)$, and since $(A^2,B^3)=1$, we conclude that each common divisor of A^2 and $4B^3$ must be 1 or 4. Hence, the possible values of (a^2,b^3) are 16 and 64.

(91) Set $d_1=(3a+5b,5a+8b)$ and $d=(a,b)$. Since $d_1|(3a+5b)$ and $d_1|(5a+8b)$, then $d_1|\big(8(3a+5b)-5(5a+8b)\big)$, that is $d_1|a$. In a similar way, we obtain that $d_1|b$ and consequently $d_1|d$.

Since $d=(a,b)$, it follows that $d|(3a+5b)$ and $d|(5a+8b)$ and therefore $d|d_1$. Since $d|d_1$ and $d_1|d$, we conclude that $d=d_1$.

The general case can he handled in a similar way, and we obtain $(ma+nb,ra+sb)=(a,b)$ when $ms-nr=1$.

(92) Let $r = (d, m)$ and set $s = d/r$. Since $r|m$, it is enough to show that $s|n$. Letting $M = m/r$, then

$$\left(\frac{d}{r}, \frac{m}{r} \right) = (s, M) = 1.$$

Since $d|mn$, there exists an integer t such that $dt = mn$, so that $rst = Mrn$, that is $s|Mn$. Since $(s, M) = 1$, it follows that $s|n$. Let $d' = (r, s)$; then $d'|r$ and $r|m$ imply that $d'|m$ and $d'|n$. We therefore have $d'|(m, n) = 1$, so that $(r, s) = 1$.

(93) We have that $d|(a + b) + (a - b) = 2a$ and therefore that $d|a$ since d is odd. Similarly, we have that $d|(a + b) - (a - b) = 2b$ and therefore that $d|b$ since d is odd. Hence, $d|(a, b)$.

(94) Since $19^2 = 361$, each composite number ≤ 360 is divisible by a prime number ≤ 17. Since there are only seven prime numbers ≤ 17, it follows by the Pigeonhole Principle that at least two of these given eight composite numbers must be divisible by the same prime number.

(95) Since $ab = r^2 = q_1^{2\alpha_1} \cdots q_k^{2\alpha_k}$ for certain prime numbers $q_1, q_2, \ldots, q_k$ and certain positive integers $\alpha_1, \alpha_2, \ldots, \alpha_k$ and since $(a, b) = 1$, it is clear that some of the $q_i^{2\alpha_i}$'s will be factors of a while the others will be factors of b, thus establishing that a and b are perfect squares.

(96) If it were true, it would follow from Problem 95 that n and $n + 1$ are two consecutive perfect squares, which is not possible.

(97) The only possible values are 1, 2, 7 and 14. Indeed, if $d = (n, n + 14)$, then $d|14$.

(98) The first statement is true because it is equivalent to $3|(n - 1)n(n + 1)$. The second statement is false: simply take $n = 4$. The third statement is true because it is easily shown to be equivalent to $8|4n(n+1)$. The fourth statement is true because $2|n(n + 1)$ and $3|n(n + 1)(n + 2)$.

(99) The answer is YES. Indeed, on the one hand, $3|n(n + 1)(n + 2)$, while on the other hand, one of the two numbers n and $n + 2$ is divisible by 4, the other by 2.

(100) Since we have $n^2 + 47 = n^2 + 48 - 1$, it is enough to show that $24|n^2 - 1$. First of all, any positive integer n is of one of the following six forms: $6k$, $6k + 1$, $6k + 2$, $6k + 3$, $6k + 4$, $6k + 5$. Since $(n, 2) = (n, 3) = 1$, it is clear that n can only be of the form $6k + 1$ or $6k + 5$, in which case it is immediate that $n^2 - 1$ is divisible by 24.

(101) Let $a = dr$ and $b = ds$ where $(r, s) = 1$. Dividing each of the integers a, $2a, \ldots, ba$ by b, we obtain the quotients

(*) $$\frac{r}{s}, \ 2\frac{r}{s}, \ \ldots, (b - 1)\frac{r}{s}, \ ds\frac{r}{s}.$$

Since $(r, s) = 1$, the only integers amongst (*) are those whose numerator is a multiple of s. Since $b = ds$, this will happen exactly d times.

(102) Part (a) is immediate. For part (b), it is sufficient to observe that $(x_0 + b)a + (y_0 - a)b = d$.

(103) Let $d = (a, mn)$; then $d|a$ and $d|mn$. Since $(m, n) = 1$, using Problem 92, we have $d = rs$, $(r, s) = 1$, $r|m$ and $s|n$. But $rs|a$ implies $r|a$ and $s|a$. It follows that $r|(a, m)$ and $s|(a, n)$, so that $d|(a, m)(a, n)$. To complete the proof, we must now show that $(a, m)(a, n)|d$. Let $d_2 = (a, m)$ and

$d_1 = (a, n)$; then $d_2|a$ and $d_2|mn$, so that $d_2|(a, mn)$. But $d_1|a$ and $d_1|mn$ imply $d_1|(a, mn)$. Since $(m, n) = 1$, it follows that $(d_1, d_2) = 1$, which allows us to conclude that $d_1 d_2|(a, mn)$.

(104) Let $d = (n^2 + 3n + 2, 6n^3 + 15n^2 + 3n - 7)$. We then have that $d|6n(n^2 + 3n + 2) - (6n^3 + 15n^2 + 3n - 7) = 3n^2 + 9n + 7$. Now since $d|n^2 + 3n + 2$, it follows that $d|3n^2 + 9n + 7 - 3(n^2 + 3n + 2) = 1$ and therefore $d = 1$.

(105) The first three problems can be solved in a similar way. For (a), we proceed as follows. Let $d = (a + b, a - b)$, so that $d|2b$ and $d|2a$. We then have $d|(2b, 2a) = 2(a, b)$ so that $d|2$, which proves that $d = 1$ or 2. Finally, for (d), it is sufficient to notice that $a^2 - 3ab + b^2 = (a + b)^2 - 5ab$.

(106) (a) Set $d = (a^3 + b^3, a^3 - b^3)$. We then have that $d|2a^3$ and $d|2b^3$, and since $(a, b) = 1$, we have $d|2$. Therefore, $d = 1$ or $d = 2$. More precisely, when a and b are of opposite parity, we find the value 1, while if a and b are of the same parity, we obtain the value 2.

(b) We have

$$(a^2 - b^2, a^3 - b^3) = (a - b)\left(a + b, a^2 - ab + b^2\right)$$
$$= (a - b)\left(a + b, (a + b)^2 - 3ab\right) = (a - b)\left(a + b, 3ab\right).$$

Let $d = (a + b, 3ab)$, so that $d|3b(a + b) - 3ab = 3b^2$ and $d|3a(a + b) - 3ab = 3a^2$, and therefore $d|3$. It follows that $(a^2 - b^2, a^3 - b^3) = a - b$ or $3(a - b)$. More precisely, the value is $a - b$ if $3 \nmid (a + b)$ and $3(a - b)$ if $3|(a + b)$.

(107) (a) False. Indeed, $(2, 3) = (2, 5) = 1$ even though $6 = [2, 3] \neq [2, 5] = 10$.

(b) True. It is enough to show that: $(a, b) = g \implies (a^2, b^2) = g^2$. We know that if $(A, B) = 1$, then $(A^2, B^2) = 1$. But by hypothesis we have $a = Ag$ and $b = Bg$ with $(A, B) = 1$. It follows that $a^2 = A^2 g^2$ and $b^2 = B^2 g^2$, which means that $(a^2, b^2) = (A^2 g^2, B^2 g^2) = g^2(A^2, B^2) = g^2$.

(c) True. Indeed, let $g = (a, b)$ and $h = (a, b, c)$. It is clear that $h|g$. Therefore, it follows that $g|h$. But $g = (a, b) = (a, c)$, which implies that $g|a$, $g|b$ and $g|c$. It follows that $g|(a, b, c) = h$, as was to be shown.

(108) (a) The statement is true. Indeed, let $(a, b) = d$, so that $a = dA$ and $b = dB$ with $(A, B) = 1$. Therefore, $(A^n, B^n) = 1$, and since $a^n|b^n$, we obtain $A^n d^n|B^n d^n$, that is $A^n|B^n$. Hence, $A^n|(A^n, B^n) = 1$, which shows that $A = 1$ and therefore that $d = a$. It follows that $b = dB = aB$, which proves the statement.

One can also prove this result by writing $a = \prod p_i^{a_i}$ and $b = \prod p_i^{b_i}$, and then using the fact that $a^n|b^n$ to obtain that $na_i \leq nb_i$; that is $a_i \leq b_i$ for each i, so that $a|b$.

(b) The statement is true because $a^m|b^n$ implies $a^n a^{m-n}|b^n$ and therefore $a^n|b^n$. From part (a), we draw the conclusion.

(c) False. Indeed, $(2^3)^2|(2^2)^3$, although $2^3 \nmid 2^2$.

(109) We have $(a, b) = 1 \iff$ there exist $x, y \in \mathbb{Z}$ such that $ax + by = 1$. Since $c|a$, there exists $q \in \mathbb{Z}$ such that $a = qc$; therefore, $ax + by = qcx + by = 1 \iff (c, b) = 1$.

(110) We have $(a, bc) = 1 \iff$ there exist $x, y \in \mathbb{Z}$ such that $ax + bcy = 1$. We therefore obtain that $(a, b) = 1$ and $(a, c) = 1$.

(111) Assuming that $(a, b) = 1$, we must show that $1 = (a + b, ab)$. Setting $d = (a + b, ab)$, we obtain that $d|a^2$ and $d|b^2$, so that $d|(a^2, b^2) = 1$. The more general case $(a, b) = d > 1$ can be obtained from the first part, using the fact that $(a/d, b/d) = 1$.

Let a and b be the integers such that $a + b = 186$ and $[a, b] = 1440$. Since $(a, b) = (a + b, [a, b])$ and since $(186, 1440) = 6$, then $a = 6A$ and $b = 6B$ where $(A, B) = 1$. This leads to $A + B = 31$ and $[A, B] = 240$, and since $(A, B) = 1$, we have $AB = 240$. We then have $A(31 - A) = 240$ and therefore $A = 15$ or $A = 16$. The other two numbers are therefore $90 = 2 \cdot 3^2 \cdot 5$ and $96 = 2^5 \cdot 3$.

(112) (a) Let $d = (a, bc)$ and $g = (a, (a, b)c)$. We have that $g|a$ and $g|(ac, bc)$ and therefore that $g|bc$. Consequently, $g|d$. But $d|a$ and $d|bc$ imply that $d|(ac, bc) = (a, b)c$ and therefore that $d|g$. Hence, $d = g$.

(b) From part (a), we have
$$(a, bc) = (a, c(a, b)) = (a, (a, c)(a, b)).$$

(113) Indeed, $(c, ab) = (c, (a, c)b) = (c, b)$.

(114) It is enough to show that the two numbers are both powers of the same prime number. Assume that p^α divides either one of the two expressions. Then, $p^\alpha|mn$, and since $(m, n) = 1$, then either $p^\alpha|m$ and $(p, n) = 1$ or else $p^\alpha|n$ and $(p, m) = 1$. Since both cases are identical, we can assume that $p^\alpha|m$, in which case we must have $(p, n) = 1$. Therefore,

$$p^\alpha|(ma + nb, mn) \iff p^\alpha|(ma + nb)$$
$$\iff p^\alpha|b \iff p^\alpha|(b, m) \iff p^\alpha|(a, n)(b, m).$$

Setting $b = a$, we obtain $(a(m + n), mn) = (a, n)(a, m)$. Since $(m, n) = 1$ also implies $(m + n, mn) = 1$, we conclude that $(a, mn) = (a, n)(a, m)$, thus also obtaining the result of Problem 103.

(115) The answer is NO. This follows from the identity

$$\binom{n}{s}\binom{s}{r} = \binom{n}{r}\binom{n-r}{s-r}$$

and from the fact that both quantities $\binom{s}{r}$ and $\binom{n-r}{n-s}$ are larger than 1.

REMARK: This problem remained unsolved until one thought about using the above identity (see Guy [16], B31). P. Erdős and G. Szekeres [11] asked if the largest prime factor of the greatest common divisor of $\binom{n}{r}$ and $\binom{n}{s}$ is always larger than r, the only counter-example with $r > 3$ being

$$\left(\binom{n}{r}, \binom{n}{s} \right) = \left(\binom{28}{5}, \binom{28}{14} \right) = 2^3 \cdot 3^3 \cdot 5 > 5.$$

(116) Let $(a, b) = d$ with $[a, b] - (a, b) = 143$. First of all, it is clear that $d|143$. We must therefore examine the possibilities $d = 1$, $d = 11$, $d = 13$ and $d = 143$. Set $a = Ad$ and $b = Bd$ with $(A, B) = 1$.

If $d = 1$, then $[a, b] - (a, b) = 143$ becomes $AB - 1 = 143$ and since $(A, B) = 1$, we have $A = a = 16$ and $B = b = 9$, as well as $A = a = 1$ and $B = b = 144$.

If $d = 11$, then $AB - 1 = 13$ and therefore $A = 2$ and $B = 7$ (which gives $a = 22$ and $b = 77$), as well as $A = 1$ and $B = 14$ (which gives $a = 11$ and $b = 154$).

If $d = 13$, we obtain $a = 39$ and $b = 52$.

If $d = 143$, we have $a = 143$ and $b = 286$.

The only six possible (ordered pairs) solutions are therefore

$$\{a,b\} = \{1,144\},\ \{9,16\},\ \{11,154\},\ \{22,77\},\ \{39,52\},\ \{143,286\}.$$

(117) For the first part we proceed as follows. Let $d = (a,b,c)$ and $d_1 = ((a,b),c)$. Since $d|a$ and $d|b$, we have $d|(a,b)$. Similarly, $d|c$; hence, $d|d_1$. On the other hand, $d_1|(a,b)$; it follows that $d_1|a$ and $d_1|b$, and since $d|c$, this shows that $d_1|d$. Since $d|d_1$ and $d_1|d$, we have $d = d_1$.

For the second part, we proceed in the following manner. Let $M = [a,b,c]$, $m = [a,b]$ and $m_1 = [m,c]$. From the definition of m_1, it follows that $m|m_1$ and $c|m_1$. Consequently, $a|m_1$, $b|m_1$ and $c|m_1$; that is $[a,b,c]|m_1$. Conversely, $M = [a,b,c]$ implies $a|M$, $b|M$ and $c|M$, and therefore $[a,b]|M$ and $c|M$. This allows us to conclude that $m_1 = [[a,b],c]|M$ and the result follows.

More generally, we have

$$(a_1,a_2,\ldots,a_n) = (a_1,(a_2,\ldots,a_n)) \quad\text{and}\quad [a_1,a_2,\ldots,a_n] = [a_1,[a_2,\ldots,a_n]].$$

By using Euclid's algorithm, we obtain

$$\begin{aligned}
132 &= 102 \cdot 1 + 30, \\
102 &= 30 \cdot 3 + 12, \\
30 &= 12 \cdot 2 + 6, \\
12 &= 6 \cdot 2.
\end{aligned}$$

It follows from this that $(132,102) = 6$ and therefore that

$$(132,102,36) = ((132,102),36) = (6,36) = 6.$$

Using the above system of equations starting at the second one from the bottom and moving up, we obtain successively

$$\begin{aligned}
6 &= 30 - 12 \cdot 2 = 30 - (102 - 3 \cdot 30) \cdot 2 = 7 \cdot 30 - 2 \cdot 102 \\
&= 7 \cdot (132 - 102) - 102 \cdot 2 = 7 \cdot 132 - 9 \cdot 102.
\end{aligned}$$

On the other hand, since $7 \cdot 6 + (-1) \cdot 36 = 6$, we obtain that

$$7 \cdot (7 \cdot 132 - 9 \cdot 102) - 36 = 6 \Rightarrow 49 \cdot 132 + (-63) \cdot 102 + (-1) \cdot 36 = 6.$$

We may thus choose $x = 49$, $y = -63$ and $z = -1$.

(118) It is easy to see that $(n, n+1, n+2) = ((n, n+1), n+2) = (1, n+2) = 1$.

Since $(n, n+1) = 1$, it follows that $[n, n+1, n+2] = [[n, n+1], n+2] = [n(n+1), n+2]$. Since $(n(n+1), n+2) = (n, n+2) = 1$ or 2, then

$$[n, n+1, n+2] = \begin{cases} n(n+1)(n+2) & \text{if } n \text{ is odd}, \\ n(n+1)(n+2)/2 & \text{if } n \text{ is even}. \end{cases}$$

(119) We know that $(*)$ $(ab,c)[ab,c] = abc$. Since $(a,b) = 1$, it follows that $[a,b] = ab$, and since $(a,c) = (b,c) = 1$, we have $(ab,c) = 1$. Therefore, $(*)$ becomes $[ab,c] = abc$, so that $[[a,b],c] = abc$. By using Problem 117, we reach the conclusion.

(120) The answer is YES. If $(a, b) = 1$, we have $(a^2, b^2) = 1$; hence, using Problem 117, we obtain

$$(a^2, ab, b^2) = ((a^2, b^2), ab) = (1, ab) = 1.$$

(121) The answer is YES. If $(a, b) = 1$, then $(a^2, b^2) = 1$, $[a^2, b^2] = a^2 b^2$ and Problem 120 allows us to obtain that $(a^2, ab, b^2) = 1$. Consequently, from Problem 117, we have

$$[a^2, ab, b^2] = [[a^2, b^2], ab] = [a^2 b^2, ab] = a^2 b^2 = [a^2, b^2].$$

For the general case $(a, b) = d$, it is enough to redo the last part with $(a/d, b/d) = 1$.

(122) The answer is YES. We set $h = (a, b, c)$ and $g = ((a, b), (a, c))$, and we easily show that $g|h$ and $h|g$.

(123) The answer is NO. It is enough to consider the counter-example provided by choosing $a = 6$, $b = 3$ and $c = 15$.

(124) *This problem was stated by the mathematician Jean-Henri Lambert (1728–1777).* Letting $(m, n) = e$ and using the fact that $[m, n](m, n) = mn$, we have

$$d^{[m,n]} - 1 = (d^m - 1 + 1)^{n/e} - 1 = (d^m - 1)^{n/e}$$
$$+ \binom{n/e}{1}(d^m - 1)^{n/e-1} + \cdots + \binom{n/e}{n/e - 1}(d^m - 1)$$

and we conclude that $a|d^{[m,n]} - 1$. Similarly,

$$d^{[m,n]} - 1 = (d^n - 1 + 1)^{m/e} - 1 = (d^n - 1)^{m/e}$$
$$+ \binom{m/e}{1}(d^n - 1)^{m/e-1} + \cdots + \binom{m/e}{m/e - 1}(d^n - 1)$$

and we obtain that $b|d^{[m,n]} - 1$. Since $(a, b) = 1$, the result follows.

(125) Assume that $(m, n) = 1$, $m > n$. We will show that

(1) $$(a^m - 1, a^n - 1) = a - 1.$$

Since $(a, b + ma) = (a, b)$, we have

$$(a^m - 1, a^n - 1) = (a^m - 1 - (a^n - 1), a^n - 1) = (a^m - a^n, a^n - 1).$$

Since $(a^n, a^n - 1) = 1$, this shows that

$$(a^m - 1, a^n - 1) = (a^n(a^{m-n} - 1), a^n - 1) = (a^{m-n} - 1, a^n - 1).$$

Without any loss in generality, we may assume that $m > n$, in which case we can write $m = nq + r$, $0 \le r < n$, so that

$$(a^m - 1, a^n - 1) = (a^r - 1, a^n - 1).$$

Then, writing $n = rs + t$, $0 \le t < r$, we obtain

$$(a^r - 1, a^n - 1) = (a^r - 1, a^t - 1),$$

and so on until we arrive at $(a - 1, a - 1) = a - 1$, which proves (1). Assume now that $d = (m, n) > 1$. Since $(m/d, n/d) = 1$, we are brought back to the first case, and we thus have

$$\left((a^d)^{m/d} - 1, (a^d)^{n/d} - 1\right) = a^d - 1,$$

which takes care of the first part of the problem. For the other cases mentioned in the second part (which by the way cover also the first part), we proceed in the following manner. First letting $(m, n) = 1$ and $u = \pm 1$, $v = \pm 1$, we have

$$
\begin{aligned}
(a^m + u, a^n + v) &= (a^m + u - uv(a^n + v), a^n + v) \\
&= (a^m - uva^n, a^n + v) = (a^{m-n} - uv, a^n + v),
\end{aligned}
$$

since $(a^n, a^n \pm 1) = 1$. Continuing this process, we obtain

$$
(a^m + 1, a^n + 1) = (a^m + 1, a^n - 1) = (a + 1, a + 1) \text{ or } (a + 1, a - 1)
$$

according to the parities of m and n. More precisely, we have the following: Since $(a + 1)|(a^k + 1)$ for k odd and since $(a + 1)|(a^k - 1)$ for k even, it follows that taking into account the fact that $a + 1$ cannot divide $a^k + 1$ if k is even unless $a = 1$, we obtain that

$$
(a^m + 1, a^n + 1) = \begin{cases} a + 1 & \text{if } mn \text{ is odd,} \\ 1 & \text{if } mn \text{ is even and } a \text{ is even,} \\ 2 & \text{if } mn \text{ is even and } a \text{ is odd} \end{cases}
$$

and that

$$
(a^m + 1, a^n - 1) = \begin{cases} 1 & \text{if } n \text{ is odd and } a \text{ is even,} \\ 2 & \text{if } n \text{ is odd and } a \text{ is odd,} \\ a + 1 & \text{if } n \text{ is even.} \end{cases}
$$

When $d = (m, n) > 1$, we can proceed essentially as we did for the first case. To find the value of $(a^m - b^m, a^n - b^n)$, we may assume that $(a, b) = 1$ and $a > b$. In this case, set $d = (m, n)$, $u = a^d - b^d$ and $v = (a^m - b^m, a^n - b^n)$. Since $d|m$, it follows that $u|(a^m - b^m)$, and since $d|n$, we also have $u|(a^n - b^n)$ and we obtain that $u|v$. Then, we only need to show that $v|u$. Choose integers $x > 0$ and $y > 0$ such that $mx - ny = d$. It is clear that

$$
a^{mx} = a^{ny+d} = a^{ny}(b^d + u),
$$

and therefore

$$
a^{mx} - b^{mx} = a^{ny}(b^d + u) - b^{ny+d} = b^d(a^{ny} - b^{ny}) + ua^{ny}.
$$

Since $v|(a^m - b^m)$, we have $v|(a^{mx} - b^{mx})$ and similarly $v|(a^{ny} - b^{ny})$, and the last equation allows one to obtain that $v|ua^{ny}$. Since $(a, b) = 1$, we have $(v, a) = 1$. Indeed, every common divisor of a and v divides a^m and $a^m - b^m$, and therefore divides b^m, and since $(a, b) = 1$, we have $(a, v) = 1$. Finally, $v|ua^{ny}$ implies $v|u$ and the result follows.

(126) Let a and b be two arbitrary integers, and set $x = 5a$ and $y = 5b$. In order to have $x + y = 5a + 5b = 40$, we must have $a + b = 8$. Moreover, to have $(x, y) = 5$, we must have $(a, b) = 1$. Therefore, it remains to show that it is possible to find infinitely many relatively prime pairs of integers a and b such that $a + b = 8$. To do so, it is enough to choose, for example, $a = 3 + 2t$ and $b = 5 - 2t$, where $t \in \mathbb{Z}$.

(127) There are four possible pairs: $a = 15$, $b = 90$; $a = 90$, $b = 15$; $a = 30$, $b = 45$; $a = 45$, $b = 30$. For the general case, we proceed in the following way. Since $(a, b) = d$, there exist integers A and B such that $a = dA$, $b = dB$, where $(A, B) = 1$. But $[a, b] = m$ implies that $[dA, dB] = d[A, B] = dAB = m$. Hence, the system of equations $(a, b) = d$, $[a, b] = m$

has solutions if and only if $d|m$. These solutions will be the same as that of $AB = m/d$ where $(A, B) = 1$. For each prime number p dividing m/d, we cannot have both $p|A$ and $p|B$. Therefore, either A contains the largest power of p which divides m/d, or else A does not have p as a divisor. Hence, for each prime factor p of m/d, we have two choices for the pair $\{A, B\}$, and therefore in total as many pairs as m/d has distinct prime factors, that is as many as $2^{\omega(m/d)}$.

(128) This follows from the fact that $3|m$ and $3|n$ while $3 \nmid 101$.

(129) (a) We observe that
$$a^{2^{m-r}} \cdot a^{2^{m-r}} = a^{2^{m-r+1}}.$$

It follows that
$$(*) \qquad a^{2^m} - 1 = (a^{2^{m-1}} + 1)(a^{2^{m-1}} - 1)$$
$$= (a^{2^{m-1}} + 1)(a^{2^{m-2}} + 1) \cdots (a^2 + 1)(a + 1)(a - 1).$$

Hence, if $m > n$, $a^{2^n} + 1$ is a divisor of $a^{2^m} - 1$, as required.

(b) Note that $a^{2^m} - 1 = (a^{2^m} + 1) - 2$ and that this integer is divisible by each of the factors on the right-hand side of $(*)$. Let $d = (a^{2^m} + 1, a^{2^n} + 1)$. We may assume that $n < m$, and therefore $a^{2^n} + 1 | (a^{2^m} + 1) - 2$, which implies $d | (a^{2^m} + 1) - 2$. Therefore, $d|2$ so that $d = 1$ or $d = 2$.

(130) (*AMM, Vol. 75, 1971, p. 201*). Let d be the greatest common divisor of the given numbers. In particular, d divides the sum of these numbers and since (see Problem 17 (b))
$$\binom{2n}{1} + \binom{2n}{3} + \binom{2n}{5} + \cdots + \binom{2n}{2n-1} = 2^{2n-1},$$
it follows that d must be of the form 2^a. If $n = 2^k r$, where r is an odd integer and k a nonnegative integer, then since $\binom{2n}{1} = 2^{k+1} r$, it follows that any common divisor of the given numbers cannot be larger than 2^{k+1}. To show that 2^{k+1} divides all these numbers, we first write, for $m = 1, 3, \ldots, 2n - 1$,
$$\binom{2n}{m} = \binom{2^{k+1} r}{m} = \frac{2^{k+1} r}{m}\binom{2^{k+1} r - 1}{m - 1}.$$
Since the binomial coefficients are integers and since m is an odd number, we have
$$\binom{2n}{m} = \binom{2^{k+1} r}{m} = 2^{k+1} M,$$
where M is an integer and $m = 1, 3, \ldots, 2n - 1$. This proves that 2^{k+1} is the greatest common divisor of the given numbers.

(131) Each a_i can be written in the form $a_i = 2^{\alpha_i} b_i$, where $\alpha_i \geq 0$ and b_i is odd. Let $B = \{b_1, b_2, \ldots, b_{n+1}\}$. We have $b_i \leq 2n$, $i = 1, 2, \ldots, n+1$. But there exist only n odd numbers $\leq 2n$; hence, there exist j, k such that $b_j = b_k$. Then, consider the two integers
$$a_j = 2^{\alpha_j} b_j \quad \text{and} \quad a_k = 2^{\alpha_k} b_k.$$
It is clear that $\alpha_j \neq \alpha_k$ (since $b_j = b_k$). If $\alpha_j < \alpha_k$, then $a_j | a_k$. If $\alpha_k < \alpha_j$, then $a_k | a_j$. In each case, the result follows.

(132) (*Contribution of Imre Kátai, Budapest*). Let

$$S := \sum_{j=1}^{n}(a_j^{(1)} + a_j^{(2)})(a_j^{(1)} + a_j^{(3)}).$$

Then,

$$S = \sum_{j=1}^{n}(a_j^{(1)})^2 + \sum_{j=1}^{n}a_j^{(1)}a_j^{(3)} + \sum_{j=1}^{n}a_j^{(2)}a_j^{(1)} + \sum_{j=1}^{n}a_j^{(2)}a_j^{(3)} = n.$$

But since it is clear that the expression $(a_j^{(1)}+a_j^{(2)})(a_j^{(1)}+a_j^{(3)})$ is a multiple of 4, the result follows.

(133) In order to show that a series made up of nonnegative real numbers converges, we only need to bound it by a series which converges. So let $\ell(n)$ be the number of digits of the positive integer n, in its decimal representation. We first observe that, for each positive integer r, we have

$$\sum_{\substack{n \in A \\ \ell(n)=r}} 1 = 8 \cdot 9^{r-1}.$$

Hence, it follows from this that

$$\sum_{n \in A}\frac{1}{n} = \sum_{r=1}^{\infty}\sum_{\substack{n \in A \\ \ell(n)=r}}\frac{1}{n} < \sum_{r=1}^{\infty}\frac{1}{10^{r-1}}\sum_{\substack{n \in A \\ \ell(n)=r}}1$$

$$= \sum_{r=1}^{\infty}\frac{8 \cdot 9^{r-1}}{10^{r-1}} = 8\sum_{r=1}^{\infty}\left(\frac{9}{10}\right)^{r-1} = 80,$$

from which the result follows.

(134) If $a > b$, it is clear that $a - b \geq (a,b)$, and we know that $(a,b)[a,b] = ab$. Hence, since

$$(u_{n+1} - u_n)[u_{n+1}, u_n] \geq (u_{n+1}, u_n)[u_{n+1}, u_n] = u_{n+1} \cdot u_n,$$

we obtain

$$\frac{1}{[u_{n+1}, u_n]} \leq \frac{u_{n+1} - u_n}{u_{n+1} \cdot u_n} = \frac{1}{u_n} - \frac{1}{u_{n+1}}.$$

Therefore the series is bounded above by a convergent series and this is why it converges.

(135) (a) With MAPLE, we have > `for i from 3 by 2 to 525 do`
   > `if isprime(2^i-1)`
   > `then print(2^i-1, ' is a prime number ')`
   > `else fi; od;`
   (b) With MAPLE, we may use
   > `nextprime(10^(100)+1);`
   We thus obtain the integer $10^{100} + 267$.

(136) With MAPLE, the program below enumerates the first N (here $N = 120$) prime numbers.
   > `for i to 120 do`
   > `p.i :=ithprime(i) od;`
   For example, $p.(1..120)$ gives the first 120 prime numbers.

(137) In order to find four consecutive integers with the same number of prime
factors, we must use the function Ω. First we type in

```
> readlib(ifactors):  with(numtheory):
```
and thereafter, we type in the following instructions:
```
> Omega:=n->sum(ifactors(n)[2][i][2],
> i=1..nops(factorset(n))):
> for n to 1000 do if Omega(n)=Omega(n+1) and
> Omega(n+1)=Omega(n+2) and Omega(n+2)=Omega(n+3)
> then print(n) else fi; od;
```
To find four consecutive integers having the same number of divisors, it
is enough to type in the instructions
```
> with(numtheory):
> for n to 1000 do
> if tau(n)=tau(n+1) and tau(n+1)=tau(n+2)
> and tau(n+2)=tau(n+3)
> then print(n) else fi; od;
```

(138) (a) With the procedure "return", the search is easily done:
```
> return:=proc(n::integer)
> local m,s;
> m:=n; s:=0;
> while m<>0 do
> s:=10*s+irem(m,10);
> m:=iquo(m,10) od; s end:
```
And for our problem, we have the following procedure:
```
> invp:=proc(N::integer) local n;
> for n from 1 to N do if isprime(n)
> and isprime(return(n)) then
> print (n) fi; od; end:
```
Without the procedure "return", we may proceed as follows:
```
> invp:=proc(N)
> for j from 169 to N do
> L:=convert(ithprime(j),base,10);# N <= 1229
> if type(1000*L[1]+100*L[2]+10*L[3]+L[4],prime)=true
> then print(ithprime(j)) else fi; od; end:
```

(b)
```
> invp:=proc(N::integer)
> local n;
> for n from 1 to N do
> if isprime(n)=false then elif
> type(sqrt(return(n)),integer)=true
> then print(n) else fi; od; end:
```

If we do not use the procedure "return", we may proceed as follows:

```
> invp:=proc(N) local j, L;
> for j from 26 to N
> do L:=convert(ithprime(j),base,10);# N <= 168
> if type(sqrt(100*L[1]+10*L[2]+L[3]),integer)=true then
> print(ithprime(j)) else fi; od; end:
```

or the following procedure:

```
> invp:=proc(N) local j, L;
> for j from 169 to N
> do L:=convert(ithprime(j),base,10);# N <= 1229
> if type(sqrt(1000*L[1]+100*L[2]+10*L[3]+L[4]),
> integer)=true
> then print(ithprime(j)) else fi; od; end:
```

(139) With MAPLE:

```
> for n from 3 by 2 to 10000 do
> if isprime(n) and isprime(n + 2) and isprime(n + 6)
> then print(n) else fi; od;
```

(140) We prove this result using induction on k. The result is immediate for $k = 2$. Assume that the result is true for a certain integer $k > 2$, that is for which we have $p_k < 2^k$. It is enough to show that $p_{k+1} < 2^{k+1}$. From Bertrand's Postulate, there exists a prime number between p_k and $2p_k$, in which case $p_{k+1} < 2p_k$, and the result is proved.

(141) It is enough to show that $d \not\equiv 2, 4 \pmod 6$. First of all, assume that $d = 2$: if $p_k \equiv 1 \pmod 3$, then $p_{k+1} = p_k + 2 \equiv 0 \pmod 3$, contradicting the fact that p_{k+1} is prime; similarly if $p_k \equiv 2 \pmod 3$, then $p_{k-1} = p_k - 2 \equiv 0 \pmod 3$, contradicting the fact that p_{k-1} is prime. The same type of contradiction emerges when we assume that $d = 4$. If $d = 6k + 2$ or $6k + 4$ with $k \geq 1$, the same argument works. For $d = 6$, it is $p_{16} = 53$; for $d = 12$, it is $p_{47} = 211$; for $d = 18$, it is $p_{2285} = 20\,201$.

REMARK: It is interesting to observe that the gap $d = 24$ is reached earlier than might be expected in the sequence of prime numbers, namely with $p_{1939} = 16\,787$.

(142) This statement follows from the fact that each of the listed numbers is a perfect square, since

$$12321 = 111^2, \quad 1234321 = 1111^2, \ldots,$$

$$12345678987654321 = 111111111^2.$$

(143) Let $k \geq 2$. Since each number $\leq n_k$ is either 1, a prime number or else a composite number, it is clear that

(1) $n_k = 1 + \pi(n_k) + k.$

By using MATHEMATICA and the program

```
n=1;Do[n=n+1;While[PrimePi[n]!=n-10^a-1,n++]; Print[10^a,
" ",n],{a,1,3}]
```

we obtain the table

10	18
100	133
1000	1197

This reveals that $n_{10} = 18$, $n_{100} = 133$ and $n_{1000} = 1197$. For values of k larger than 1000, and to accelerate the computations, one can use the approximation (guaranteed by the Prime Number Theorem) $\pi(x) \approx \frac{x}{\log x} + \frac{x}{\log^2 x}$, so that (1) gives

$$n_k \approx \frac{n_k}{\log n_k} + \frac{n_k}{\log^2 n_k} + k$$

and therefore that

(2) $$n_k \left(1 - \frac{1}{\log n_k} - \frac{1}{\log^2 n_k} \right) \approx k,$$

which in particular means that

(3) $$\log n_k \approx \log k.$$

Combining (2) and (3), we obtain the approximation

(4) $$n_k \approx k \cdot \left(1 - \frac{1}{\log k} - \frac{1}{\log^2 k} \right)^{-1}.$$

Setting $s_k(n) := 1 + \pi(n) + k - n$, it follows that if a number n satisfies $s_k(n) = 0$, then $n = n_k$.

First consider the case $k = 10^4$. From (4), we have as a first approximation $n_{10^4} \approx 11\,369$. By using MATHEMATICA and the program

```
n=11369;While[(a=s[n])!=0,n=n+a];Print[n]
```

where $s(n) = s_{1000}(n)$, we obtain that $n_{10000} = 11\,374$. Similarly, with the approximation $n_{10^5} \approx 110\,425$, we obtain that $n_{10^5} = 110\,487$. The following is the table giving the values of n_{10^α} for $1 \le \alpha \le 10$.

α	n_{10^α}	α	n_{10^α}
1	18	6	1 084 605
2	133	7	10 708 555
3	1 197	8	106 091 745
4	11 374	9	1 053 422 339
5	110 487	10	10 475 688 327

(144) Let $k \ge 2$. Setting $r = [\log n_k / \log 2]$, it is clear that the number n_k satisfies

$$\sum_{i=1}^{r} \pi(n_k^{1/i}) = k.$$

From this relation and the approximation $\pi(x) \approx \dfrac{x}{\log x}$ (guaranteed by the Prime Number Theorem), it follows that

$$\frac{n_k}{\log n_k} \approx k,$$

so that $\log n_k \approx \log k + \log \log n_k \approx \log k$ and therefore that

$$n_k \approx k \log n_k \approx k \log k,$$

which gives a starting point for the computation of the exact value of n_k.

Using MATHEMATICA and the program

```
Do[k = 10^j; n = Floor[N[k*Log[k]]];
While[r = Floor[N[Log[n]/Log[2]]];
s=Sum[PrimePi[n^(1/i)],{i,1,r}];(a=k-s)!=0,n=n+a];
Print[j,"->", n,"=",FactorInteger[n]],{j,3,10}]
```

we finally obtain the following table:

α	n_{10^α}	α	n_{10^α}
1	16	6	15 474 787
2	419	7	179 390 821
3	7 517	8	2 037 968 761
4	103 511	9	22 801 415 981
5	1 295 953	10	252 096 677 813

(145) If n is even, then $2^n + n^2$ is also even and therefore not a prime. It follows that $n \equiv 1, 3$ or 5 modulo 6. If $n = 6k + 1$ for a certain nonnegative integer k, then $2^n = 2^{6k+1} \equiv 2 \pmod 3$ and $n^2 \equiv 1 \pmod 3$; in this case, we have that $2^n + n^2 \equiv 2 + 1 \equiv 0 \pmod 3$. Similarly, if $n = 6k + 5$ for a certain nonnegative integer k, we easily show that $3 | 2^n + n^2$. Therefore, the only way that $2^n + n^2$ can be a prime number is that $n \equiv 3 \pmod 6$.

Thus, by considering all the positive integers $n < 100$ of the form $n = 6k + 3$ and using a computer, we easily find that the only prime numbers of the form $2^n + n^2$, with $n < 100$, are those corresponding to $n = 1, 9, 15, 21, 33$.

(146) We will show that if n is of the form $n = 3k + 1$ or $n = 3k + 2$ with $k \geq 1$, then $7 | a_n$. Moreover, we will show that if n is of the form $n = 3(3k + 1)$ or $n = 3(3k + 2)$ with $k \geq 1$, then $73 | a_n$. Finally, since $3 | a_n$ if n is even, it will follow that, for a_n to be prime, n must be an odd multiple of 9.

So let $n = 3k + a$, with $a = 1$ or 2. Since $8^k \equiv 1 \pmod 7$ for each integer $k \geq 1$, we have

$$a_n = 2^n(2^n + 1) + 1 = 2^{3k+a}(2^{3k+a} + 1) + 1$$
$$= 8^k 2^a(8^k 2^a + 1) + 1 \equiv 2^a(2^a + 1) + 1 \pmod 7.$$

But

$$2^a(2^a + 1) + 1 = \begin{cases} 7 \equiv 0 \pmod 7 & \text{if } a = 1, \\ 21 \equiv 0 \pmod 7 & \text{if } a = 2, \end{cases}$$

which establishes our first statement.

Let us now assume that $n = 3(3k + a)$ with $a = 1$ or 2. Since $2^9 \equiv 1 \pmod{73}$, we have

$$a_n = 2^n(2^n + 1) + 1 = 2^{9k+3a}(2^{9k+3a} + 1) + 1$$
$$= (2^9)^k 2^{3a}((2^9)^k 2^{3a} + 1) + 1 \equiv 2^{3a}(2^{3a} + 1) + 1 \pmod{73}.$$

But

$$2^{3a}(2^{3a} + 1) + 1 = \begin{cases} 2^3(2^3 + 1) + 1 = 73 \equiv 0 \pmod{73} & \text{if } a = 1, \\ 2^6(2^{12} + 1) + 1 = 4161 \equiv 0 \pmod{73} & \text{if } a = 2, \end{cases}$$

which establishes our second statement.

Having observed that a_n is prime for $n = 1, 3$ and 9, and then considering all the numbers of the form $n = 9(2k + 1)$, we obtain using a computer that a_n is composite for each integer n, $10 \leq n < 1000$.

(147) That number is $n = 9$; we then have $a_9 = 326981 = 79 \cdot 4139$.

(148) First observe that it follows from Wilson's Theorem that

$$\left[\frac{(j - 1)! + 1}{j} - \left[\frac{(j - 1)!}{j} \right] \right] = \begin{cases} 1 & \text{if } j \text{ is prime}, \\ \\ 0 & \text{if } j \text{ is composite}. \end{cases}$$

Hence, to obtain the formula of Mináč and Willans, we only need to prove that

$$p_n = 2 + \sum_{m=2}^{2^n} \left[\left[\frac{n}{1 + \pi(m)} \right]^{1/n} \right].$$

But we easily prove that

$$\left[\left[\frac{n}{1 + \pi(m)} \right]^{1/n} \right] = \begin{cases} 1 & \text{if } \pi(m) \leq n - 1, \\ \\ 0 & \text{otherwise.} \end{cases}$$

Now, as m varies from 2 to 2^n, we have that $\pi(m) \leq n - 1$ for $m = 2, 3, \ldots, p_n - 1$, that is a total of $p_n - 2$ numbers. Therefore,

$$2 + \sum_{m=2}^{2^n} \left[\left[\frac{n}{1 + \pi(m)} \right]^{1/n} \right] = 2 + p_n - 2 = p_n,$$

as was to be shown.

(149) We use an induction argument. The result is true for $n = 1$ and for $n = 2$. So let $n \geq 3$. Assume that the result is true for all natural numbers $\leq n - 1$ and let us show that it implies that it must be true for n. Let $P_n = \prod_{p \leq n} p$. First of all, if n is even, then $P_n = P_{n-1}$, so that the result is true for n. Let us examine the case where n is odd, that is $n = 2k + 1$ for a certain positive integer k. It follows that each prime number p such that $k + 2 \leq p \leq 2k + 1$ is a divisor of the number

$$(*) \qquad \binom{2k + 1}{k} = \frac{(2k + 1)(2k)(2k - 1)(2k - 2) \cdots (k + 2)}{1 \cdot 2 \cdot 3 \cdots k}.$$

Since

$$2^{2k+1} = (1 + 1)^{2k+1} > \binom{2k + 1}{k} + \binom{2k + 1}{k + 1} = 2 \binom{2k + 1}{k},$$

we obtain

$$\binom{2k + 1}{k} < 4^k.$$

It follows that the product of all the prime numbers p such that $k + 2 \leq p \leq 2k + 1$ is a divisor of $\binom{2k + 1}{k}$ and therefore smaller than 4^k. On the other hand, using the induction hypothesis, we have that $P_{k+1} \leq 4^{k+1}$. This is why

$$P_n = P_{2k+1} = \prod_{p \leq k+1} p \cdot \prod_{k+2 \leq p \leq 2k+1} p < 4^{k+1} \cdot 4^k = 4^{2k+1} = 4^n,$$

as was to be shown.

(150) Let $m = (a, c)$. Then, there exist two integers u and v such that $(u, v) = 1$ and such that $a = mu$ and $c = mv$. Hence, since $ab = cd$, we have $mub = mvd$ and therefore $ub = vd$. Since $(u, v) = 1$, we have $u|d$ and this is why there exists an integer n such that $d = nu$. Since $ub = vnu$, we therefore have $b = nv$. It follows from these relations that

$$a^2 + b^2 + c^2 + d^2 = m^2u^2 + n^2v^2 + m^2v^2 + n^2u^2 = m^2(u^2 + v^2)$$
$$+ n^2(u^2 + v^2) = (u^2 + v^2)(m^2 + n^2),$$

a product of two integers larger than 1.

(151) Since

$$4n^3 + 6n^2 + 4n + 1 = n^4 + 4n^3 + 6n^2 + 4n + 1 - n^4 = (n+1)^4 - n^4$$
$$= ((n+1)^2 - n^2)((n+1)^2 + n^2) = (2n+1)(2n^2 + 2n + 1),$$

the product of two integers larger than 1, the result follows.

(152) First of all, since $p + q$ is even, we can write

$$(*) \qquad\qquad p + q = 2 \cdot \frac{p+q}{2}.$$

Since $\frac{p+q}{2}$ is an integer located between the two consecutive prime numbers p and q, it must be composite, that is the product of at least two prime numbers, and this is why the right-hand side of $(*)$ has at least three prime factors.

(153) The answer is YES. We look for positive integers n, a, b and c such that

$$\frac{n}{2} = a^2, \quad \frac{n}{3} = b^3, \quad \frac{n}{5} = c^5.$$

It is sufficient to find integers a, b and c such that

$$2a^2 = 3b^3 = 5c^5.$$

The task is therefore to find integers α_i, β_i and γ_i $(i = 1, 2, 3)$ such that

$$2\left(2^{\alpha_1} 3^{\beta_1} 5^{\gamma_1}\right)^2 = 3\left(2^{\alpha_2} 3^{\beta_2} 5^{\gamma_2}\right)^3 = 5\left(2^{\alpha_3} 3^{\beta_3} 5^{\gamma_3}\right)^5.$$

To do so, we must find integers α_i, β_i and γ_i $(i = 1, 2, 3)$ such that

$$2\alpha_1 + 1 = 3\alpha_2 = 5\alpha_3, \quad 2\beta_1 = 3\beta_2 + 1 = 5\beta_3, \quad 2\gamma_1 = 3\gamma_2 = 5\gamma_3 + 1.$$

We easily find

$$\alpha_1 = 7, \alpha_2 = 5, \alpha_3 = 3, \quad \beta_1 = 5, \beta_2 = 3, \beta_3 = 2, \quad \gamma_1 = 3, \gamma_2 = 2, \gamma_3 = 1.$$

We then obtain that $n = 2(2^7 \cdot 3^5 \cdot 5^3)^2 = 30\,233\,088\,000\,000$ serves our purpose.

(154) This follows from the identity

$$n^{42} - 27 = (n^{14})^3 - 3^3 = (n^{14} - 3)(n^{28} + 3n^{14} + 3^2).$$

(155) We proceed by contradiction by assuming that there does not exist any prime number in the interval $]x, 2x]$, in which case we have $\theta(2x) = \theta(x)$. By using the inequalities $0.73x < \theta(x) < 1.12x$, we would then have

$$1.46x = 2(0.73)x < \theta(2x) = \theta(x) < 1.12x,$$

a contradiction.

(156) We proceed by induction. First of all, for $n = 4$, the result is true, since $121 = 11^2 = p_5^2 < p_1 p_2 p_3 p_4 = 210$. Assume that the inequality $p_k^2 < p_1 p_2 \cdots p_{k-1}$ is true for a certain integer $k \geq 5$. By using Bertrand's Postulate in the form $p_{k+1} < 2p_k$, we then have

$$p_{k+1}^2 < 4p_k^2 < 4p_1 p_2 \cdots p_{k-1} < p_1 p_2 \cdots p_k,$$

and the result then follows by induction.

(157) If there exist $q, r, a \in \mathbb{N}$ such that $q^r = (q^{r/2})^2 = a^2$, where r is even and $q^r = p + m^2$ with p prime and $m \in \mathbb{N}$, then $a^2 - m^2 = p$, so that $(a-m)(a+m) = p$. Since p is prime, we must have $a-m = 1$ and $a+m = p$, and therefore $m = a - 1$ and $p = 2a - 1$. Hence, if $2a - 1 = 2q^{r/2} - 1$ is composite, q^r cannot be written as $p + m^2$, as was to be shown.

(158) For $p = 3$, the result is immediate. Assume that $p \geq 5$. If $p = 3k + 1$ for a certain positive integer k, then $8k + 1 = 24k + 9$, a multiple of 3. Otherwise, that is if $p = 3k - 1$ for a certain positive integer k, then $8p - 1 = 24k - 9$, a multiple of 3, which contradicts the fact that $8p - 1$ is prime. In both cases, the result is proved.

(159) If a positive integer of the form $3k + 2$ has no prime factor of the form $3k + 2$, then all its prime factors are of the form $3k + 1$. Since the product of two integers of the form $3k + 1$ is of the form $3k + 1$, the result follows.

Since each product of prime numbers of the form $4k + 1$ is of the same form and since each product of prime numbers of the form $6k + 1$ is of the same form, the result follows.

(160) (a) We have $23 = 3 \cdot 3! + 2 \cdot 2! + 1 \cdot 1!$ and $57 = 2 \cdot 4! + 1 \cdot 3! + 1 \cdot 2! + 1 \cdot 1!$.

(b) To find the Cantor expansion of a positive integer n, we proceed as follows. Let m be the largest positive integer such that $m! \leq n$ and let a_m be the largest positive integer such that $a_m \cdot m! \leq n$. It is clear that $0 < a_m \leq m$; otherwise, this would contradict the maximal choice of m. If $a_m \cdot m! = n$, then the Cantor expansion is given by $n = a_m \cdot m!$. Otherwise, that is if $a_m \cdot m! < n$, let $d_1 = n - a_m \cdot m! > 0$, let m_1 be the largest positive integer such that $m_1! \leq d_1$ and let a_{m_1} be the largest positive integer such that $a_{m_1} \cdot m_1! \leq d_i$. As above, we have $a < a_{m_1} \leq m_1$. If $a_{m_1} \cdot m_1! = d_1$; then the Cantor expansion is given by $n = a_m \cdot m! + a_{m_1} \cdot m_1!$, where $0 < a_{m_1} \leq m_1 < m$. If $a_{m_1} \cdot m_1! < d_1$, then we set $d_2 = d_1 - a_{m_1} \cdot m_1!$ and we let m_2 be the largest positive integer such that $m_2! \leq d_2$. And so on. We thus build a sequence of positive integers $m > m_1 > m_2 > \ldots$ with the corresponding integers $0 < a_{m_i} \leq m_i$. Since the sequence of m_i's is decreasing, it must have an end. Let us show the uniqueness of this representation. Assume that for $0 \leq a_j, b_j \leq j$, we have

$$n = a_m m! + \cdots + a_1 1! = b_m m! + \cdots + b_1 1!,$$

that is $(a_m - b_m)m! + \cdots + (a_1 - b_1)1! = 0$. If both expansions are different, then there exists a smaller integer j such that $1 \leq j < m$ and $a_j \neq b_j$. Hence,

$$j! \left((a_m - b_m)\frac{m!}{j!} + \cdots + (a_{j+1} - b_{j+1})(j + 1) + (a_j - b_j) \right) = 0$$

and therefore

$$b_j - a_j = (a_m - b_m)\frac{m!}{j!} + \cdots + (a_{j+1} - b_{j+1})(j + 1)$$

$$= (j + 1) \left((a_m - b_m)\frac{m!}{(j + 1)!} + \cdots + (a_{j+1} - b_{j+1}) \right),$$

which implies that $(j + 1)|(b_j - a_j)$. Since $0 \leq a_j, b_j \leq j$, it follows that $a_j = b_j$, a contradiction.

(161) (*TYCM, Vol. 19, 1988, p. 191*). The expression in the statement can be written as

$$\frac{(p-1)!+1}{p} + \frac{(-1)^d d!(p-1)!+1}{p+d}.$$

Since $(p+d-1)! = (p+d-1)(p+d-2)\cdots(p+d-d)(p-1)!$, we have $(p+d-1)! \equiv (-1)^d d!(p-1)! \pmod{p+d}$, and it follows that the expression in the statement is an integer if and only if

$$(1) \qquad \frac{(p-1)!+1}{p} + \frac{(p+d-1)!+1}{p+d}$$

is an integer. From Wilson's Theorem, if p and $p+d$ are two prime numbers, then each of the terms of (1) is an integer, which proves the necessary condition.

Conversely, assume that expression (1) is an integer. If p or $p+d$ is not a prime, then by Wilson's Theorem, at least one of the terms of (1) is not an integer. This implies that none of the terms of (1) is an integer or equivalently neither of p and $p+d$ is prime. It follows that both fractions of (1) are in reduced form.

It is easy to see that if a/b and a'/b' are reduced fractions such that $a/b + a'/b' = (ab' + a'b)/(bb')$ is an integer, then $b|b'$ and $b'|b$.

Applying this result to (1), we obtain that $(p+d)|p$, which is impossible. We may therefore conclude that if (1) is an integer, then both p and $p+d$ must be primes.

(162) If $p = 3$, then $p+2 = 5$ is prime and $p^2 + 2p - 8 = 7$ is prime. It is the only number with this property. Indeed, $p = 2$ does not have this property, while if $p > 3$, then

$$p^2 + 2p - 8 \equiv 1 + 2p - 2 \equiv 2(p+1) \equiv 0 \pmod 3 \iff 3|(p+1).$$

But for $p > 3$, $p = 3k \pm 1$, and in each of the cases it is easily seen that at least one of the two numbers $p+2$ and $p^2 + 2p - 8$ is not a prime.

(163) The answer is YES. If $p = 3$, then $p^2 + 8 = 17$ is prime and $p^3 + 4 = 31$ is prime. It is the only prime number with this property. Indeed, $p = 2$ does not have this property, while if $p > 3$, then $p \equiv \pm 1 \pmod 3$, in which case $p^2 \equiv 1 \pmod 3$, that is $p^2 + 8 \equiv 9 \equiv 0 \pmod 3$, so that $p^2 + 8$ is not a prime. Thus the result.

(164) (Ribenboim [**28**], p. 145). First assume that the congruence is satisfied. Then $n \neq 2, 4$ and $(n-1)! + 1 \equiv 0 \pmod n$. Thus, using Wilson's Theorem, n is prime. Moreover, $4(n-1)! + 2 \equiv 0 \pmod{n+2}$; thus, multiplying by $n(n+1)$ we obtain

$$4[(n+1)!+1] + 2n^2 + 2n - 4 \equiv 0 \pmod{n+2}$$

and therefore

$$4[(n+1)!+1] + (n+2)(2n-2) \equiv 0 \pmod{n+2};$$

hence, $4[(n+1)!+1] \equiv 0 \pmod{n+2}$. This is why, using Wilson's Theorem, $n+2$ is also prime.

Conversely, if n and $n+2$ are prime, then $n \neq 2$ and

$$(n-1)! + 1 \equiv 0 \pmod n,$$
$$(n+1)! + 1 \equiv 0 \pmod{n+2}.$$

But $n(n+1) = (n+2)(n-1)+2$, and this is why $2(n-1)!+1 = k(n+2)$, where k is an integer. From the relation $(n-1)! \equiv -1 \pmod{n}$, we obtain $2k+1 \equiv 0 \pmod{n}$. Now, $2(n-1)!+1 = k(n+2)$ is equivalent to $4(n-1)!+2 \equiv 0 \equiv -(n+2) \pmod{n+2}$. Moreover, $4(n-1)!+2 \equiv 4k \equiv -2 \equiv -(n+2) \pmod{n}$. Hence, $4(n-1)!+2 \equiv -(n+2) \pmod{n(n+2)}$; that is $4\big((n-1)!+1\big) + n \equiv 0 \pmod{n(n+2)}$.

(165) The prime number $p = 3$ is the only one with this property, because if $p > 3$, then $p = 2k+1$ for a certain integer $k \geq 2$, in which case

$$2^p = 2^{2k+1} = 4^k \cdot 2 \equiv 2 \pmod{3}$$

while

$$p^2 \equiv 1 \pmod{3},$$

so that

$$2^p + p^2 \equiv 0 \pmod{3}.$$

(166) The answer is $p = 19$. Indeed, $17p+1 = a^2 \Rightarrow 17p = (a-1)(a+1)$. We then have $17 = a-1$ and $p = a+1$ or $17 = a+1$ and $p = a-1$. The first case yields $a = 18$ and $p = 19$, while the second case yields $a = 16$ and $p = 15$, which is to be rejected. Thus the result.

(167) (a) The possible values of (a^2, b) are p and p^2.
   (b) The only possible value of (a^2, b^2) is p^2.
   (c) The possible values of (a^3, b) are p, p^2 and p^3.
   (d) The possible values of (a^3, b^2) are p^2 and p^3.

(168) (a) The only possible value is p^3.
   (b) The only possible value is p.
   (c) The only possible value is p.
   (d) The possible values are p^2, p^3, p^4 and p^5.

(169) We have $(a^2b^2, p^4) = p^4$ and $(a^2 + b^2, p^4) = p^2$.

(170) (a) True.   (b) True.   (c) True.
   (d) False. Indeed, we have $13|2^2 + 3^2$ and $13|3^2 + 2^2$, while $13 \nmid 2^2 + 2^2 = 8$.

(171) It is an immediate consequence of Theorem 12.

(172) Let

$$\begin{cases} a = p_1^{\alpha_1} \cdots p_r^{\alpha_r}, \\ b = p_1^{\beta_1} \cdots p_r^{\beta_r}, \\ c = p_1^{\gamma_1} \cdots p_r^{\gamma_r}. \end{cases}$$

From Theorem 12,

$$(a, b, c) = p_1^{\min(\alpha_1, \beta_1, \gamma_1)} \cdots p_r^{\min(\alpha_r, \beta_r, \gamma_r)}$$

and

$$[a, b, c] = p_1^{\max(\alpha_1, \beta_1, \gamma_1)} \cdots p_r^{\max(\alpha_r, \beta_r, \gamma_r)}.$$

To prove the result, we proceed by contradiction. Assume for example that $(a, b) > 1$. Using the fact that $(a, b, c)[a, b, c] = abc$, it follows, using the above notation, that

$$\min(\alpha_i, \beta_i, \gamma_i) + \max(\alpha_i, \beta_i, \gamma_i) = \alpha_i + \beta_i + \gamma_i \qquad (i = 1, 2, \ldots, r).$$

But it is easy to prove that for the sum of three nonnegative integers to be equal to the sum of the smallest and of the largest of these same three numbers, at least two of these numbers must be 0. But this contradicts the fact that $(a, b) > 1$, an inequality which means that there exists an i_0 $(1 \leq i_0 \leq r)$ for which $\min(\alpha_{i_0}, \beta_{i_0}) \geq 1$. Hence, the result.

(173) We use Theorem 12 and the fact that

$$\min\{\alpha_i, \beta_i, \gamma_i\} = \min\{\alpha_i, \beta_i\} + \min\{\beta_i, \gamma_i\} + \min\{\alpha, \gamma_i\}$$
$$- \min\{\alpha_i + \beta_i, \beta_i + \gamma_i, \alpha_i + \gamma_i\}.$$

The second part follows from the first part and Problem 171.

(174) Let

$$\begin{cases} a = p_1^{\alpha_1} \cdots p_r^{\alpha_r}, \\ b = p_1^{\beta_1} \cdots p_r^{\beta_r}, \\ c = p_1^{\gamma_1} \cdots p_r^{\gamma_r}. \end{cases}$$

Since $[a, b] = \prod_{i=1}^{r} p_i^{\max\{\alpha_i, \beta_i\}}$ and $(a, b) = \prod_{i=1}^{r} p_i^{\min\{\alpha_i, \beta_i\}}$, it is enough to show that, for each i,

$$2\max\{\alpha_i, \beta_i, \gamma_i\} - \max\{\alpha_i, \beta_i\} - \max\{\beta_i, \gamma_i\} - \max\{\gamma_i, \alpha_i\}$$
$$= 2\min\{\alpha_i, \beta_i, \gamma_i\} - \min\{\alpha_i, \beta_i\} - \min\{\beta_i, \gamma_i\} - \min\{\gamma_i, \alpha_i\}.$$

Without loosing in generality, we may assume that, for a given i, $\alpha_i \geq \beta_i \geq \gamma_i$, from which the result easily follows.

(175) Let $a = \prod_{i=1}^{r} p_i^{a_i}$, $b = \prod_{i=1}^{r} p_i^{b_i}$, $c = \prod_{i=1}^{r} p_i^{c_i}$. Without loosing in generality, we may assume that $a_i \leq b_i \leq c_i$. The equation of the statement allows one to conclude that $c_i + a_i = \frac{1}{2}(a_i + b_i + c_i)$ and therefore that $a_i + c_i = b_i$, which implies, since $c_i \geq b_i$, that $b_i = c_i$ and $a_i = 0$. This means that in order for the relation to be true, for the same prime number, two of the exponents must be equal while the third one should be 0. Hence, we can choose $a = 2^1 \cdot 3^1 \cdot 5^0 = 6$, $b = 2^1 \cdot 3^0 \cdot 5^1 = 10$ and $c = 2^0 \cdot 3^1 \cdot 5^1 = 15$. Note that the numbers 42, 70, 15 will also do.

(176) The left inequality is obvious. To prove the right equality, first observe that

$$\#n = [1, 2, \ldots, n] = \prod_{p \leq n} p^{\delta_p},$$

where p^{δ_p} is the largest power of p not exceeding n. In other words, δ_p is defined implicitly by the inequalities $p^{\delta_p} \leq n < p^{\delta_p + 1}$. It follows successively that

$$\delta_p \log p \leq \log n < (\delta_p + 1) \log p,$$

$$\delta_p \leq \frac{\log n}{\log p} < \delta_p + 1,$$

$$\delta_p = \left[\frac{\log n}{\log p} \right].$$

We have thus established that

$$\#n = \prod_{p \leq n} p^{[\log n / \log p]},$$

which was to be shown.

(177) It is easy to see that

$$(p, p + r) = \begin{cases} p & \text{if } p | r, \\ 1 & \text{if } p \nmid r \end{cases}$$

and that

$$[p, p + r] = \begin{cases} p + r & \text{if } p | r, \\ p(p + r) & \text{if } p \nmid r. \end{cases}$$

(178) We have that $p | (8ad - bd) - (8bc - bd) = 8(ad - bc)$, and therefore that $p | (ad - bc)$, since p is odd.

(179) Since p is odd, it is clear that $p + p + 2 = 2(p + 1)$ is a multiple of 4. On the other hand, since $p + 2$ is prime, the prime number p must be of the form $3k + 2$. It follows that $p + p + 2 = 2p + 2 = 2(3k + 2) + 2 = 6k + 6$, a multiple of 3. Hence, the result.

(180) Set $n = pr$, where p is a prime number. If $p \neq r$, then p and r show up as factors in the product $(n - 1)!$ and therefore n divides $(n - 1)!$. If $r = p$, then $n = p^2$ and

$$(n - 1)! = (p^2 - 1)(p^2 - 2) \cdots p \cdots 1.$$

Hence, in order for $(n-1)!$ to be divisible by p^2, we must have that $(n-1)!$ contains the factors p and $2p$; that is we must have $p^2 - \alpha = 2p$ for a certain integer $\alpha \geq 2$. But this is possible only if we choose $\alpha = p(p-2)$ (provided that $p > 2$). If $p = 2$, that is $n = 4$, it is clear that the result does not hold.

(181) If it is the case, we will have

$$\frac{n(n + 1)}{2} \, \Big| \, n!, \quad \text{that is} \quad \frac{n + 1}{2} \, \Big| \, (n - 1)!.$$

This means that we are looking for the positive integers n for which there exists a positive integer $M = M(n)$ such that

$$(n - 1)! = M \frac{(n + 1)}{2}, \quad \text{that is} \quad 2 \frac{(n - 1)!}{(n + 1)} = M.$$

If $n + 1 = p$, with p prime, then $n = p - 1$, in which case M is not an integer. Therefore, $n + 1$ must be composite; that is $n + 1 = pr$, where p is prime. If p and r can be chosen in such a way that $p \neq r$, then p and r will show up as factors in the product $(n - 1)!$, implying that M is an integer. If the only possible choice for p and r is $p = r$, then we have $n + 1 = r^2$ and

$$2(n - 1)! = 2(p^2 - 2)! = 2(p^2 - 2)(p^2 - 3) \cdots p \cdots 1.$$

Hence, in order for $2(n - 1)!$ to be divisible by p^2, we must have that $2(n-1)!$ contains the factors p and $2p$; that is we must have $(p^2 - \alpha) = 2p$ for a certain integer $\alpha \geq 2$. But this is possible only if $\alpha = p(p - 2)$ (provided $p > 2$). It follows that the result is true for all integers n such that $n \neq p - 1$, p being an odd prime number.

(182) (*AMM, Vol. 81, 1974, p. 778*). From the solution of Problem 181, we have that if $n > 5$ is a composite integer, $(n - 2)!/n$ is an even integer and therefore that

$$\sin \frac{\pi}{2} n \left(\frac{(n - 2)!}{n} \right) = 0.$$

On the other hand, if $n = p$ is prime, then by Wilson's Theorem, $(p-2)! \equiv -(p-1)! \equiv 1 \pmod{p}$, in which case there exists an integer k such that $(p-2)! = kp + 1$ and therefore

$$\left(\frac{(p-2)!}{p}\right) = k = \frac{(p-2)! - 1}{p}.$$

Hence, if $p > 5$, then $4 | (p-2)!$ and therefore

$$\sin \frac{\pi}{2} p \left(\frac{(p-2)!}{p}\right) = \sin \frac{\pi}{2}((p-2)! - 1) = \sin \left(\frac{\pi}{2}(p-2)! - \frac{\pi}{2}\right) = -1.$$

These two cases allow us to conclude that for $n > 5$, the term indexed by n in the sum is 0 if n or $n+2$ is composite and is $(-1)(-1) = 1$ if n and $n+2$ are prime. Note that the term "2" is necessary in order to count the pairs of twin primes $(3,5)$ and $(5,7)$.

(183) Assume that there does not exist any prime p between n and $n!$. Then, consider the number $N = n! - 1$. If N is prime, we have found a prime number between n and $n!$, a contradiction. If N is composite, then there exists a prime number p such that $p \le n$ and $p | (n! - 1)$; but since $p | n!$, we must have $p | 1$, again a contradiction. Thus, the result.

(184) The answer is NO. Since $1 + 2 + 2^2 + \cdots + 2^n = 2^{n+1} - 1$, it is easy to check that $1 + 2 + \cdots + 2^8$ can be written as

$$2^9 - 1 = (2^3)^3 - 1 = 8^3 - 1 = (8 - 1)(8^2 + 8 + 1),$$

a composite number, while the preceding number, that is 255, is also composite.

REMARK: The prime numbers of the form $2^k - 1$ are called Mersenne primes, and it is not known if there exist infinitely many of them. See the next problem.

(185) It is easy to see that

$$a^n - 1 = (a - 1)(a^{n-1} + a^{n-2} + \cdots + a + 1)$$

where the second factor is larger than 1. This implies that $a - 1 = 1$; that is $a = 2$. Moreover, if n is composite, then there exist integers r and s such that $n = rs$, $r > 1$, $s > 1$, and therefore

$$a^n - 1 = 2^{rs} - 1 = (2^r - 1)(2^{r(s-1)} + \cdots + 2^r + 1),$$

where each factor is larger than 1, which contradicts the fact that $a^n - 1$ is prime.

(186) If a is odd, then $a^n + 1 \ge 4$ and is an even integer, hence not prime. On the other hand, if n has an odd factor $r > 1$, then there exists a positive integer m such that $n = mr$, in which case

$$a^n + 1 = (a^m + 1)(a^{m(r-1)} - a^{m(r-2)} + \cdots - a^m + 1).$$

Since $r \ge 3$, both factors are larger than 1, and this contradicts the fact that $a^n + 1$ is prime. Hence, n has no odd factor larger than 1 and n must be of the form 2^r.

(187) (*TYCM, Vol. 13, 1982, p. 208*). We reduce these expressions modulo 3. Since $2^{2^z} + 1$ with $z > 0$ is of the form $2^{2t} + 1$ with $t > 0$, it follows that

$$2^{2^z} + 1 = 2^{2t} + 1 = (2^2)^t + 1 \equiv 2 \pmod{3}.$$

But $2^x - 1 \equiv 0$ or $1 \pmod 3$ depending whether x is even or odd. Hence, $(2^x - 1)(2^y - 1) \equiv 0$ or $1 \pmod 3$, and since $2^{2^z} + 1 \equiv 2 \pmod 3$, the result follows.

(188) The result is true for $n = 1$, since in this case it is easy to check that $F_0 = 2^{2^0} + 1 = 3$ and $F_1 - 2 = 2^{2^1} + 1 - 2 = 5 - 2 = 3$. Assume that the result is true for $n = k$ and let us show that it implies that the result is then true for $n = k + 1$. Indeed, by the induction hypothesis, we have

$$
\begin{aligned}
F_0 F_1 F_2 \cdots F_{k-1} F_k &= (F_k - 2) F_k = \left(2^{2^k} - 1\right)\left(2^{2^k} + 1\right) \\
&= 2^{2^{k+1}} - 1 = \left(2^{2^{k+1}} + 1\right) - 2 = F_{k+1} - 2,
\end{aligned}
$$

as required.

(189) Assume the contrary, that is that there exist two integers $m > n \geq 0$ such that $(F_m, F_n) = d > 1$. Then, using Problem 188, we have

$$(*) \qquad\qquad F_0 F_1 \cdots F_{m-1} = F_m - 2.$$

Since F_n is one of the factors on the left of $(*)$, it follows that $d | 2$. But since each Fermat number is odd, it is impossible to have $d = 2$. Hence, $d = 1$, and the result follows.

(190) With the help of a computer, we find that this number is $29\,341$.

(191) From Problem 189, all Fermat numbers are pairwise relatively prime. Each Fermat number therefore introduces in its factorization at least one new prime number. As a consequence, the Fermat numbers generate infinitely many prime numbers.

(192) To prove part (a), we proceed by induction. First of all, it is clear that $3^2 | 2^{3^1} + 1$. Assuming that $3^k | f_{k-1}$ for some $k \geq 2$, we will show that this implies that $3^{k+1} | f_k$. Using the fact that $a^3 + 1 = (a + 1)(a^2 - a + 1)$, we have

$$
\begin{aligned}
(*) \qquad f_k &= 2^{3^k} + 1 = \left(2^{3^{k-1}}\right)^3 + 1 \\
&= \left(2^{3^{k-1}} + 1\right)\left(\left(2^{3^{k-1}}\right)^2 - \left(2^{3^{k-1}}\right) + 1\right) \\
&= A \cdot B,
\end{aligned}
$$

say. The expression A is divisible by 3^k because of the induction hypothesis. It therefore remains to show that $3 | B$. But $B = a^2 - a + 1$, where $a = 2^{3^{k-1}} \equiv 2 \pmod 3$. It follows that

$$B = a^2 - a + 1 \equiv 2^2 - 2 + 1 \equiv 0 \pmod 3,$$

as required.

To prove part (b), we only need to observe that $f_{n-1} | f_n$, as is implied by the second line of $(*)$.

(193) (This problem can be found on page 64 of the book of D.J. Newman [**24**]). Consider the arithmetic progression $15k + 7$, $k = 1, 2, \ldots$, which by Dirichlet's Theorem contains infinitely many prime numbers. If $p = 15k + 7$, it is clear that $p - 2 = 15k + 5$ is a multiple of 5 and that $p + 2 = 15k + 9$ is a multiple of 3, which proves the result.

(194) Since

$$2^{16} = 2^{2^4} = 65536 \equiv 154 \pmod{641},$$

we have

$$2^{32} = (2^{16})^2 \equiv (154)^2 = 23716 \equiv 640 \equiv -1 \pmod{641},$$

and the result follows.

(195) First of all, it is clear that $F_2 = 2^{2^2} + 1 = 17 \equiv 7 \pmod{10}$. Therefore, it is enough to show that if $F_k \equiv 7 \pmod{10}$ for a certain $k \geq 2$, then $F_{k+1} \equiv 7 \pmod{10}$. Indeed, we have by the induction hypothesis that

$$F_{k+1} = 2^{2^{k+1}} + 1 = \left(2^{2^k}\right)^2 + 1$$

$$= \left((2^{2^k} + 1) - 1\right)^2 + 1 \equiv (7 - 1)^2 + 1 = 37 \equiv 7 \pmod{10},$$

as required.

(196) We proceed by contradiction in assuming that 2^k divides an integer $m \in E \setminus \{2^k\}$, in which case $m = 2^k r$ for a certain integer $r > 1$, implying that 2^{k+1} is in the set E, which contradicts the minimal choice of 2^k.

For the second part of the problem, assume that the given sum is an integer M. The smallest common multiple of the elements of E must be of the form $2^k m$, where m is an odd number. Multiplying the sum by $m2^{k-1}$, we obtain

$$m2^{k-1} \sum_{j=1}^n \frac{1}{j} = m2^{k-1}M.$$

But when the left-hand side is expanded, one of the n terms is equal to $m/2$ while all the others are integers, which yields a contradiction since m is odd.

(197) We have

$$2^{5n} - 1 = (2^5)^n - 1 = (2^5 - 1)(2^{5(n-1)} + 2^{5(n-2)} + \cdots + 2^5 + 1),$$

which implies that the number $2^{5n} - 1$ is divisible by 31 for any positive integer n. Hence, $p = 31$ will serve our purpose.

(198) We have $M_1 = 3$, $M_2 = 7$, $M_3 = 31$, $M_4 = 211$, $M_5 = 2\,311$, which are all prime numbers, while $M_6 = 59 \cdot 509$ and $M_7 = 19 \cdot 97 \cdot 277$ are composite. REMARK: Using the MAPLE program

```
> for k from 8 to 10 do print(
> M(k) = ifactor(product(ithprime(i), i=1..k)+1)) od;
```

we obtain $M_8 = 347 \cdot 27953$, $M_9 = 317 \cdot 703763$ and $M_{10} = 331 \cdot 571 \cdot 34231$.

As of 2006, we still don't know if the sequence $\{M_k\}$ contains infinitely many prime numbers; with the help of a computer, we can nevertheless easily establish that the only values of $k < 1000$ for which M_k is a prime number are: 1, 2, 3, 4, 5, 11, 75, 171, 172, 384, 457, 616 and 643.

(199) Any prime number dividing $p_1 p_2 \cdots p_r + 1$ is distinct from any of the primes $p_1, p_2, \ldots, p_r$; hence, it follows that

$$p_{r+1} \leq p_1 p_2 \cdots p_r + 1,$$

and using an induction argument, we obtain that

$$p_{r+1} \leq 2^{2^0} \cdot 2^{2^1} \cdot 2^{2^2} \cdots 2^{2^{r-1}} + 1 < 2^{2^r},$$

which proves the result.

(200) Let $x \geq 3$. Choose $r \in \mathbb{N}$ such that

(1) $$e^{e^{r-1}} < x \leq e^{e^r}.$$

We easily observe that such a choice of r is unique. The left inequality of (1), the fact that $\pi(x)$ is a nondecreasing function and the relation $p_r \leq 2^{2^{r-1}}$ allow us to write

(2) $$\pi(x) \geq \pi(e^{e^{r-1}}) \geq \pi(2^{2^{r-1}}) \geq \pi(p_r) = r.$$

The right inequality of (1) guarantees that

(3) $$r \geq \log \log x.$$

Combining (2) and (3), we obtain the required inequality.

(201) Assume that there is only a finite number of prime numbers of the form $4n + 3$. Denote them by

$$q_1 < q_2 < \ldots < q_k$$

and consider the number

(∗) $$N = 4q_1 q_2 \cdots q_k - 1 = 4(q_1 q_2 \cdots q_k - 1) + 3,$$

which is clearly of the form $4n + 3$. If N is prime, then we have found a prime number of the form $4n + 3$ larger then q_k, thereby yielding a contradiction. If N is composite, then N cannot be the product of only prime numbers of the form $4n + 1$ (since N would then also be of the form $4n + 1$). Therefore there exists a prime number q of the form $4n + 3$ which divides N. If q is equal to one of the q_i's, that would mean, in light of relation (∗), that $q | 1$, again a contradiction. Hence, $q > q_k$ and the result is proved.

(202) Assume that there is only a finite number of prime numbers of the form $6n + 5$. Denote them by

$$q_1 < q_2 < \ldots < q_k$$

and consider the number

(∗) $$N = 6q_1 q_2 \cdots q_k - 1 = 6(q_1 q_2 \cdots q_k - 1) + 5,$$

which is surely of the form $6n + 5$. If N is prime, then we have found a prime number of the form $6n + 5$ larger than q_k, thereby yielding a contradiction. If N is composite, then N cannot be the product of only prime numbers of the form $6n + 1$ (since N would also be of the form $6n + 1$). Therefore there exists a prime number q of the form $6n + 5$ which divides N. If q is equal to one of the q_i's, that would mean, in light of (∗), that $q | 1$, again a contradiction. Hence, $q > q_k$ and the result is proved.

(203) It is enough to consider the polynomial

$$f(x) = (x - p_1)(x - p_2) \cdots (x - p_r) + x,$$

where p_k stands for the k-th prime number.

(204) The answer is NO. Consider such a number N with $2k + 1$ digits, $k \geq 2$. We first notice that, for each integer $k > 1$,

$$(1 + 10^2 + \cdots + 10^{2k})(10^2 - 1) = (10^{2k+2} - 1) = (10^{k+1} - 1)(10^{k+1} + 1).$$

Hence,

$$1 + 10^2 + \cdots + 10^{2k} = \frac{(10^{k+1} - 1)(10^{k+1} + 1)}{9 \cdot 11}.$$

Since $k > 1$, both factors on the right-hand side, after dividing by 99, have two factors larger than 1, so that the number N is composite. On the other hand, in the particular case $k = 1$, we find the prime number 101.

(205) Let $G_n = 2^{2^n} + 5$. First of all, $G_0 = 2^1 + 5 = 7$, which is prime. We will show that all the other G_n's, that is those with $n \geq 1$, are divisible by 3. To do so, it is enough to prove that $2^{2^n} \equiv 1 \pmod{3}$. But this is true since

$$2^{2^n} = \left(2^2\right)^{2^{n-1}} = 4^{2^{n-1}} \equiv 1^{2^{n-1}} = 1 \pmod{3}.$$

Clearly, we could have obtained the same result if instead of 5 we would have used a number of the form $3k+2$, except that in this case, one should first check whether $2 + 3k + 2 = 3k + 4$ is prime or not.

(206) The answer is NO. Indeed, the next gap in the list is 14; it occurs when $p_{r+1} - p_r = 127 - 113 = 14$, while the first gaps of 10 and 12 occur respectively when $p_{r+1} - p_r = 149 - 139 = 10$ and $p_{r+1} - p_r = 211 - 199 = 12$.

(207) Let S be this series; then

$$S = \sum_p \left(\frac{1}{p^2} + \frac{1}{p^3} + \frac{1}{p^4} + \cdots\right) = \sum_p \frac{1}{p^2}\left(1 + \frac{1}{p} + \frac{1}{p^2} + \frac{1}{p^3} + \cdots\right)$$

$$= \sum_p \frac{1}{p^2}\frac{1}{1 - \frac{1}{p}} = \sum_p \frac{1}{p(p-1)} < \sum_{n=2}^{\infty} \frac{1}{n(n-1)} = 1.$$

In fact, the exact value of S is $0.773156669\ldots$.

(208) We have successively

$$\sum_{n=1}^{\infty} \frac{\mu(n)}{n} f(x^{1/n}) = \sum_{n=1}^{\infty} \left(\frac{\mu(n)}{n} \sum_{m=1}^{\infty} \frac{1}{m}\pi(x^{1/mn})\right)$$

$$= \sum_{n=1}^{\infty}\sum_{m=1}^{\infty} \frac{\mu(\frac{mn}{m})}{mn}\pi(x^{1/mn})$$

$$= \sum_{r=1}^{\infty} \left(\frac{\pi(x^{1/r})}{r} \sum_{d|r} \mu(d)\right) = \pi(x),$$

where we used the fact that $\sum_{d|r} \mu(d) = 0$ if $r > 1$ and 1 if $r = 1$.

(209) The result is immediate for $2 \leq n \leq 6$. On the other hand, since

$$\sum_{n \leq m^2 \leq 2n} 1 = \left[\sqrt{2n}\right] - \left[\sqrt{n}\right] + 1 \geq \sqrt{2n} - 1 - \sqrt{n} + 1 = (\sqrt{2} - 1)\sqrt{n} > 1,$$

for $n > 6$, the result follows for each integer $n \geq 2$.

(210) Assume the contrary, that is that $n^3 = p + m^3$. We then have $n^3 - m^3 = p$, which implies that $(n-m)(n^2 + mn + m^2) = p$ and therefore that $n - m = 1$ and $n^2 + mn + m^2 = p$. This shows that $n^2 + n(n - 1) + (n - 1)^2 = p$,

that is $3n^2 - 3n + 1 = p$, which contradicts the fact that $3n^2 - 3n + 1$ is composite.

(211) The prime numbers $p < 10\,000$ of the indicated form are 2, 5, 17, 37, 101, 197, 257, 401, 577, 677, 1297, 1601, 2917, 3137, 4357, 5477, 7057, 8101 and 8837. Given $n \geq 1$, we have $n \equiv 0, 1, 2, 3, 4, 5, 6, 7, 8, 9 \pmod{10}$, in which case $n^2 \equiv 0, 1, 4, 9, 6, 5, 6, 9, 4, 1 \pmod{10}$ and therefore $n^2 + 1 \equiv 1, 2, 5, 0, 7, 6, 7, 0, 5, 2 \pmod{10}$. Since the numbers $n^2 + 1$, congruent to 0, 2, 6 or 5 modulo 10, are composite, we are left with the numbers n for which $n^2 + 1 \equiv 1, 7 \pmod{10}$. Finally, since the numbers $n \equiv 4, 6 \pmod{10}$ are such that $n^2 + 1 \equiv 7 \pmod{10}$ while only the numbers $n \equiv 0 \pmod{10}$ are such that $n^2 + 1 \equiv 1 \pmod{10}$, this explains why the digit 7 seems to appear twice as often.

(212) This follows from the fact that, from Theorem 27, we have

$$n! = \prod_{p \leq n} p^{\sum_{j=1}^{\infty} \left[\frac{n}{p^j}\right]} \leq \prod_{p \leq n} p^{n \sum_{j=1}^{\infty} \frac{1}{p^j}} = \prod_{p \leq n} p^{\frac{n}{p-1}},$$

where we used the fact that

$$\frac{1}{p} + \frac{1}{p^2} + \frac{1}{p^3} + \cdots = \frac{1}{p}\left(1 + \frac{1}{p} + \frac{1}{p^2} + \cdots\right) = \frac{1}{p-1}.$$

(213) Every positive integer $n \geq 6$ can be written as $n = 6k$, $n = 6k + 1$, $n = 6k + 2$, $n = 6k + 3$, $n = 6k + 4$ or $n = 6k + 5$, in which case the corresponding values of $n^2 + 2$ are respectively multiples of 2, multiples of 3, multiples of 2, of the form $6K + 5$, multiples of 2 and multiples of 3. It follows that only those $n = 6k + 3$ (with $n^2 + 2 = 6K + 5$) are possible candidates for ensuring that $n^2 + 2$ is prime, thus the result.

(214) Let $N + i$ be one of these numbers. If i is prime, then $i | N + i$ and $N + i$ is composite. While if i is not prime, then i is divisible by a prime number $p_0 < i < p$, in which case $p_0 | N + i$ and $N + i$ is composite.

(215) We write n as

$$n = a_k 10^k + a_{k-1} 10^{k-1} + \cdots + a_2 10^2 + a_1 10 + a_0,$$

where $k \geq 2$ and where the a_i's are integers satisfying $0 \leq a_i \leq 9$, $a_k \neq 0$. Part (a) is trivial. To prove (b), it is enough to observe that, since $4 | 10^j = 5^j \cdot 2^j$ for each integer $j \geq 2$, it follows that $4 | a_1 10 + a_0$ if and only if $4 | n$. To prove (c), it is enough to observe that, since $8 | 10^j = 5^j \cdot 2^j$ for each integer $j \geq 3$, it follows that $8 | a_2 100 + a_1 10 + a_0 \Leftrightarrow 8 | n$. Therefore it becomes clear that one can generalize this result as follows: *If $n > 1$ is an integer having at least k digits, then $2^k | n$ if and only if the number made up of the last k digits of n is divisible by 2^k.*

(216) (Hlawka [**19**]). Let $n > e^e$ and set $f(n) = \displaystyle\sum_{p|n, p > \log n} 1$. It follows that

$$n \geq \prod_{\substack{p|n \\ p > \log n}} p \geq (\log n)^{f(n)}$$

and therefore that $\log n \geq f(n) \log \log n$; that is $f(n) \leq (\log n)/\log \log n$. On the other hand, for n sufficiently large, we have

$$\log\left(1 - \frac{1}{\log n}\right) \geq -\frac{2}{\log n}.$$

It follows that

$$0 \geq \log P(n) = \sum_{p|n, p > \log n} \log\left(1 - \frac{1}{p}\right) \geq f(n) \log\left(1 - \frac{1}{\log n}\right)$$

$$\geq -2\frac{f(n)}{\log n} \geq -\frac{2}{\log \log n}.$$

Hence, $\lim_{n \to \infty} \log P(n) = 0$, and the result is proved.

(217) (*MMAG, April 1992, p. 130*). Assume the contrary, that is that each interval $[n^2, (n+1)^2]$ contains less than 1000 prime numbers. We know that the sum of the reciprocals of the prime numbers diverges. Hence, according to our hypothesis, we have

$$+\infty = \sum_p \frac{1}{p} = \sum_{n=1}^{\infty} \sum_{n^2 < p \leq (n+1)^2} \frac{1}{p} < \sum_{n=1}^{\infty} \frac{1}{n^2} \sum_{n^2 < p \leq (n+1)^2} 1$$

$$= \sum_{n=1}^{\infty} \frac{1}{n^2}\left(\pi((n+1)^2) - \pi(n^2)\right) < 1000 \sum_{n=1}^{\infty} \frac{1}{n^2} < +\infty,$$

a contradiction.

(218) We only need to show that if $x < y$ are any two positive real numbers, then there exist two prime numbers p and q such that $x < p/q < y$. It is obvious that

$$(1) \qquad \pi(qy) - \pi(qx) = \pi(qx)\left(\frac{\pi(qy)}{\pi(qx)} - 1\right).$$

On the other hand, using the Prime Number Theorem, we have

$$(2) \qquad \lim_{q \to \infty} \frac{\pi(qy)}{\pi(qx)} = \frac{y}{x} > 1.$$

It follows from (1) and (2) that $\lim_{q \to \infty}(\pi(qy) - \pi(qx)) = +\infty$. This means that for q sufficiently large, say $q = q_0$, there exists at least one prime number p_0 such that $q_0 x < p_0 < q_0 y$, in which case we have

$$x < \frac{p_0}{q_0} < y,$$

as required.

(219) Assume that $q_1 < q_2 < \ldots < q_m$ are prime numbers such that

$$\frac{1}{q_1} + \cdots + \frac{1}{q_m} = n$$

for a certain integer $n \geq 1$. Then,

$$\frac{1}{q_1} = n - \frac{1}{q_2} - \cdots - \frac{1}{q_m} = \frac{r}{q_2 \cdots q_m},$$

where r is an integer. In this case, the product $q_2 \cdots q_m$ is divisible by q_1, which is impossible.

(220) We have successively

$$\sum_{\substack{n \le x \\ P(n) > \sqrt{x}}} 1 = \sum_{\sqrt{x} < p \le x} \sum_{\substack{n \le x \\ P(n) = p}} 1 = \sum_{\sqrt{x} < p \le x} \sum_{\substack{mp \le x \\ P(m) \le p}} 1 = \sum_{\sqrt{x} < p \le x} \sum_{\substack{m \le x/p \\ P(m) \le p}} 1$$

$$= \sum_{\sqrt{x} < p \le x} \left[\frac{x}{p}\right] = \sum_{\sqrt{x} < p \le x} \frac{x}{p} + \sum_{\sqrt{x} < p \le x} \left(\left[\frac{x}{p}\right] - \frac{x}{p}\right)$$

$$= x \log 2 + x R(x) - x R(\sqrt{x}) + S(x),$$

say, where $|S(x)| \le \pi(x)$. Relation (3) then follows from (1) and (2). To show the last part, it is sufficient to observe that, since $\log 2 = 0.69\ldots$, then if x is sufficiently large,

$$\frac{1}{x} \sum_{\substack{n \le x \\ P(n) > \sqrt{n}}} 1 > \frac{1}{x} \sum_{\substack{n \le x \\ P(n) > \sqrt{x}}} 1 = \log 2 + T(x) > \frac{2}{3}.$$

(221) Let $p_1, p_2, \ldots, p_r$ be the prime numbers $\le \sqrt{x}$. Then, all odd integers $\le x$ which are not divisible by $p_1, p_2, \ldots, p_r$ are prime numbers. Consequently, $\pi(x) - \pi(\sqrt{x})$ counts the number of prime numbers $> \sqrt{x}$. But the number of positive integers $\le x$ which are divisible by none of the primes $p_1, p_2, \ldots, p_r$ is equal to

$$(*) \qquad [x] - \sum_{1 \le i \le r} \left[\frac{x}{p_i}\right] + \sum_{1 \le i < j \le r} \left[\frac{x}{p_i p_j}\right] - \sum_{1 \le i < j < k \le r} \left[\frac{x}{p_i p_j p_k}\right]$$

$$+ \cdots + (-1)^r \left[\frac{x}{p_1 \cdots p_r}\right].$$

Indeed, let n be an integer $\le x$ which is divisible only by the prime numbers $p_1, p_2, \ldots, p_r$; in this case, the sum is equal to

$$1 - \binom{r}{1} + \binom{r}{2} - \cdots + (-1)^r \binom{r}{r} = (1-1)^r = 0,$$

while if n is not divisible by any of the primes $p_1, \ldots, p_r$, then its contribution to the right-hand side is obviously 1.

REMARK: Observe that expression $(*)$ can also be written as

$$\sum_{n | p_1 p_2 \cdots p_r} \mu(n) \left[\frac{x}{n}\right].$$

(222) Let $n > 5$. From Conjecture A, if n is even, there exist two prime numbers p and q such that $n - 2 = p + q$, that is $n = p + q + 2$; while if n is odd, there exist two prime numbers p and q such that $n - 3 = p + q$, so that $n = p + q + 3$. In both cases, Conjecture B follows.

Let n be an even integer ≥ 4. From Conjecture B, there exist three prime numbers p, q, r such that $n + 2 = p + q + r$. Since $n + 2$ is even, it is clear that one of the three prime numbers p, q, r must be even, that is equal to 2. Assume that $r = 2$. It follows that $n = p + q$, which establishes Conjecture A.

(223) (This result is attributed to Mináč; see P. Ribenboim [29]). First of all, we observe that if $n \ne 4$ is not a prime number, then $n | (n-1)!$. Indeed, either $n = ab$, with $2 \le a < b \le n - 1$, in which case $n | (n-1)!$, or

$n = p^2 \neq 4$, in which case $2 < p \leq n - 1 = p^2 - 1$ and $2p \leq p^2 - 1$, which implies that $n | 2p^2 = p \cdot 2p$, an expression which divides $(n - 1)!$.

To prove the stated relation, we analyze separately the cases "j prime" and "j composite".

If j is prime, then by Wilson's Theorem, there exists $k \in \mathbb{N}$ such that $(j - 1)! + 1 = kj$ so that

$$\left[\frac{(j-1)! + 1}{j} - \left[\frac{(j-1)!}{j} \right] \right] = \left[k - \left[k - \frac{1}{j} \right] \right] = [k - (k - 1)] = 1.$$

If j is composite, $j \geq 6$, then $j | (j - 1)!$ in light of the above observation. Therefore, there exists an integer k such that $(j - 1)! = kj$. It follows that

$$\left[\frac{(j-1)! + 1}{j} - \left[\frac{(j-1)!}{j} \right] \right] = \left[k + \frac{1}{j} - k \right] = 0.$$

Finally, if $j = 4$, we have $\left[\frac{3! + 1}{4} - \left[\frac{3!}{4} \right] \right] = 0$, which completes the proof of Mináč's formula.

(224) (a) We have

$$A(n) = \sum_{\substack{m \leq n \\ a | m}} 1 = \sum_{ar \leq n} 1 = \sum_{r \leq \frac{n}{a}} 1 = \left[\frac{n}{a} \right].$$

It is therefore easy to see that the quotient $A(n)/n$ tends to $1/a$ as $n \to \infty$.

(b) We have

$$A(n) = \sum_{\substack{m \leq n \\ a | m}} 1 - \sum_{\substack{m \leq n \\ a | m, \, a_0 | m}} 1 = \sum_{ar \leq n} 1 - \sum_{\substack{m \leq n \\ [a, a_0] | m}} 1 = \sum_{r \leq \frac{n}{a}} 1$$

$$- \sum_{r \leq \frac{n}{[a, a_0]}} 1 = \left[\frac{n}{a} \right] - \left[\frac{n}{[a, a_0]} \right].$$

It is therefore easy to see that the quotient $A(n)/n$ tends to $\frac{1}{a} - \frac{1}{[a, a_0]}$ as $n \to \infty$.

(c) Using the inclusion-exclusion principle, we have

$$n - A(n) = \sum_{1 \leq i \leq r} \sum_{\substack{m \leq n \\ q_i | m}} 1 - \sum_{1 \leq i < j \leq r} \sum_{\substack{m \leq n \\ q_i q_j | m}} 1$$

$$+ \sum_{1 \leq i < j < k \leq r} \sum_{\substack{m \leq n \\ q_i q_j q_k | m}} 1 - \cdots + (-1)^{r+1} \sum_{\substack{m \leq n \\ q_1 q_2 \cdots q_r | m}} 1$$

$$= \sum_{1 \leq i \leq r} \left[\frac{n}{q_i} \right] - \sum_{1 \leq i < j \leq r} \left[\frac{n}{q_i q_j} \right] + \sum_{1 \leq i < j < k \leq r} \left[\frac{n}{q_i q_j q_r} \right]$$

$$- \cdots + (-1)^{r+1} \left[\frac{n}{q_1 q_2 \cdots q_r} \right].$$

It follows that

$$1 - \frac{A(n)}{n} = \sum_{1 \le i \le r} \frac{1}{n}\left[\frac{n}{q_i}\right] - \sum_{1 \le i < j \le r} \frac{1}{n}\left[\frac{n}{q_i q_j}\right]$$
$$+ \sum_{1 \le i < j < k \le r} \frac{1}{n}\left[\frac{n}{q_i q_j q_r}\right] - \cdots + (-1)^{r+1}\frac{1}{n}\left[\frac{n}{q_1 q_2 \cdots q_r}\right].$$

But, as $n \to \infty$, this last expression tends to

$$\sum_{1 \le i \le r} \frac{1}{q_i} - \sum_{1 \le i < j \le r} \frac{1}{q_i q_j} + \sum_{1 \le i < j < k \le r} \frac{1}{q_i q_j q_r} - \cdots$$
$$+ (-1)^{r+1}\frac{1}{q_1 q_2 \cdots q_r} = -\prod_{i=1}^{r}\left(1 - \frac{1}{q_i}\right) + 1,$$

as required.

(225) We will show that

$$\underline{\mathbf{d}}\,\mathcal{A} = \frac{1}{3} < \frac{2}{3} = \overline{\mathbf{d}}\,\mathcal{A}.$$

To do so, we prove that

$$\lim_{k \to \infty} \frac{A(2^{2k+1})}{2^{2k+1}} = \frac{2}{3}, \quad \text{while} \quad \lim_{k \to \infty} \frac{A(2^{2k})}{2^{2k}} = \frac{1}{3}.$$

Indeed,

$$A(2^{2k+1}) = \sum_{1 \le n < 2} 1 + \sum_{2^2 \le n < 2^3} 1 + \sum_{2^4 \le n < 2^5} 1 + \cdots + \sum_{2^{2k} \le n < 2^{2k+1}} 1$$
$$= 1 + (2^3 - 2^2) + (2^5 - 2^4) + \cdots + (2^{2k+1} - 2^{2k})$$
$$= 1 + 2^2 + 2^4 + \cdots + 2^{2k}$$
$$= 1 + 4 + 4^2 + \cdots + 4^k = \frac{4^{k+1} - 1}{4 - 1} = \frac{1}{3}\left(4^{k+1} - 1\right).$$

It follows from this that

$$\lim_{k \to \infty} \frac{A(2^{2k+1})}{2^{2k+1}} = \lim_{k \to \infty} \frac{1}{3}\frac{4^{k+1} - 1}{2 \cdot 4^k} = \frac{2}{3}.$$

On the other hand,

$$A(2^{2k}) = \sum_{\substack{n \in \mathcal{A} \\ n \le 2^{2k}}} 1 = \sum_{\substack{n \in \mathcal{A} \\ n < 2^{2k-1}}} 1 = \sum_{\substack{n \in \mathcal{A} \\ n \le 2^{2(k-1)+1}}} 1 = \frac{1}{3}(4^k - 1),$$

which implies that

$$\lim_{k \to \infty} \frac{A(2^{2k})}{2^{2k}} = \lim_{k \to \infty} \frac{1}{3}\frac{4^k - 1}{4^k} = \frac{1}{3}.$$

(226) To each element $a_i \in \mathcal{A}$, we associate its largest odd divisor d_i. It is clear that all the d_i's are distinct; indeed, if $d_i = d_j$ for two positive integers $i \ne j$, then $a_i | a_j$ or $a_j | a_i$, which is possible only if $i = j$. It follows that $A(2n) \le n$, since there are no more than n odd numbers $\le 2n$. Hence, the result.

(227) To each element $a_i \in \mathcal{A}$, we associate its largest prime factor q_i. Let $B_i = \{n \in \mathbb{N} : p(n) > q_i\}$, where $p(n)$ stands for the smallest prime factor of n. Let also $C_i = a_i B_i = \{a_i \cdot b : b \in B_i\}$. The sets C_i are disjoint; indeed, if $a_i r = a_j s$ (with $q_i \leq q_j$), where $r \in B_i$ and $s \in B_j$, then $a_i | a_j$, which is possible only if $i = j$. It follows that, for each positive integer k, we have

$$\sum_{i=1}^{k} \mathbf{d}\, C_i \leq 1.$$

From (a) and (c) of Problem 224, we have

$$\mathbf{d}\, C_i = \frac{1}{a_i} \mathbf{d}\, B_i = \frac{1}{a_i} \prod_{p \leq q_i} \left(1 - \frac{1}{p}\right),$$

so that

(1)
$$\sum_{i=1}^{\infty} \frac{1}{a_i} \prod_{p \leq q_i} \left(1 - \frac{1}{p}\right) \leq 1.$$

But from Mertens' Theorem, we have that

(2)
$$\prod_{p \leq q_i} \left(1 - \frac{1}{p}\right) \gg \frac{1}{\log q_i} \geq \frac{1}{\log a_i}.$$

The result then follows by combining (1) and (2).

(228) Part (a) is obvious. To prove (b), first observe that the norm of every element of E is always ≥ 5. Assume that $3 = (a+b\sqrt{-5})(c+d\sqrt{-5})$; taking the norm, we have $9 = (a^2 + 5b^2)(c^2 + 5d^2)$. This is however impossible since both factors on the right-hand side are larger than 5. Hence, 3 is a prime belonging to E. We easily obtain that $29 = (3+2\sqrt{-5})(3-2\sqrt{-5})$ and is therefore a composite number in E. Part (c) follows from the fact that

$$9 = (3 + 0\sqrt{-5}) \cdot (3 + 0\sqrt{-5}) = (2 + \sqrt{-5})(2 - \sqrt{-5}).$$

(229) Since

$$A(n) - A(n-1) = \begin{cases} 1 & \text{if } n \in A, \\ 0 & \text{otherwise}, \end{cases}$$

we have that

$$\sum_{\substack{n \leq x \\ n \in A}} \frac{1}{n} = \sum_{2 \leq n \leq x} \frac{A(n) - A(n-1)}{n}$$

$$= \sum_{2 \leq n \leq x} A(n) \left(\frac{1}{n} - \frac{1}{n+1}\right) + \frac{A(x)}{[x]+1}$$

$$= \sum_{n \leq x} \frac{A(n)}{n(n+1)} + \frac{A(x)}{[x]+1}.$$

(230) Let n be a palindrome with $2r$ digits. Hence, there exist digits $d_1 > 0$, $d_2, \ldots, d_r$ such that

$$
\begin{aligned}
n &= d_1 10^{2r-1} + d_2 10^{2r-2} + \cdots + d_r 10^r + d_r 10^{r-1} + \cdots \\
&\qquad\qquad\qquad\qquad\qquad\qquad\qquad + d_3 10^2 + d_2 10 + d_1 \\
&= d_1 (10^{2r-1} + 1) + d_2 (10^{2r-2} + 10) + \cdots + d_r (10^r + 10^{r-1}) \\
&= d_1 (10^{2r-1} + 1) + 10 d_2 (10^{2r-3} + 1) + \cdots + 10^{r-2} d_{r-1} (10^3 + 1) \\
&\qquad\qquad\qquad\qquad\qquad\qquad\qquad\qquad + 10^{r-1} d_r (10 + 1).
\end{aligned}
$$

The numbers $10^{2i+1} + 1$ for $i = 0, 1, \ldots, r-1$ are all multiples of 11. Indeed, using the Binomial Theorem, we obtain that

$$
10^{2i+1} + 1 = 1 + (11 - 1)^{2i+1} = 1 + \sum_{k=0}^{2i+1} \binom{2i+1}{k} (-1)^k 11^{2i+1-k} = 11M
$$

for some positive integer M, in which case the number n is divisible by 11.

(231) It is enough to consider the next number, since

$$
1442 = [6, 6, 2]_{15} = 6! + 6! + 2! \, .
$$

(232) Assume that two such representations exist, that is that there also exist positive integers $e_1 < e_2 < \ldots < e_r$ such that

$$
n = d_1! + d_2! + \cdots + d_r! = e_1! + e_2! + \cdots + e_r!.
$$

We proceed by induction. If $r = 1$, then it is clear that $d_1 = e_1$, in which case the result is proved. Assume that the result is true for $r - 1$, and let us show that it is true for r. Without any loss in generality, we may assume that $d_r \geq e_r$. If $d_r = e_r$, then the conclusion follows by an induction argument. We may therefore assume that $d_r > e_r$, in which case $d_r \geq e_r + 1$. We then have

$$
(e_r + 1)! \leq d_r! < d_1! + d_2! + \cdots + d_r! = e_1! + e_2! + \cdots + e_r! \leq r e_r!,
$$

so that $e_r + 1 < r$ and $e_r \leq r - 2$. But the e_i's being distinct, we must have $e_r \geq r$, a contradiction, and the result follows.

(233) (Sierpinski [**39**], Problem #194) Assume that an integer solution $\{x, y\}$ does exist. We then have

$$
(*) \quad x^2 + 1 = y^3 + (2c)^3 = (y + 2c)(y^2 - 2cy + 4c^2) = (y + 2c)((y - c)^2 + 3c^2).
$$

Since $c^2 \equiv 1 \pmod 8$, we have $3c^2 \equiv 3 \pmod 8$. First of all, if y is even, then

$$
x^2 = y^3 + (2c)^3 - 1 \equiv 7 \pmod 8,
$$

a contradiction.

On the other hand, if y is odd, then $y - c$ is even and $(y - c)^2 + 3c^2$ is of the form $4k + 3$. It follows that $(y - c)^2 + 3c^2$ has a prime divisor of the form $4k + 3$, which is then itself a divisor of $x^2 + 1$. But this is impossible because we would then have that the congruence $x^2 \equiv -1$ $\pmod p$ is solvable, which is false since $\left(\frac{-1}{p}\right) = -1$, where $\left(\frac{a}{p}\right)$ stands for the Legendre Symbol (see Definition 21).

As for the second part of the problem, the infinite set $\{8c^3 - 1 \mid c = 1, 2, 3, \ldots\}$ will do.

(234) (Sierpiński [**39**], Problem #211) We observe that, for $n \geq 4$, we have
$$5^{n+4} - 5^n = 5^n(5^4 - 1) = 5^n \cdot 2^4 \cdot 39 = 5^4 \cdot 2^4 \cdot 5^{n-4} \cdot 39$$
$$= 10^4 \cdot 5^{n-4} \cdot 39.$$

Therefore, $5^{n+4} \equiv 5^n \pmod{10\,000}$. It follows that the block of the four last digits of 5^n repeats itself for each number 5^{n+4}. We easily verify that these "periodic" blocks are 0625, 3125, 5625 and 8125.

(235) It is clear that it is enough to consider the cases of two and three cubes. For $a \in \mathbb{Z}$, we have $a^3 \equiv 0, 1, 8 \pmod 9$. From this it follows that, given any integers x and y, we have
$$x^3 + y^3 \equiv 0, 1, 2, 7, 8 \equiv 0, \pm 1, \pm 2 \pmod 9.$$

This implies that not all integers $n \equiv \pm 3, \pm 4 \pmod 9$ can be written as the sum of two cubes. Similarly, given integers x, y and z, we have
$$x^3 + y^3 + z^3 \equiv 0, 1, 2, 3, 6, 7, 8 \equiv 0, \pm 1, \pm 2, \pm 3 \pmod 9.$$

This implies that not all integers $n \equiv \pm 4 \pmod 9$ can be written as the sum of three cubes.

(236) (Anglin [**2**], p. 194). Let m be this integer. First of all, it is clear that $k := \dfrac{m - m^3}{6}$ is an integer. It is then easy to see that
$$m = m^3 + (k+1)^3 + (k-1)^3 + (-k)^3 + (-k)^3,$$

which proves the result.

(237) (*The College Mathematics Journal, March 99, p. 144; solution by T. Amdeberban*). Let (m, n) be such a pair. We write $m = d_1 d_2 \cdots d_r$, where $d_1, d_2, \ldots, d_r$ stand for the digits of m. We then have
$$10^{r-1} \leq m < n = (d_1 + d_2 + \cdots + d_r)^2 \leq (9r)^2,$$

an inequality which can hold only if $r \leq 4$. It follows that $m < n \leq (9 \cdot 4)^2 = 36^2$. Using a computer, one only needs to check all perfect squares ≤ 36. We then find that there is only one solution to this problem, namely $(m, n) = (13^2, 16^2) = (169, 256)$.

(238) (Anglin [**2**], p. 4).

(a) We use induction. Assume that the result is true for each fraction whose numerator is smaller than n (with $n > 1$), and consider the fraction n/m, $n < m$, $(n, m) = 1$. Let $r \geq 2$ be the unique integer such that
$$\frac{1}{r} < \frac{n}{m} < \frac{1}{r-1}.$$
We then have $0 < nr - m < n$. But
$$\frac{n}{m} = \frac{1}{r} + \frac{nr - m}{mr}.$$

Hence, using the induction hypothesis, the fraction $\dfrac{nr - m}{mr}$ is the sum of unitary fractions. Moreover, none of these unitary fractions is equal to $1/2$, because $\dfrac{1}{r} > \dfrac{nr - m}{mr}$. This completes the proof.

(b) Here indeed is a counter-example:
$$\frac{3}{7} = \frac{1}{3} + \frac{1}{11} + \frac{1}{231} = \frac{1}{4} + \frac{1}{7} + \frac{1}{28}.$$

(c) The result follows immediately from the identity
$$\frac{4}{4m+3} = \frac{1}{m+2} + \frac{1}{(m+1)(m+2)} + \frac{1}{(m+1)(4m+3)}.$$

REMARK: Paul Erdős made the following conjecture: *For each integer $n > 4$, the fraction $4/n$ can be written as the sum of three distinct unitary fractions.*

(239) First of all, it is clear that if n is complete, we have $\sqrt{10} \cdot 10^4 < n < 10^5$. To obtain the stated result, we can use the following MATHEMATICA program:

```
w={ };Do[If[Sort[IntegerDigits[n²]]=={0,1,2,3,4,5,6,7,8,9},
w=Append[w,n]],{n,32000,100000}];Print[w]
```

We then obtain the 87 numbers: 32043, 32286, 33144, 35172, 35337, 35757, 35853, 37176, 37905, 38772, 39147, 39336, 40545, 42744, 43902, 44016, 45567, 45624, 46587, 48852, 49314, 49353, 50706, 53976, 54918, 55446, 55524, 55581, 55626, 56532, 57321, 58413, 58455, 58554, 59403, 60984, 61575, 61866, 62679, 62961, 63051, 63129, 65634, 65637, 66105, 66276, 67677, 68763, 68781, 69513, 71433, 72621, 75759, 76047, 76182, 77346, 78072, 78453, 80361, 80445, 81222, 81945, 83919, 84648, 85353, 85743, 85803, 86073, 87639, 88623, 89079, 89145, 89355, 89523, 90144, 90153, 90198, 91248, 91605, 92214, 94695, 95154, 96702, 97779, 98055, 98802, 99066.

Therefore, there exist only 87 complete numbers. The smallest complete number is therefore 32 043.

As for the second question, it is easy to see that if n is complete, then by definition, the sum of the digits of n^2 is
$$0 + 1 + 2 + 3 + \cdots + 9 = \frac{9 \cdot 10}{2} = 45,$$

a number divisible by 9. It follows that n is a multiple of 3 and therefore not prime.

(240) Assume that $p = x^3 + y^3$ with $x, y \in \mathbb{N}$. Then,
$$p = (x + y)(x^2 - xy + y^2) = (x + y)((x - y)^2 + xy).$$

Since $x + y \geq 2$, we must have
$$p = x + y \qquad \text{and} \qquad (x - y)^2 + xy = 1,$$

which proves that $x = y$ and $xy = 1$ and therefore that $x = y = 1$ and $p = 2$.

(241) First let $p = x^3 - y^3$, with $x, y \in \mathbb{N}$. Then,
$$p = (x - y)(x^2 + xy + y^2),$$

so that $x - y = 1$ and
$$p = x^2 + xy + y^2 = (y + 1)^2 + (y + 1)y + y^2 = 3y^2 + 3y + 1$$
$$= 3y(y + 1) + 1,$$

which proves the first implication.

Conversely, assume that the prime number p is of the form $p = 3k(k + 1) + 1$ with $k \in \mathbb{N}$. Then, letting $x = k + 1$ and $y = k$, we obtain

$$
\begin{aligned}
p &= 3k(k+1) + 1 = 3k^2 + 3k + 1 = (k+1)^2 + (k+1)k + k^2 \\
&= x^2 + xy + y^2 = (x - y)(x^2 + xy + y^2) = x^3 - y^3,
\end{aligned}
$$

and the result is established.

The ten smallest prime numbers of this form are 7, 19, 37, 61, 127, 271, 331, 397, 547 and 631.

REMARK: As of 2006, we still do not know if there exist infinitely many prime numbers of this form, a result which is actually a consequence of Schinzel's Hypothesis H (stated on page 12).

(242) First of all, if a and b are two even integers, it is easy to see that $a^2 - b^2$ is of the form $4k$, while if a and b are two odd integers, $a^2 - b^2$ is of the form $8k$. On the other hand, if one of a or b is even and the other odd, $a^2 - b^2$ is odd. These observations prove that the condition is necessary.

Assume now that n is not of the form $4k + 2$. This means that either n is odd or it is of the form $4k$. If it is odd, then $n - 1$ and $n + 1$ are both even, in which case $\frac{n-1}{2}$ and $\frac{n+1}{2}$ are integers. It follows that

$$
n = \left(\frac{n+1}{2}\right)^2 - \left(\frac{n-1}{2}\right)^2.
$$

On the other hand, if n is divisible by 4, then

$$
n = \left(\frac{n}{4} + 1\right)^2 - \left(\frac{n}{4} - 1\right)^2.
$$

We have thus proved that the condition is sufficient, and the result is proved.

REMARK: It is interesting to observe that this result allows one to directly solve Problem 2, that is without using the expression for $\sum_{k=1}^{n} k^3$.

(243) (*Contribution of Nicolas Doyon*) It is easy to notice that the decimal representation of an automorphic number ends with a 5 or a 6. We will display an algorithm which allows one to construct infinitely many automorphic numbers ending with the digit 5. So let n be an automorphic number with r digits ending with the digit 5. We now show that the number m made up of the $r + 1$ last digits of n^2 is also automorphic. Since n is automorphic, we can write $m = d \cdot 10^r + n$ for a certain nonnegative integer d. We then have

$$
m^2 = d^2 \cdot 10^{2r} + 2nd \cdot 10^r + n^2.
$$

Since n ends with the digit 5, the number $2n \cdot 10^r$ ends with $r+1$ zeros. The $r+1$ last digits of m^2 are therefore equal to the $r+1$ last digits of n^2, which are in fact equal to the digits of m. Hence, m is automorphic. Iterating this process, we conclude that there exist infinitely many automorphic numbers.

This last algorithm allows one to build the sequence of automorphic numbers, whose first terms are as follows : 5, 25, 625, (0625), 90625, 890625, 2890625, 12890625, 212890625, ...

A variation of this algorithm allows one to build infinitely many automorphic numbers ending with the digit 6.

(244) (*Contribution of Nicolas Doyon*) In the case where one of the digits of n is equal to 0, the result is immediate. We may therefore assume that $d_i \neq 0$ for $1 \leq i \leq r$.

We will show that the difference $\Delta(n) := n - d_1 d_2 \cdots d_r$ is minimal when $d_1 = d_2 = \ldots = d_{r-1} = 1$. Let us however start by showing that the difference $n - d_1 d_2 \cdots d_r$ attains its minimal value when $d_1 = 1$, that is if the digits $d_2, \ldots, d_r$ are fixed; then the number $n = [d_1, d_2, \ldots, d_r]$ for which $\Delta(n)$ is minimal is the one with $d_1 = 1$.

So let $n = d_1 d_2 \cdots d_r$ and $n' = (d_1 + 1) d_2 \cdots d_r$. Since $n' = n + 10^{r-1}$, we have

$$n' - (d_1 + 1) d_2 \cdots d_r = n - d_1 d_2 \cdots d_r + 10^{r-1} - d_2 d_3 \cdots d_r.$$

Since $10^{r-1} - d_2 d_3 \cdots d_r \geq 10^{r-1} - 9^{r-1} > 0$, it follows that

$$n' - (d_1 + 1) d_2 \cdots d_r > n - d_1 d_2 \cdots d_r,$$

as we wanted to show.

Assume now that $n = [\underbrace{1, 1, \ldots, 1}_{k}, d_{k+1}, d_{k+2}, \ldots, d_r]$. Let us show that the difference $n - d_{k+1} d_{k+2} \cdots d_r$ is minimal when $d_{k+1} = 1$.

Set $n' = [\underbrace{1, 1, \ldots 1}_{k}, d_{k+1} + 1, d_{k+2}, \ldots, d_r]$. Since $n' = n + 10^{r-k-1}$, we have

$$n' - (d_{k+1} + 1) d_{k+2} \cdots d_r$$
$$= n - d_{k+1} d_{k+2} \cdots d_r + 10^{r-k-1} - d_{k+2} d_{k+3} \cdots d_r.$$

Since $10^{r-k-1} - d_{k+2} d_{k+3} \cdots d_r \geq 10^{r-k-1} - 9^{r-k-1} > 0$, it follows that

$$n' - (d_{k+1} + 1) d_{k+2} \cdots d_r > n - d_{k+1} d_{k+2} \cdots d_r$$

and therefore that the difference $n - d_{k+1} d_{k+2} \cdots d_r$ is minimal when $d_{k+1} = 1$. By induction, we obtain that the difference $n - d_1 d_2 \cdots d_r$ is minimal when $n = [\underbrace{1, 1, \ldots, 1}_{r-1}, d_r]$. We conclude from this that if all the digits of n are different from 0, $n - d_1 d_2 \cdots d_r \geq \underbrace{111...1}_{r-1} 0 > 10^{r-1}$, thus completing the proof.

(245) (*Contribution of Nicolas Doyon*) We will show that this number is 97247. Let n be a number having this property and whose digits are $d_1, d_2, \ldots, d_r$. We must then have

$$n = d_1^5 + d_2^5 + \cdots + d_r^5 + d_1 d_2 \cdots d_r.$$

From the preceding problem, and since $n > 9$, we have

$$10^{r-1} \leq n - d_1 d_2 \cdots d_r.$$

We then have the inequalities

$$10^{r-1} \leq d_1^5 + d_2^5 + \cdots + d_r^5 \leq r \cdot 9^5.$$

But $10^{r-1} < r \cdot 9^5$ does not hold if $r \geq 7$. It follows that n cannot have more than 6 digits and is therefore smaller than 10^6. Using a computer, we easily find five numbers satisfying the given property, namely the numbers 1324, 4150, 16363, 93084 and 97247.

(246) Let $a_1 = 2$, $a_2 = 12$, $a_3 = 112$ and $a_4 = 2112$. Having determined $a_{k-1} = 2^{k-1}b_{k-1}$, $k \geq 3$, here is the algorithm allowing us to determine a_k:

- if b_{k-1} is odd, set $a_k = 10^{k-1} + a_{k-1}$;
- if b_{k-1} is even, set $a_k = 2 \cdot 10^{k-1} + a_{k-1}$.

It is then easy to check that the number a_k thus chosen satisfies the conditions (i), (ii) and (iii). Using induction, we easily prove that this sequence is unique. The first 14 terms of the sequence are 2, 12, 112, 2112, 22112, 122112, 2122112, 12122112, 212122112, 1212122112, 11212122112, 111212122112, 1111212122112 and 11111212122112.

(247) In fact, it is enough to prove that the smallest number n such that $s(n) = k$ is $n = (a + 1)10^b - 1$, where $b = [k/9]$ and $a = k - 9b$. First of all, clearly, in order for n to be as small as possible, the digit 9 must appear as often as possible and at the end of the number n. But the maximal number b of 9's that one can place at the end of n is $b = [k/9]$, and to obtain $s(n) = k$, the first digit of n must be $a = k - 9b$. The number n is therefore $n = a99\ldots9$. Hence, $n = (a + 1)10^b - 1$, as was to be shown.

(248) (This is part of the problem proposed in Problem #10605 [1997,p. 567] in AMM by J.W. Borwein and C.G. Pinner; solution by D. Bradley, **106** (1999), p. 173.) For m fixed, set $f(n) = n(n - m)/(n^2 - mn + m^2)$. We then have

$$P(m) = \prod_{n \neq m} \frac{f(n)}{f(n + m)}.$$

It is clear that the product is then telescopic, in which case we obtain

$$P(m) = f(2m) \prod_{n=1}^{m-1} f(n) = \frac{2}{3} \prod_{n=1}^{m-1} f(n).$$

Since

$$f(n) = \frac{n(n^2 - m^2)}{n^3 + m^3},$$

we have

$$\prod_{n=1}^{m-1} f(n) = \prod_{n=1}^{m-1} n \cdot \prod_{n=1}^{m-1} (n - m) \cdot \prod_{n=1}^{m-1} \frac{n + m}{n^3 + m^3}$$

$$= (-1)^{m+1}((m - 1)!)^2 \cdot \prod_{n=1}^{m-1} \frac{n + m}{n^3 + m^3}$$

$$= (-1)^{m+1}((m - 1)!)^2 \cdot m^2 \cdot \prod_{n=1}^{m} \frac{n + m}{n^3 + m^3}$$

$$= (-1)^{m+1}(m!)^2 \cdot \prod_{n=1}^{m} \frac{n + m}{n^3 + m^3},$$

as required.

(249) We will show that the numbers 1, 2, 145 and 40585 are the only ones with this property. Indeed, let n be such a number. Then,

$$(*) \qquad n = d_1! + d_2! + \cdots + d_r!,$$

where $d_1, d_2, \ldots, d_r$ are the digits of n. By trial and error, we end up seeing that $n = 1$, $n = 2$ and $n = 145$ satisfy relation (*). On the other hand, we can show that the number of positive integers n satisfying (*) must be finite. Indeed, it is clear that such a number n must satisfy the double inequality

$$10^{r-1} < n = d_1! + d_2! + \cdots + d_r! \leq r \cdot 9! = 362\,880r.$$

But $10^{r-1} < 362\,880r$ does not hold if $r \geq 8$, that is if $n \geq 10^7$. Therefore, any solution of (*) must be smaller than 10^7. Hence, using a computer, it remains to examine each number $n < 10^7$ to check if it satisfies the condition (*), a process that reveals that the additional number is indeed $n = 40\,585$.

(250) We will see that this number is $n = 2592$, because we have $2592 = 2^5 \cdot 9^2$. Assume that the number $n = d_1 d_2 \cdots d_{2k}$ satisfies the property

(*) $$n = d_1^{d_2} \cdot d_3^{d_4} \cdots d_{2k-1}^{d_{2k}}.$$

Then, it is clear that

$$10^{2k-1} \leq n \leq k \cdot 9^9,$$

an inequality which can only be satisfied if $k \leq 5$. This proves that there exist only finitely many numbers with this property. But this also means that it is also possible, theoretically, that such a number n could have as many as 10 digits. We should therefore find a way to limit the number of computations in order not to have to check each number $n < 10^{10}$. Such a task is possible, because the only possible factors of any n satisfying (*) are 2, 3, 5 and 7 (since its digits belong to the set $\{0, 1, 2, \ldots, 9\}$). It follows that n is of the form

$$n = 2^\alpha \cdot 3^\beta \cdot 5^\gamma \cdot 7^\delta,$$

for certain nonnegative integers $\alpha, \beta, \gamma, \delta$. Since $2^\alpha < 10^{10}$ only if $\alpha \leq 33$, $3^\beta < 10^{10}$ only if $\beta \leq 20$, $5^\gamma < 10^{10}$ only if $\gamma \leq 14$, and $7^\delta < 10^{10}$ only if $\delta \leq 11$, it is easy to see that the number of possible candidates is at most $34 \cdot 21 \cdot 15 \cdot 12 = 128\,520$.

Hence, using MATHEMATICA, we can write the following program:
```
Do[n=2^a*3^b*5^c*7^d;v=IntegerDigits[n];r=Length[v];
If[EvenQ[r]&&(Apply[Times,Table[v[[2*i-1]],{i,1,r/2}]]!
=0) &&(Product[v[[2*i-1]]^v[[2*i]],{i,1,r/2}]==n),
Print[n," ",FactorInteger[n]]],{a,0,33},{b,0,20},
{c,0,14},{d,0,11 }]
```
a program that reveals only the number $n = 2592 = 2^5 \cdot 3^4$ (see Dudeney [**9**]).

(251) Such a number n must satisfy

$$10^{r-1} \leq n \leq r \cdot 9 + r \cdot 9^3 = 738r,$$

a double inequality which does not hold if $r \geq 6$. We must therefore have $r \leq 5$, that is $n < 10^5$. Using a computer, we then find that the only numbers having this property are 12, 30, 666, 870, 960 and 1998.

(252) It follows from the relation

(*) $$3^k = q2^k + r \qquad (1 \leq r < 2^k)$$

that the number $n = q2^k - 1$ trivially satisfies the inequality $n < 3^k$. Hence, the only admissible positive values for the x_i's in the representation

$$n = x_1^k + x_2^k + \cdots + x_r^k$$

are $x_i = 1$ and $x_i = 2$. It follows that

$$n = \underbrace{2^k + 2^k + \cdots + 2^k}_{q-1} + \underbrace{1 + 1 + \cdots + 1}_{2^k - 1} = (q-1) \cdot 2^k + (2^k - 1) \cdot 1.$$

Since from $(*)$ we have $q = [(3/2)^k]$, we have thus established that

$$g(k) \geq q - 1 + 2^k - 1 = 2^k - 2 + q = 2^k - 2 + \left[\left(\frac{3}{2}\right)^k\right],$$

as required.

(253) The required number is 69: indeed, we have $69^2 = 4761$ and $69^3 = 328\,509$. To show uniqueness, we first observe that $47 \leq n \leq 99$, since $46^2 = 2116$ and $46^3 = 97336$ gather together only 9 digits, while 100^2 and 100^3 already have 11 digits. Finally, using a computer and examining the candidates 47, 48, ..., 99, we easily check that 69 is the only number with this property.

(254) The required number is 6534: indeed, we have $6534^2 = 42693156$ and $6534^3 = 278957081304$. To show uniqueness, we first observe that $4642 \leq n \leq 9999$, since $4641^2 = 21538881$ and $4641^3 = 99961946721$ gather together only 19 digits, while 10000^2 and 10000^3 already have 21 digits. Finally, using a computer and examining the candidates 4642, 4643, ..., 9999, we easily check that 69 is the only number with this property.

(255) We will show that there exist exactly 10 integers having this property, namely 497375, 539019, 543447, 586476, 589629, 601575, 646479, 858609, 895688 and 959097. To show that these numbers are the only ones with this property, we first observe that $464159 \leq n \leq 999999$. Finally, using a computer and examining the candidates 464159, ..., 999999, we easily check that 497375, 539019, 543447, 586476, 589629, 601575, 646479, 858609, 895688 and 959097 are the only numbers with this property.

(256) The 4-digit vampire numbers are $1260 = 21 \cdot 60$, $1395 = 15 \cdot 93$, $1435 = 35 \cdot 41$, $1530 = 30 \cdot 51$, $1827 = 21 \cdot 87$, $2187 = 27 \cdot 81$ and $6880 = 80 \cdot 86$. The following is a MATHEMATICA program which generates all 4-digit vampire numbers:

```
Do[r=Length[v=Select[Divisors[n], (n/#<100)&&(#<100)
&&(#<N[Sqrt[n]])&]]; Do[d=v[[i]];
If[Sort[Join[IntegerDigits[d],IntegerDigits[n/d]]]
== Sort[IntegerDigits[n]],
Print[n," = ",d," x ",n/d]],{i,1,r}],{n,1000,9999}]
```

REMARK: Using a similar program, we find 155 vampire numbers with six digits and 3382 with eight digits (see Weisstein [40]).

(257) Using a computer, we quickly notice that the numbers 1, 136, 153, 244, 370, 371, 407, 919 and 1459 are the only solutions $n < 10^4$ of equation $(*)$ $g_3(g_3(n)) = n$. To prove that there is no other solution, we will show that if n satisfies $(*)$, then $n < 10^4$, which will be enough to establish that the

above nine numbers are indeed the only solutions of $(*)$. Since, for each number n with r digits, we have

$$10^{r-1} \le n < 10^r,$$

it is clear that $r \le \log_{10} n + 1$, so that if n is a solution of $(*)$, we shall have

$$g_3(n) \le 9^3 \cdot r = 729r \le 729(\log_{10} n + 1) < n,$$

this last inequality holding if $n \ge 10^4$. Therefore, to find all the solutions n of $(*)$, we only need to examine one by one each number $n \le 9\,999$.

(258) First of all, we observe that if $f(n) > n$ for a certain positive integer $n = d_1 d_2 \cdots d_r$, then

$$10^{r-1} < n = d_1 d_2 \cdots d_r < f(n) = d_1! + d_2! + \cdots + d_r! \le r \cdot 9!\,.$$

But $10^{r-1} < r \cdot 9!$ is only possible if $r \le 7$, that is if $n < 10^7$. Hence, it remains to find the largest integer $n < 10^7$ such that $f(n) > n$ (since if $n \ge 10^7$, we necessarily have $f(n) \le n$). The following MATHEMATICA program does that:

```
k=1;While[f[n=10^7-k]<n,k++];Print[n," ",f[n]]
```

and we then obtain that with $n = 1\,999\,999$, we have $f(n) = 2\,177\,181$. Again using a computer, we obtain that each number $n \le 1\,999\,999$ is such that the corresponding sequence $f_1(n), f_2(n), f_3(n), \ldots$ eventually enters one of the six loops mentioned in the statement.

(259) In fact, much more is true, namely: *Given any integer $k \ge 2$, there exists a polynomial $p(x)$ of degree k with integer coefficients and a positive integer m such that $n = p(m)$.* Indeed, first let $m = [n^{1/k}]$. By writing n in basis m, we obtain

$$n = c_k m^k + c_{k-1} m^{k-1} + \cdots + c_1 m + c_0,$$

where $c_k = 1$, $c_{k-1}, c_{k-2}, \ldots, c_1, c_0$ are the digits of n in basis m, with of course $0 \le c_i < m$ for $i = 0, 1, \ldots, k$. By choosing

$$p(x) = x^k + c_{k-1} x^{k-1} + \cdots + c_1 x + c_0,$$

we have thus found the required polynomial $p(x)$ for which $p(m) = n$.
REMARK: This observation is at the very basis of the *Number Field Sieve Factorization Method* (see Pomerance [**27**]).

(260) Assume the contrary, that is that k is not prime. Then, $k = ab$ with $2 \le a \le b < k$, in which case

$$111\ldots1 = \frac{10^k - 1}{9} = \frac{10^{ab} - 1}{9} = \frac{10^{ab} - 1}{10^b - 1} \cdot \frac{10^b - 1}{9},$$

that is the product of two integers larger than 1.
REMARK: Only seven numbers k for which the corresponding number $\underbrace{11\ldots1}_{k}$ is prime are known: these are 2, 19, 23, 317, 1031, 49081 and 86453 (see Weisstein [**40**]).

(261) Let $n = q_1 q_2 q_3$. First of all, it is clear that

$$(3) \qquad\qquad q_1 < q_2 < q_3 < n^{1/3},$$

in which case

$$n^{1/3} > q_3 > q_2 = \frac{n}{q_3 q_1} > \frac{n}{n^{1/3} q_1} = \frac{n^{2/3}}{q_1},$$

which implies that

$$q_1 > \frac{n^{2/3}}{n^{1/3}} = n^{1/3},$$

which contradicts (3).

If the condition is instead given by (2), then if $3|n$, there exist prime numbers p and q both different from 3 such that

$$0 \equiv n = 9 + p^2 + q^2 \equiv 1 + 1 = 2 \pmod 3,$$

which contradicts the fact that $n \equiv 0 \pmod 3$; while if 3 does not divide n, there exist prime numbers p, q and r all different from 3 such that

$$0 \not\equiv n = p^2 + q^2 + r^2 \equiv 1 + 1 + 1 \equiv 0 \pmod 3,$$

again a contradiction.

REMARK: There exist at least eight numbers n (with $n \neq p^3$) such that $n = \sum_{p|n} p^3$, namely $n = 378, 2548, 2\,836\,295, 4\,473\,671\,462, 23\,040\,925\,705,$ $13\,579\,716\,377\,989, 21\,467\,102\,506\,955$ and $119\,429\,556\,097\,859$.

(262) (*Contribution of Jean-Lou De Carufel*) We will show that $n = 3$ is the only number with this property, so let n be such a number. There exists a positive integer r such that $10^{r-1} < n \leq 10^r$. We must have

$$\sum_{i=1}^{\infty} n \left(\frac{1}{10^r} \right)^i = \frac{1}{n},$$

a relation which is successively equivalent to $\frac{n/10^r}{1-1/10^r} = \frac{1}{n}$, $\frac{n^2}{10^r} = 1 - \frac{1}{10^r}$ and $n^2 = 10^r - 1$.

Taking $r \geq 2$, we find, since $n^2 \equiv 0, 1 \pmod 4$, $0 \equiv 0 - 1 \pmod 4$ or $1 \equiv 0 - 1 \pmod 4$, so that $0 \equiv 3 \pmod 4$ or $1 \equiv 3 \pmod 4$, which is impossible. Hence, we must have $r = 1$, in which case $n^2 = 10^1 - 1 = 9$, and $n = 3$, as required.

(263) These numbers are

$$
\begin{aligned}
42 &= 2 \cdot 3 \cdot 7 = 2^5 + 3 + 7, \\
140 &= 2^2 \cdot 5 \cdot 7 = 2^7 + 5 + 7, \\
290 &= 2 \cdot 5 \cdot 29 = 2^8 + 5 + 29, \\
618 &= 2 \cdot 3 \cdot 103 = 2^9 + 3 + 103, \\
2\,058 &= 2 \cdot 3 \cdot 7^3 = 2^{11} + 3 + 7, \\
6\,747 &= 3 \cdot 13 \cdot 173 = 3^8 + 13 + 173, \\
131\,430 &= 2 \cdot 3 \cdot 5 \cdot 13 \cdot 337 = 2^{17} + 3 + 5 + 13 + 337, \\
531\,531 &= 3^2 \cdot 7 \cdot 11 \cdot 13 \cdot 59 = 3^{12} + 7 + 11 + 13 + 59.
\end{aligned}
$$

To find these numbers, we first observe that the number r of prime factors of n satisfying (*) must be odd. Indeed, assume that r is even. If $q_1 = 2$, then n is even and $q_2 + \cdots + q_r$ is odd, but $n = 2^a + q_2 + \cdots + q_r$ is odd, a contradiction. On the other hand, if $q_1 \geq 3$, then n is odd and $q_2 + \cdots + q_r$ is odd, in which case $n = q_1^a + q_2 + \cdots + q_r$ is even, a contradiction. We must therefore have $r > 1$, r odd. On the other hand, since for any integer n satisfying the given property, we must have

$$q_1^a = n - (q_2 + \cdots + q_r),$$

it is enough to verify that

$$a = \log_{q_1} (n - (q_2 + \cdots + q_r)) \text{ is an integer}.$$

The following MATHEMATICA program (which generates the numbers n in question, the corresponding exponent a and the factorization of n) is therefore quite efficient (below, `factors[n]` stands for the prime factors of n):

```
Do[w=factors[n];
If[OddQ[r=Length[w]]&&(r>1)&&
IntegerQ[a=Log[w[[1]],n-Apply[Plus,Take[w,-r+1]]]],
Print["n=",n," a=",a," ",FactorInteger[n]]],
{n,2,1000000}]
```

REMARK: The other numbers $n < 4 \cdot 10^8$ with this property are

$$
\begin{aligned}
n &= 5\,124\,615 = 3 \cdot 5 \cdot 341641 = 3^{14} + 5 + 341641, \\
n &= 14\,356\,161 = 3^2 \cdot 227 \cdot 7027 = 3^{15} + 227 + 7027, \\
n &= 34\,797\,196 = 2^2 \cdot 7 \cdot 1242757 = 2^{25} + 7 + 1242757, \\
n &= 40\,265\,322 = 2 \cdot 3 \cdot 6710887 = 2^{25} + 3 + 6710887, \\
n &= 67\,239\,938 = 2 \cdot 257 \cdot 130817 = 2^{26} + 257 + 130817.
\end{aligned}
$$

(264) (*MMAG, Vol. 63, no. 2, p. 129*). First of all, it is clear that if n is a product of Mersenne primes, then $n = \prod_{p_j} (2^{p_j} - 1)$, where the product runs over certain prime numbers p_j, in which case we have

$$\sigma(n) = \prod_{p_j} 2^{p_j} = 2^{\sum_{p_j} p_j},$$

which proves that the condition is sufficient. Assume now that there exist prime numbers $q_1 < \ldots < q_r$ and positive integers $\alpha_1, \ldots, \alpha_r, \alpha$ such that

$$n = q_1^{\alpha_1} q_2^{\alpha_2} \cdots q_r^{\alpha_r} \text{ with } \sigma(n) = 2^\alpha.$$

It then follows that for each integer i, $1 \le i \le r$, there exists a positive integer β_i such that

$$(*) \qquad \sigma(q_i^{\alpha_i}) = 1 + q_i + q_i^2 + \cdots + q_i^{\alpha_i} = 2^{\beta_i}.$$

Let i be fixed, $1 \le i \le r$. Relation $(*)$ implies in particular that q_i and α_i are odd. We therefore have that $\alpha_i = 2k_i + 1$ for a certain integer k_i, so that

$$\sigma(q_i^{\alpha_i}) = (1 + q_i)(1 + q_i^2 + q_i^4 + \cdots + q_i^{2k_i}) = 2^{\beta_i}.$$

But then there must exist a positive integer δ_i such that $1 + q_i = 2^{\delta_i}$, which means that $q_i = 2^{\delta_i} - 1$ is a Mersenne prime. Since this is true for each integer i, $1 \le i \le r$, the result follows.

(265) (*MMAG, Vol. 64, no. 5, p. 351*). We will show that these are the positive integers N which are not a power of 2. Indeed let N be an integer of the form

$$N = \binom{k}{2} + kn = \frac{k(k-1)}{2} + nk \qquad (k > 1, n \geq 1).$$

Since $2N = k(k + 2n - 1)$, it follows that $2N$ must have an odd factor larger than 2, and therefore similarly for N. It follows that N cannot be a power of 2.

Conversely, let N be a positive integer which has an odd factor larger than 2. Consider the factorization of $2N$ as a product of two positive integers of which one is odd. Let A be the smallest of these two factors and B the largest. Setting $k = A$ and $n = \frac{B+1-A}{2}$, it follows that

$$\binom{k}{2} + kn = \frac{A(A-1)}{2} + A\frac{B+1-A}{2} = \frac{AB}{2} = N,$$

which gives the result.
REMARK: Since

$$\binom{k}{2} + kn = 1 + 2 + \cdots + (k-1) + kn = n + (n+1) + \cdots + (n+k-1),$$

the problem is equivalent to the one that consists of searching for the positive integers which can be written as the sum of consecutive integers.

(266) They are the integers n of the form $n = 4m + 2$, $m = 0, 1, 2 \ldots$, since $3^{4m+2} \equiv 9 \equiv -1 \pmod{10}$, while $3^{4m} \equiv 1 \pmod{10}$, $3^{4m+1} \equiv 3 \pmod{10}$ and $3^{4m+3} \equiv 7 \pmod{10}$.

(267) It is the number 5. Indeed, since $n! \equiv 0 \pmod 7$ as soon as $n \geq 7$, we have

$$1! + 2! + \cdots + 50! \equiv 1! + 2! + 3! + 4! + 5! + 6!$$
$$\equiv 1 + 2 + 6 + 3 + 1 + 6 \equiv 5 \pmod 7.$$

(268) Since for $i \geq 4$, $12|i!$, the remainder is $1 + 2 + 6 = 9$.

(269) For n odd, $10 \cdot 32^n + 1 \equiv 0 \pmod 3$, while for each even integer n, $10 \cdot 32^n + 1 \equiv 0 \pmod{11}$.

(270) The answer is YES. Since $n^6 \equiv 1 \pmod 9$ for each integer n such that $(n,3) = 1$ and since $n^2 \equiv 4 \pmod 9$ for $n \equiv 2 \pmod 9$, it follows that if $n \equiv 2 \pmod 9$, we have $n^6 + n^2 + 4 \equiv 0 \pmod 9$. On the other hand, since $n^6 + n^2 + 4 \equiv 0 \pmod 4$ for all even n, we may conclude that $36|n^6 + n^2 + 4$ for $n = 18k + 2$, $k = 0, 1, 2, \ldots$.

(271) If the equation $3k - 1 = x^2 + 3y^2$ had a solution, then we would have $x^2 \equiv -1 \equiv 2 \pmod 3$, which is impossible because $x^2 \equiv 0, 1 \pmod 3$.

(272) We know that

$$m = \sum_{i=1}^{[\log n / \log p]} \left[\frac{n}{p^i}\right].$$

Amongst the integers $1, 2, \ldots, n$, those which are divisible by p are: $p, 2p, \ldots,$ $k_1 p$, where $k_1 = [n/p]$. Since

$$n! = 1 \cdot 2 \cdots (p-1)(\mathbf{p})(p+1)(p+2) \cdots (2p-1)(\mathbf{2p})$$
$$\cdot (2p+1)(2p+2) \cdots (3p-1)(\mathbf{3p})((k_1-1)p+1)((k_1-1)p+2) \cdot$$
$$\cdots (k_1 p - 1)(\mathbf{k_1 p})(k_1 p + 1)(k_1 p + 2) \cdots n$$

and since from Wilson's Theorem, the product of the integers in each set $\{1, 2, \ldots, p-1\}$, $\{p+1, p+2, \ldots, 2p-1\}, \ldots, \{(k_1-1)p+1, (k_1-1)p+2, \ldots, k_1 p - 1\}$ is congruent modulo p to -1, it follows that

$$\frac{n!}{p^{k_1}} \equiv (-1)^{k_1} k_1! \left(n - \left[\frac{n}{p} \right] p \right)! \pmod{p}.$$

Now, amongst the integers $1, 2, \ldots, k_1$, those which are divisible by p are: $p, 2p, \ldots, k_2 p$, where $k_2 = [k_1/p] = [n/p^2]$. It follows that

$$\frac{n!}{p^{k_1 + k_2}} \equiv (-1)^{k_1 + k_2} k_2! \left(n - \left[\frac{n}{p} \right] p \right)! \left(\left[\frac{n}{p} \right] - \left[\frac{n}{p^2} \right] p \right)! \pmod{p},$$

where $1 \le k_2 < k_1$. Continuing this process, the result follows.

(273) We must show that $n^{13} - n \equiv 0 \pmod{10}$ or equivalently that $n^{13} - n \equiv 0 \pmod{2}$ and $n^{13} - n \equiv 0 \pmod{5}$. Using Fermat's Little Theorem, $n^2 \equiv n \pmod{2}$ which implies $n^{13} \equiv n \pmod{2}$. Similarly, $n^5 \equiv n \pmod{5}$ implies $n^{13} \equiv n \pmod{5}$.

(274) Since n must be divisible by 7 and by 11, it can be written as $n = 7^a \cdot 11^b$. But $n/7 = 7^{a-1} \cdot 11^b$ must be the 7-th power of an integer, in which case $a \equiv 1 \pmod{7}$ and $b \equiv 0 \pmod{7}$. Moreover, $n/11 = 7^a \cdot 11^{b-1}$ must be the 11-th power of an integer, so that $a \equiv 0 \pmod{11}$ and $b \equiv 1 \pmod{11}$. Solving this system of congruences gives $a \equiv 22 \pmod{77}$ and $b \equiv 56 \pmod{77}$. Hence, the smallest positive integer satisfying the given constraints is $n = 7^{22} \cdot 11^{56}$.

(275) Consider the system of congruences $x + j - 1 \equiv 0 \pmod{p_j^2}$, $j = 1, 2, \ldots, k$, where p_j stands for the j-th prime number. From the Chinese Remainder Theorem, this system has one solution; that is there exists an integer n which verifies these k congruences. Therefore, each of the k integers $n, n+1, \ldots, n+k-1$ is divisible by a perfect square, as required.

(276) Since $x \equiv a \pmod{m}$, there exists $k \in \mathbb{Z}$ such that $x = a + km$ and therefore $a + km \equiv b \pmod{n}$. Hence, there exists $j \in \mathbb{Z}$ such that $a + km = b + jn$, that is $km - jn = -(a - b)$. Since $(m, n) | m$ and $(m, n) | n$, it follows that $(m, n) | (a - b)$.

Reciprocally, assume that $(m, n) | (a - b)$. Then, there exists $M \in \mathbb{Z}$ such that $a - b = M(m, n)$ and since $(m, n) = k_1 m + k_2 n$, $k_1, k_2 \in \mathbb{Z}$, it follows that there exist integers j and k such that $a - b = -km + jn$, $k = -k_1 M$, $j = k_2 M$. Therefore, we have $a + km = b + jn$. Setting $x = a + km$, we obtain $x \equiv a \pmod{m}$ and moreover $x = a + km = b + jn$, that is $x \equiv b \pmod{n}$.

(277) Letting $N = \binom{p}{k}$, then

$$k! N = p(p-1) \cdots (p - k + 1) \equiv 0 \pmod{p},$$

and since $(k!, p) = 1$ then $N \equiv 0 \pmod{p}$.

(278) (a) This follows from Problem 277 and induction on n.

(b) Since $a^p \equiv b^p \pmod{p}$, then by Fermat's Little Theorem, we have $a \equiv b \pmod{p}$ and therefore there exists an integer k such that $a = b + kp$. Hence, by the Binomial Theorem, there exists an integer K such that

$$a^p = (b + kp)^p$$
$$= b^p + \binom{p}{1} b^{p-1} kp + \binom{p}{2} b^{p-2} k^2 p^2 + \cdots + k^p p^p = b^p + Kp^2,$$

where we used the result of Problem 277, thus completing the proof of part (b).

(279) Let $N = \binom{p-1}{k} = \dfrac{(p-1)(p-2)\cdots(p-k)}{k!}$. We then have

$$k! \, N \equiv (-1)^k k! \pmod{p}$$

and since $(k!, p) = 1$, we conclude that $N \equiv (-1)^k \pmod{p}$.

(280) From Wilson's Theorem,

$$(p-1)! = (p-1)(p-2)\cdots(p-r)(p-r-1)!$$
$$\equiv (-1)^r r!(p-r-1)! \equiv -1 \pmod{p}.$$

Since $(-1)^r r! \equiv 1 \pmod{p}$, we obtain the result.

For the second part, it is enough to notice that $(-1)^9 9! \equiv 1 \pmod{269}$ and that $(-1)^{15} 15! \equiv 1 \pmod{479}$.

(281) Assume that a solution exists. First, if β is odd,

$$2^\beta - 1 \equiv (-1)^\beta - 1 \equiv -2 \equiv 1 \not\equiv 0 \pmod{3},$$

which contradicts the given equation. Similarly, if β is even,

$$2^\beta - 1 = \left(2^{\beta/2} - 1\right)\left(2^{\beta/2} + 1\right),$$

which means that $3|(2^{\beta/2} - 1) > 3$ or $3|(2^{\beta/2} + 1) > 3$, and this is why we must have that $p|(2^{\beta/2} - 1)$ and $p|(2^{\beta/2} + 1)$, implying that $p|2$, which is not possible.

(282) (P. Giblin [**14**]) Assume that q is a prime factor of n. Since n is odd, it follows that q is odd. We will first prove that $p|(q - 1)$. Observe that $4^p = 2^{n-1} \equiv 1 \pmod{n}$, so that $4^p \equiv 1 \pmod{q}$. It follows that r, the order of 4 modulo q, is a factor of p; we therefore have that $r = 1$ or $r = p$. If $r = 1$, then $4 \equiv 1 \pmod{q}$, which implies that $q = 3$, in which case $3|n$, which contradicts the fact that n is not a multiple of 3. Hence, $r = p$, which implies that $p|q - 1$, as required. We shall finally show that $n = q$. Since $q - 1 > p$, we have $q \geq p - 1 > n/2 \geq \sqrt{n}$, because $n \geq 4$. We have thus shown that each prime factor q of n is larger than $\sqrt{n}$, which is impossible unless n itself is a prime number.

(283) (Francesco Sica) Assume that $p^k \| a - b$. Then there exists a positive integer c which is not divisible by p and such that

$$b = a + cp^k.$$

We then have

$$b^p = (a + cp^k)^p = \sum_{i=0}^{p} \binom{p}{i} a^i c^{p-i} p^{k(p-i)}$$

$$\equiv a^p + pa^{p-1}cp^k + \frac{p(p-1)}{2}a^{p-2}c^2p^{2k} \pmod{p^{k+2}}$$

$$\equiv a^p + a^{p-1}cp^{k+1} \pmod{p^{k+2}}.$$

We have thus established that

$$(**)\qquad\qquad a^p - b^p \equiv a^{p-1}cp^{k+1} \pmod{p^{k+2}},$$

hence, in particular (*). Moreover, it follows from (**) that p^{k+2} divides $a^p - b^p + a^{p-1}cp^{k+1}$, but, since $p \nmid a^{p-1}c$, it follows that p^{k+2} divides exactly $a^p - b^p$, as required.

(284) The answer is NO. If $p = 2$, then $p|1$, a contradiction. Hence, $p \geq 3$. If δ is even, then $p^\delta + 1 \equiv 1 + 1 = 2 \pmod 4$ while $2^\nu \equiv 0 \pmod 4$, a contradiction, while if δ is odd, then

$$2^\nu = p^\delta + 1 = (p+1)(p^{\delta-1} - p^{\delta-2} + \cdots - p + 1) = (p+1)Q,$$

where $Q > 1$ is odd, which is nonsense.

(285) If a solution $\{m, n\}$ exists, then it is clear that $n > 1$ and that $m > n > 1$, in which case

$$1 + n = m^2 - n^2 = (m - n)(m + n) \geq m + n > 1 + n,$$

which is nonsense.

   Second solution. Assume that $1 + n + n^2 = m^2$ with $n > 1$, $m > 1$. We then have $4 + 4n + 4n^2 = 4m^2$ and therefore $(2n + 1)^2 + 3 = (2m)^2$. But, the only squares which differ by 3 are 1 and 4. This implies that $n = 0$, which contradicts the fact that $n > 1$.

(286) Let (1) be the equation for which we seek the solutions and let $\{p, q\}$ be a solution. First of all, it is clear that

$$(2)\qquad\qquad p^2 + 1 < q < p^2 + p.$$

Indeed, these inequalities are consequences of the following two inequalities:

$$(p^2 + 1)^2 = p^4 + 2p^2 + 1 < p^4 + p^3 + p^2 + p + 1 = q^2$$
$$(p^2 + p)^2 = p^4 + 2p^3 + p^2 > p^4 + p^3 + p^2 + p + 1 = q^2.$$

But it follows from (1) that $p(1 + p + p^2 + p^3) = q^2 - 1 = (q-1)(q+1)$ and this shows that $p|(q-1)(q+1)$. It follows that $p|(q-1)$ or $p|(q+1)$.

   If $p|(q-1)$, then it follows from (2) that

$$p^2 < q - 1 < p^2 + p - 1, \text{ and therefore } p^2 + 1 \leq q - 1 \leq p^2 + p - 2.$$

Observing that the interval $[p^2 + 1, p^2 + p - 2]$ contains no multiple of p, it is therefore impossible that $p|(q-1)$.

   If $p|(q+1)$, then, from (2), we have

$$p^2 + 2 < q + 1 < p^2 + p + 1, \text{ and therefore } p^2 + 3 \leq q + 1 \leq p^2 + p.$$

The fact that the only multiple of p in the interval $[p^2+3, p^2+p]$ is p^2+p implies that $q+1 = p^2+p$; that is $q = p^2+p-1$. Substituting this value of q in (1), we obtain

$$
\begin{aligned}
1+p+p^2+p^3+p^4 &= (p^2+p-1)^2, \\
p^3-2p^2-3p &= 0, \\
p^2-2p-3 &= 0, \\
(p-3)(p+1) &= 0,
\end{aligned}
$$

an equation that implies that $p = 3$, which gives $q = 11$.

(287) It is easy to establish that for each integer $m \not\equiv 0 \pmod 7$, we have $m^3 \equiv +1$ or $-1 \pmod 7$. On the other hand, by hypothesis we have

$$(*) \qquad\qquad x_1^3 + x_2^3 + x_3^3 + x_4^3 + x_5^3 \equiv 0 \pmod 7.$$

Therefore, none of the x_i's is divisible by 7, and the congruence $(*)$ is impossible. Thus the result.

(288) (*AMM, Vol. 81, 1974, p. 172*). If $p = 2$, then $2^2+3^2 = 13$ is not a power of an integer larger than 1. Assume that p is odd; then using Problem 8,

$$2^p + 3^p = (2+3)\sum_{k=0}^{p-1}(-1)^k 2^{p-1-k}3^k,$$

and since $3 \equiv -2 \pmod 5$, we have that

$$\sum_{k=0}^{p-1}(-1)^k 2^{p-1-k}3^k \equiv \sum_{k=0}^{p-1} 2^{p-1} = 2^{p-1}p \pmod 5.$$

If $p \neq 5$, then $2^{p-1}p \not\equiv 0 \pmod p$ and therefore $2^p+3^p = 5k$, for $k \not\equiv 0 \pmod 5$. Hence, 2^p+3^p is never the power of an integer. On the other hand, for $p = 5$, $2^5+3^5 = 275$ is obviously not a power of an integer. Hence, the result.

(289) Letting $n = 6k+r$, $k \in \mathbb{N}$, $0 \leq r \leq 5$, then

$$1^n+2^n+3^n+4^n+5^n+6^n \equiv 1^r+2^r+3^r+4^r+5^r+6^r \pmod 7.$$

Hence, if $r = 0$, we have $1^n+2^n+3^n+4^n+5^n+6^n \equiv 6 \pmod 7$, while for $r = 1, 2, 3, 4, 5$ we have $1^n+2^n+3^n+4^n+5^n+6^n \equiv 0 \pmod 7$. Hence, the result.

(290) The answer is YES. Indeed, by hypothesis $(n, 100) = 1$; we may therefore use Euler's Theorem and obtain that $n^{\phi(100)} \equiv 1 \pmod{100}$. Hence, $n^{40} \equiv 1 \pmod{100}$, which means that the last two digits of $n^{400} = (n^{40})^{10}$ are indeed 0 and 1.

(291) We need to examine to what values the quantities $4^0, 4^1, 4^2, \ldots$ are congruent modulo 10. But we easily verify that each of these numbers is congruent to 1, 4 or 6.

(292) We must first show that $(n+1)^3 - n^3 \not\equiv 0 \pmod 3$ for each integer $n \geq 1$. But this quantity is equal to $3n^2+3n+1$, which is congruent to 1 modulo 3, thus the result. Similarly, we prove that $(n+1)^3 - n^3 \not\equiv 0 \pmod 5$, for each integer $n \geq 1$. Indeed, it is enough to consider $n = 5m+r$, $r = 0, 1, 2, 3, 4$.

(293) This is true since

$$2(32)^n + 5^2 5^n \equiv 2 \cdot 5^n - 2 \cdot 5^n \equiv 0 \pmod{27}.$$

(294) Since $98^2 \equiv -169 \equiv -13^2 \pmod{337}$, the result is immediate.

(295) Since $19^{19} \equiv 9 \pmod{10}$, then $19^{19^{19}} = 19^{9+10k}$ for a certain integer k. We thus obtain

$$19^{9+10k} \equiv 79 \pmod{100},$$

which implies that the last two digits are 7 and 9.

(296) We have $280 = 2^3 \cdot 5 \cdot 7$ and since both a and b are odd, then $a^2 \equiv 1 \pmod 8$ and $b^2 \equiv 1 \pmod 8$. Therefore, $a^{12} \equiv 1 \equiv b^{12} \pmod 8$. Using Fermat's Little Theorem, $a^4 \equiv b^4 \equiv 1 \pmod 5$ and therefore $a^{12} \equiv b^{12} \pmod 5$. Similarly, Fermat's Little Theorem allows one to obtain $a^{12} \equiv b^{12} \pmod 7$. The result then follows by combining these congruences.

(297) We only need to observe that $2730 = 2 \cdot 3 \cdot 5 \cdot 7 \cdot 13$ and use Fermat's Little Theorem five times.

(298) The required integer is 21424. Indeed, we must solve the congruences $n \equiv 4 \pmod{12}$, $n \equiv 4 \pmod{17}$, $n \equiv 4 \pmod{45}$, $n \equiv 4 \pmod{70}$. The first two are equivalent to $n \equiv 4 \pmod{204}$, while the last two give $n \equiv 4 \pmod{1530}$. Finally, the solution of these last two congruences is given by $n \equiv 4 \pmod{21420}$, which gives the result.

(299) The answer is YES. From Fermat's Little Theorem, $n^{13} \equiv n^7 \equiv n \pmod 7$, $n^{11} \equiv n^5 \pmod 7$ and $n^7 \equiv n \pmod 7$, so that the polynomial is congruent to $3n + 4n^5 + n + 3n^5 + 3n = 7n + 7n^5 \equiv 0 \pmod 7$. Thus the result.

(300) Since

$$\binom{2p}{p} = \frac{2p(2p-1)(2p-2) \cdots (2p-(p-1))}{p!}$$

and since

$$(2p-1)(2p-2) \cdots (2p-(p-1)) \equiv (p-1)! \pmod p,$$

it is clear that

$$\binom{2p}{p} \equiv 2\frac{(p-1)!}{(p-1)!} \equiv 2 \pmod p.$$

(301) It is clear that $7|n = $ "abc" if and only if

$$n = 100a + 10b + c \equiv 2a + 3b + c \equiv 0 \pmod 7,$$

and the result follows.

(302) It is clear that "$abcabc$" $=$ "abc"$\cdot 1001$. But $13|1001$, thus obtaining the result.

(303) Since $2^{561} \equiv (-1)^{561} = -1 \equiv 2 \pmod 3$ and since from Fermat's Little Theorem, $2^{561} = (2^{10})^{56} \cdot 2 \equiv 2 \pmod{11}$ and $2^{561} = (2^{16})^3 \cdot 5 \cdot 2 \equiv 2 \pmod{17}$, we conclude that $2^{561} \equiv 2 \pmod{561}$. The second part can be obtained in a similar manner.

(304) Since

$$\frac{n^{13}}{5} + \frac{n^{13}}{7} = \frac{12}{35} n^{13}$$

and since we have $n^{13} \equiv n \pmod 5$ and $n^{13} \equiv n \pmod 7$, then

$$\frac{12}{35}n^{13} + \frac{23}{35}n = \frac{n^{13}}{5} + \frac{n^{13}}{7} + \frac{23}{35}n = \frac{n^{13} - n}{5} + \frac{n^{13} - n}{7} + n,$$

a number which is an integer for each $n \in \mathbb{N}$.

(305) The answer is YES. The case $n = 1$ implies that we can choose $r = 431/481$. We will show that for this rational number r, the number in the statement is an integer for each $n \in \mathbb{N}$, $(n, 481) = 1$. But this number is an integer when $(n, 481) = 1$ if and only if

$$n \left(\frac{50}{481}n^{36} + \frac{431}{481} \right) = \frac{50}{481}n^{37} + \frac{431}{481}n$$

is an integer. Since $481 = 13 \cdot 37$ and since for each $n \in \mathbb{N}$,

$$n^{37} \equiv n \pmod{37} \quad \text{and} \quad n^{13} \equiv n \pmod{13},$$

we conclude that the number $\frac{50}{481}n^{36} + r$ is an integer for all $n \in \mathbb{N}$ when $r = 431/481$.

(306) If $p = 3$, then considering the numbers 111, $111\,111$, $111\,111\,111$, ..., that is all the numbers containing $3, 6, 9, \ldots$ times the digit "1", we obtain infinitely many numbers of the required form. Let $p \geq 7$, p prime. An integer N made up entirely of "1" can be written as $N = (10^n - 1)/9$. But from Fermat's Little Theorem, $10^{p-1} \equiv 1 \pmod p$, which means that $10^{m(p-1)} \equiv 1 \pmod p$ for $m = 1, 2, 3, \ldots$. Since $p \neq 3$, this means that $p \mid (10^{m(p-1)} - 1)/9$, for $m = 0, 1, 2, 3, \ldots$, and the result follows.

(307) Indeed, we easily check that $2^{340} \equiv 1 \pmod{341}$, while $n = 341 = 11 \cdot 31$ is not prime.

(308) (*AMM, Vol. 67, 1960, p. 923*). From Fermat's Little Theorem, it follows that $b^3 \equiv b \pmod 3$ and $b^3 \equiv b \pmod 2$ and therefore that $b^3 \equiv b \pmod 6$. Since $b^3 - b = b(b^2 - 1)$, we have

$$b^{p-1} - 1 = (b^2 - 1)(b^{p-3} + b^{p-5} + \cdots + b^2 + 1)$$

and therefore $b^3 - b$ is a factor of $b^p - b$, in which case $b^p - b \equiv 0 \pmod 6$. Fermat's Little Theorem allows one to write $b^p - b \equiv 0 \pmod p$, and since $(6, p) = 1$, we have $b^p - b \equiv 0 \pmod{6p}$. Similarly, we obtain $a^p - a \equiv 0 \pmod{6p}$. Combining the congruences $ab^p - ab \equiv 0 \pmod{6p}$ and $-ba^p + ab \equiv 0 \pmod{6p}$ then yields the result.

(309) The answer is NOT ALWAYS. Assume that n is an odd integer. Since $1 + 2 + \cdots + (n - 1) = n(n - 1)/2$ and since n is odd, it follows that $(n - 1)/2$ is an integer and consequently the congruence is true.

Assume that n is an even integer. Letting $n = 2m$, then

$$1 + 2 + \cdots + (n - 1) = m(2m - 1) \not\equiv 0 \pmod{2m}.$$

(310) Using the formula $\sum_{i=1}^{k} i^2 = \frac{k(k+1)(2k+1)}{6}$ (see Problem 1), with $k = n-1$, we obtain that n must satisfy $n \equiv \pm 1 \pmod 6$.

(311) The answer is YES. Since $1^3 + 2^3 + \cdots + (n-1)^3 = n \cdot n(n-1)^2/4$ (see Problem 1), it follows that the congruence is true if $n^3 - 2n^2 + n \equiv 0 \pmod 4$. Setting $n = 4m + r$, $0 \leq r \leq 3$, we obtain that the congruence is true except in the case $n = 4m + 2$.

(312) Since

$$5^n = (4+1)^n$$

$$= 4^n + \binom{n}{1}4^{n-1} + \cdots + \binom{n}{n-3}4^3 + \binom{n}{n-2}4^2 + \binom{n}{n-1}4 + 1,$$

it follows that

$$5^n \equiv 4n + 1 \pmod{16}$$

and that

$$5^n \equiv 1 + 4n + 8n(n-1) \pmod{64}.$$

(313) If we can show that, for each integer $k \geq 1$, we have

$$5^{2^k} \equiv 1 + 2^{k+2} \pmod{2^{k+3}},$$

then the result will follow. But this last congruence can easily be obtained by induction on k. For $k = 1$, the result is immediate. Assuming that the congruence is true for k, that is that $5^{2^k} = 1 + 2^{k+2} + M2^{k+3}$ for a certain positive integer M, then squaring each side of this last equation, we obtain

$$5^{2^{k+1}} \equiv 1 + 2^{k+3} \pmod{2^{k+4}}.$$

The general case can be handled essentially in the same manner.

(314) This follows from the fact that the given expression is equal to

$$\frac{n^5 - n}{5} + \frac{n^3 - n}{3} + n,$$

which using Fermat's Little Theorem is easily seen to be an integer.

(315) It is clear that $x \equiv 0 \pmod{13}$ is not a solution. So let $1 \leq x \leq 12$. Then, from Fermat's Little Theorem, we have that $x^{12} \equiv 1 \pmod{13}$ and this is why $x^{24} \equiv 1 \pmod{13}$. The congruence to be solved can therefore be reduced to $7x \equiv 1 \pmod{13}$, which leads to the solution $x \equiv 2 \pmod{13}$.

(316) The seven pairs are $\{2, 9\}$, $\{3, 6\}$, $\{4, 13\}$, $\{5, 7\}$, $\{10, 12\}$, $\{11, 14\}$ and $\{8, 15\}$.

(317) Since $(m_i, m_j) = 1$ for $i \neq j$, it follows from Euler's Theorem that $m_i^{\phi(m_j)} \equiv 1 \pmod{m_j}$. Since the function ϕ is a multiplicative function, we have $m_i^{\phi(m)/\phi(m_i)} \equiv 1 \pmod{m_j}$ for $i \neq j$. On the other hand, $m_j^{\phi(m)/\phi(m_j)} \equiv 0 \pmod{m_j}$, so that for $j = 1, 2, \ldots, r$, we obtain

$$m_1^{\phi(m)/\phi(m_1)} + m_2^{\phi(m)/\phi(m_2)} + \cdots + m_r^{\phi(m)/\phi(m_r)} \equiv r - 1 \pmod{m_j}.$$

Since the integers m_j are relatively prime, the result follows.

(318) From Wilson's Theorem,

$$(p-1)! = (p-1)\cdots(p-(k-1))(p-k)!$$

$$\equiv (-1)^{k-1}(k-1)!(p-k)! \equiv -1 \pmod{p},$$

and multiplying by $(-1)^{k-1}$, the result follows.

(319) The answer is YES to both questions. We first use Fermat's Little Theorem for p and then for q, in which case we obtain

$$p^{q-1} + q^{p-1} \equiv 1 \pmod{p}, \quad p^{q-1} + q^{p-1} \equiv 1 \pmod{q},$$

since $(p, q) = 1$, and the result follows.

To prove the second part, we call upon Euler's Theorem.

(320) We have

$$3^{2n+2} = 9(9^n) = 9(8+1)^n = 9(8^n + n8^{n-1} + \cdots + 8n + 1)$$
$$= 9\left(8^n + n8^{n-1} + \cdots + 8^2 \frac{n(n-1)}{2}\right) + 9(8n+1),$$

and this is why

$$3^{2n+2} \equiv 72n + 9 \equiv 8n + 9 \pmod{64}.$$

(321) We will prove that the required GCD is equal to p. First of all, from Wilson's Theorem, it follows that for p prime, $(p-1)! \equiv -1 \pmod p$, a congruence which can be written as $(p-2)!\,(p-1) \equiv -1 \pmod p$, implying that $(p-2)! \equiv 1 \pmod p$ and therefore that $p|\,((p-2)! - 1)$. It remains to show that if $2 \le k \le p-1$, then k does not divide $(p-2)! - 1$. But if $2 \le k \le p-2$ and $k|(p-2)!-1$, we obtain that $k|1$, a contradiction. It remains to consider the case when $(p-1)|\,((p-2)! - 1)$. Since p is a prime number, $p-1$ is an even number, and therefore, using Problem 180, $(p-1)|(p-2)!$ except for $p-1 = 4$, that is when $p = 5$. Hence, $(p-1) \nmid ((p-2)! - 1)$ for $p \ge 5$.

(322) This follows from the fact that dividing by 7 the number 5^{6614} leaves 4 as a remainder, while dividing by 7 the number 12^{857} leaves 3 as a remainder. Indeed,

$$5^{6614} \equiv (-2)^{6614} = 2^{6614} = 2^{3\cdot2204+2} = 8^{2204}\cdot4 \equiv 4 \pmod 7,$$

$$12^{857} \equiv 5^{857} = 5^{6\cdot142+5} \equiv 1^{142}\cdot5^5 \equiv (-2)^5 = -32 \equiv 3 \pmod 7.$$

(323) (a) Since $10 \equiv 1 \pmod 3$, we have

$$3|N \iff a_n10^n + \cdots + a_110 + a_0 \equiv 0 \pmod 3$$
$$\iff a_n + \cdots + a_1 + a_0 \equiv 0 \pmod 3.$$

   (b) We have

$$4|N \iff a_n10^n + \cdots + a_110 + a_0 \equiv 0 \pmod 4$$
$$\iff 10a_1 + a_0 \equiv 0 \pmod 4,$$

since $10^j \equiv 0 \pmod 4$ for each $j \ge 2$.

   (c) We have

$$6|N \iff a_n10^n + \cdots + a_110 + a_0 \equiv 0 \pmod 6$$
$$\iff 4(a_n + \cdots + a_2 + a_1) + a_0 \equiv 0 \pmod 6,$$
$$\iff 4(a_n + \cdots + a_2 + a_1 + a_0) \equiv 3a_0 \pmod 6,$$

since $10^j - 4 \equiv 0 \pmod 6$ for each $j \ge 1$; indeed, $10^j - 4 = 999\ldots96$, a number which is even and divisible by 3.

   (d) If N has three digits (that is $n = 2$), then the result is obvious. We examine the case $n = 3$, so that $N = 1000a_3 + 100a_2 + 10a_1 + a_0$. We must prove that

$$1000a_3 + 100a_2 + 10a_1 + a_0 \equiv 0 \pmod 7$$
$$\iff 100a_2 + 10a_1 + a_0 - a_3 \equiv 0 \pmod 7.$$

This boils down to proving that

$$1001a_3 + 100a_2 + 10a_1 + a_0 - a_3 \equiv 0 \quad (\text{mod } 7)$$
$$\Longleftrightarrow 100a_2 + 10a_1 + a_0 - a_3 \equiv 0 \quad (\text{mod } 7),$$

an equivalence which is easily verified since $7|1001$.

To prove the case $n = 4$, we proceed essentially in the same manner, this time using the identity

$$10^4a_4 + 10^3a_3 + 10^2a_2 + 10a_1 + a_0$$
$$= 10\,010a_4 + 1001a_3 + 100a_2 + 10a_1 + a_0 - (10a_4 + a_3)$$

and by observing that $7|10\,010$. The same argument works also for the case $n = 5$.

If $n \geq 6$, we use the same argument by also observing that $10^6 - 1 = (10^3 - 1)(10^3 + 1)$, where $7|10^3 + 1$; that $10^7 - 10 = 10(10^6 - 1)$; that $10^8 - 100 = 10^2(10^6 - 1)$; and so on.

(e) We have

$$8|N \quad \Longleftrightarrow \quad a_n10^n + \cdots + a_110 + a_0 \equiv 0 \quad (\text{mod } 8)$$
$$\Longleftrightarrow \quad 100a_2 + 10a_1 + a_0 \equiv 0 \quad (\text{mod } 8),$$

since $10^j \equiv 0 \ (\text{mod } 8)$ for each integer $j \geq 3$.

(f) Since $10 \equiv 1 \ (\text{mod } 9)$, it follows that

$$9|N \quad \Longleftrightarrow \quad a_n10^n + \cdots + a_110 + a_0 \equiv 0 \quad (\text{mod } 9)$$
$$\Longleftrightarrow \quad a_n + \cdots + a_1 + a_0 \equiv 0 \quad (\text{mod } 9).$$

(g) We have

$$11|N \quad \Longleftrightarrow \quad a_n10^n + \cdots + a_110 + a_0 \equiv 0 \quad (\text{mod } 11)$$
$$\Longleftrightarrow \quad a_n(11 - 1)^n + a_{n-1}(11 - 1)^{n-1} + \cdots$$
$$+ a_1(11 - 1) + a_0 \equiv 0 \quad (\text{mod } 11)$$
$$\Longleftrightarrow \quad (-1)^n a_n + (-1)^{n-1}a_{n-1} + \cdots$$
$$+ a_2 - a_1 + a_0 \equiv 0 \quad (\text{mod } 11)$$
$$\Longleftrightarrow \quad (-1)^n \left\{ (-1)^n a_n + (-1)^{n-1}a_{n-1} + \cdots \right.$$
$$\left. + a_2 - a_1 + a_0 \right\} \equiv 0 \quad (\text{mod } 11)$$
$$\Longleftrightarrow \quad a_n - a_{n-1} + \cdots + (-1)^{n+1}a_1 + (-1)^n a_0$$
$$\equiv 0 \quad (\text{mod } 11),$$

and the result follows.

(324) Observe that $168 = 8 \cdot 3 \cdot 7$. Since $8|\text{``}770ab45c\text{''}$, it follows that using Problem 215 we have $8|\text{``}45c\text{''}$ and then $c = 6$. Similarly, $3|\text{``}770ab456\text{''}$ implies $a + b \equiv 1 \ (\text{mod } 3)$, and $7|\text{``}770ab456\text{''}$ implies (using Problem 323 (e)) that $456 - (10a + b) + 77 \equiv 0 \ (\text{mod } 7)$, that is $3a + b \equiv 1 \ (\text{mod } 7)$. Therefore, $a + b = 1$ and $3a + b = 1$, which allows us to conclude that $a = 0$ and $b = 1$. The three required numbers are therefore $a = 0$, $b = 1$ and $c = 6$.

(325) Since $(a, m) = 1$, using Euler's Theorem, we have

$$a^{\phi(m)} - 1 \equiv 0 \quad (\text{mod } m).$$

But

$$a^{\phi(m)} - 1 = (a-1)(a^{\phi(m)-1} + a^{\phi(m)-2} + \cdots + a + 1)$$

and since $(a-1, m) = 1$, the result follows.

(326) If $p|a$, then $a^{(p-1)!+1} = a \cdot a^{(p-1)!} \equiv 0 \equiv a \pmod{p}$. If $p \nmid a$, then $(a, p) = 1$, and it follows from Fermat's Little Theorem that $a^{p-1} \equiv 1 \pmod{p}$ and therefore that $a^{(p-1)!} = \left(a^{p-1}\right)^{(p-2)!} \equiv 1 \pmod{p}$, in which case $a^{(p-1)!+1} \equiv a \pmod{p}$, as required.

(327) Using Fermat's Little Theorem,

$$1^{p-1} + \cdots + (p-1)^{p-1} \equiv \underbrace{1 + \cdots + 1}_{p-1} = p - 1 \equiv -1 \pmod{p}.$$

(328) Using Fermat's Little Theorem, we have $a^p \equiv a \pmod{p}$ for each positive integer a. Hence,

$$1^p + 2^p + \cdots + (p-1)^p \equiv 1 + 2 + \cdots + p = p(p+1)/2 \equiv 0 \pmod{p},$$

since $p + 1$ is an even number.

(329) This is a consequence of the congruence $(k-1)!(p-k)! \equiv (-1)^k \pmod{p}$ (see Problem 318) and Fermat's Little Theorem, because

$$\sum_{k=1}^{p-1} (k-1)!(p-k)!k^{p-1}$$

$$\equiv -1^{p-1} + 2^{p-1} - 3^{p-1} + \cdots - (p-2)^{p-1} + (p-1)^{p-1}$$
$$\equiv -1 + 1 - 1 + \cdots - 1 + 1 = 0 \pmod{p}.$$

(330) From Wilson's Theorem, we have $(4n)! \equiv -1 \pmod{p}$, in which case

$$(4n)(4n-1)\cdots[4n-(2n-1)](2n)! \equiv -1 \pmod{p}.$$

Since $4n = p - 1 \equiv -1 \pmod{p}$, we have $4n - 1 \equiv -2 \pmod{p}$ and therefore $4n - 2 = p - 3 \equiv -2 \pmod{p}$, so that $4n - (2n-1) \equiv -2n \pmod{p}$, and the result follows.

For the generalization, we have from Wilson's Theorem $(m+n)! \equiv -1 \pmod{p}$, and therefore

$$(*) \qquad (m+n)(m+n-1)\cdots[m+n-(n-1)]m! \equiv -1 \pmod{p}.$$

We have $m + n = p - 1 \equiv -1 \pmod{p}$ and $m + n - 1 \equiv -2 \pmod{p}$, and so on, until we obtain $m + n - (n-1) \equiv -n \pmod{p}$. Then, substituting in $(*)$, we find

$$(**) \qquad\qquad (-1)^n m! n! \equiv -1 \pmod{p}.$$

Since $m + n$ is even, the second relation of the problem is proved. Finally, the last congruence can be obtained by setting $m = n = \frac{p-1}{2}$ in $(**)$.

(331) From Wilson's Theorem, n is prime if and only if $(n-1)! \equiv -1 \pmod{n}$. Therefore,

$$-1 \equiv (n-1)! = (n-1)(n-2)(n-3)! \equiv 2(n-3)! \pmod{n},$$

and the result follows.

(332) This follows immediately from Fermat's Little Theorem and Wilson's Theorem. Indeed, $a^p \equiv a \pmod{p}$ and $a(p-1)! \equiv -a \pmod{p}$, allowing us to conclude that $a^p + a(p-1)! \equiv 0 \pmod{p}$.

(333) From Wilson's Theorem, $\dfrac{(n-1)!+1}{n}$ is an integer if and only if n is a prime number, in which case the sum appearing in the statement is equal to

$$\sum_{p\le x}(\pm 1)^2 = \sum_{p\le x}1 = \pi(x),$$

as required.

(334) If $d=(r,s)$, then $r=dr_1$ and $s=ds_1$. It is clear that

$$(a^d)^{r/d}\equiv 1\pmod{m_1}\quad\text{and}\quad (a^d)^{s/d}\equiv 1\pmod{m_2}.$$

Therefore,

$$a^{[r,s]}=(a^d)^{(r/d)(s/d)}\equiv 1^{(s/d)}\equiv 1\pmod{m_1},$$

$$a^{[r,s]}=(a^d)^{(s/d)(r/d)}\equiv 1^{(r/d)}\equiv 1\pmod{m_2},$$

and the result follows.

(335) Let $m=q_1^{\alpha_1}q_2^{\alpha_2}\cdots q_r^{\alpha_r}$. If $(a,q_i)=1$, $1\le i\le r$, then $a^{\phi(q_i^{\alpha_i})}\equiv 1\pmod{q_i^{\alpha_i}}$. Now, since $q_i^{\alpha_i}|m$ implies $\phi(q_i^{\alpha_i})|\phi(m)$, then $a^{\phi(m)}\equiv 1\pmod{q_i^{\alpha_i}}$. If $a>0$ and $q\ge 2$ are positive integers, then on the one hand, we have $q^{a-1}\ge a$ (we can prove this by induction on a) and on the other hand, for $i=1,2,\dots,r$, we have $q_i^{\alpha_i-1}|m$ and $q_i^{\alpha_i-1}|\phi(m)$. Therefore,

$$(*)\qquad\qquad q_i^{\alpha_i-1}|m-\phi(m).$$

Since $m-\phi(m)>0$, then for $m>1$, it follows from $(*)$ that

$$m-\phi(m)\ge q_i^{\alpha_i-1}\ge \alpha_i.$$

Therefore, in the case $(a,q_i)>1$, that is $q_i|a$, we have

$$q_i^{\alpha_i}|q_i^{m-\phi(m)}|a^{m-\phi(m)}.$$

It follows that for each positive integer a, the relation

$$q_i^{\alpha_i}|a^{m-\phi(m)}(a^{\phi(m)}-1)$$

is true for $i=1,2,\dots,r$ and therefore that $m|a^{m-\phi(m)}$.

(336) Let $a_1,a_2,\dots,a_m$ be a complete residue system modulo m. Since $(m+1)/2$ is a positive integer, say $(m+1)/2=k$, it follows that

$$\sum_{i=1}^{m}a_i\equiv\sum_{i=1}^{m}i\equiv\frac{m(m+1)}{2}\equiv mk\equiv 0\pmod{m},$$

as was to be shown.

(337) Let $E=\{x_1,x_2,\dots,x_n\}$ be a complete residue system. The set E' contains the same number of elements as E and for $x_i,x_j\in E$, $i\ne j$, we have $x_i\ne x_j$. If

$$ax_i+b\equiv ax_j+b\pmod{m},$$

then $ax_i\equiv ax_j\pmod{m}$ and therefore $x_i\equiv x_j\pmod{m}$, which contradicts our hypothesis.

(338) The answer is YES. Indeed, the set $\{6,12,18,24,30,36\}$ is a reduced residue system modulo 7.

(339) We must show that

$$\sum_{\substack{k \leq m \\ (k,m)=1}} k \equiv 0 \pmod{m}.$$

Let $a_1, a_2, \ldots, a_{\phi(m)}$ be integers smaller than m and relatively prime to m. Since $(k,m) = 1 \iff (m-k, m) = 1$, we have

$$\begin{aligned} a_1 + a_2 + \cdots + a_{\phi(m)} &= (m - a_1) + (m - a_2) + \cdots + (m - a_{\phi(m)}) \\ &= m\phi(m) - (a_1 + a_2 + \cdots + a_{\phi(m)}). \end{aligned}$$

Since $\phi(m)$ is an even integer when $m > 2$, we then have

$$\sum_{\substack{k \leq m \\ (k,m)=1}} k = \frac{\phi(m)}{2} m \equiv 0 \pmod{m}.$$

(340) The result follows immediately from Wilson's Theorem since $r_1 r_2 \cdots r_{p-1} \equiv (p-1)! \pmod{p}$.

(341) The set $\{1, 3, 7, 9\}$ is a reduced residue system modulo 10. However,

$$E' = \{3x + 2 \mid x \in E\} = \{5, 11, 23, 29\}$$

is not a reduced residue system modulo 10, since $(5, 10) \neq 1$.

(342) (*MMAG, Vol. 64, 1991, p. 63*). The only solution is $(x, y, z) = (2, 3, 5)$. First of all, we observe that $(x, y) = (x, z) = (y, z) = 1$. Then, $2 \leq x < y < z$, and combining the three given congruences we obtain

$$xy + xz + yz - 1 \equiv 0 \pmod{x, y \text{ and } z}.$$

Since x, y and z are pairwise coprime, we have

$$xy + xz + yz - 1 \equiv 0 \pmod{xyz}.$$

It follows that $xy + xz + yz - 1 = k(xyz)$ for some integer $k \geq 1$. Dividing by xyz, we obtain that

$$\frac{1}{z} + \frac{1}{y} + \frac{1}{x} = \frac{1}{xyz} + k > 1.$$

Since $x < y < z$, it follows that

(*) $$1 < \frac{1}{x} + \frac{1}{y} + \frac{1}{z} < \frac{3}{x}$$

and this is why $x = 2$. In this case, the inequalities give

$$\frac{1}{2} < \frac{1}{y} + \frac{1}{z} < \frac{2}{y},$$

which implies that $y = 3$. It follows that the only possible values of z are 4 and 5. Hence, for $2 \leq x < y < z$, the solutions are $(x, y, z) = (2, 3, 4)$ and $(2, 3, 5)$. Since 2 and 4 are not relatively prime, the only solution is $(x, y, z) = (2, 3, 5)$.

(343) Let p_r stand for the r–th prime number. For each integer i, $1 \le i \le n$, let $m_i = p_{(i-1)k+1} \cdot p_{(i-1)k+2} \cdot p_{(i-1)k+3} \cdots p_{ik-1} \cdot p_{ik}$, and consider the system of congruences

$$\begin{cases} x \equiv -1 \pmod{m_1}, \\ x \equiv -2 \pmod{m_2}, \\ \quad \vdots \\ x \equiv -n \pmod{m_n}. \end{cases}$$

Since the m_i's are pairwise coprime, the Chinese Remainder Theorem guarantees a solution x_0. Then, $m_1|(x_0+1), \ldots, m_n|(x_0+n)$. Therefore, $x_0+1, x_0+2, \ldots, x_0+n$ is a sequence of n consecutive integers which are divisible by at least k prime numbers.

For the second part ($n = 4$ and $k = 1$), we must solve

$$\begin{cases} x \equiv -1 \pmod 3, \\ x \equiv -2 \pmod 5, \\ x \equiv -3 \pmod 7, \\ x \equiv -4 \pmod{11}. \end{cases}$$

In this case, $x \equiv 788 \pmod{1155}$ and therefore $x_0 = 788$. The four numbers are therefore 789, 790, 791 and 792.

(344) We must solve the system

$$\begin{cases} x \equiv 1 \pmod 3, \\ x \equiv 2 \pmod 4, \\ x \equiv 3 \pmod 5. \end{cases}$$

Using the Chinese Remainder Theorem, we find that $x \equiv 58 \pmod{60}$. The required positive integers are therefore the numbers $60j + 58$, with $j = 0, 1, 2, \ldots$.

(345) We must solve the system

$$\begin{cases} a \equiv 0 \pmod 2, \\ a+1 \equiv 0 \pmod 3, \\ a+2 \equiv 0 \pmod 4, \\ a+3 \equiv 0 \pmod 5, \\ a+4 \equiv 0 \pmod 6. \end{cases}$$

This system is equivalent to:

$$\begin{cases} a \equiv 2 \pmod 2, \\ a \equiv 2 \pmod 3, \\ a \equiv 2 \pmod 4, \\ a \equiv 2 \pmod 5, \\ a \equiv 2 \pmod 6. \end{cases}$$

Since $[2,3,4,5,6] = 60$, we have $a \equiv 2 \pmod{60}$. Hence, the smallest integer a is 62.

(346) We obtain

$$\frac{1}{3} = 0.\overline{3}, \quad \frac{1}{3^2} = 0.\overline{1} \quad \text{of period 1,}$$

$1/3^3 = 0.\overline{037}$ of period 3, $1/3^4 = 0.\overline{012345679}$ of period 9. However,

$$\frac{1}{7} = 0.\overline{142857} \quad \text{is of period 6,}$$

and

$$\frac{1}{7^2} = 0.\overline{020408163265306122448979591836734693877551}$$

is of period 42 ($= 6 \cdot 7$). On the other hand, the period of $1/7^3$ is $6 \cdot 7 \cdot 7$. It seems reasonable to make the following conjectures:

- Let p be a prime number such that $(p, 30) = 1$; if $1/p$ is of period m, then $1/p^n$ is of period mp^{n-1}.
- For $n \geq 2$, $1/3^n$ is of period 3^{n-2}.

(347) (*TYCM, Vol. 28, no. 4, 1997, p. 320*). Assume that the decimal expansion of a/b is formed by the repetition of the block $B = ab$ of length $n \geq 1$. Then,

$$\frac{a}{b} = 0.BBB\ldots = \frac{B}{10^n - 1} = \frac{ab}{10^n - 1},$$

so that $b^2 = 10^n - 1$. Hence, for $n \geq 1$, b must be an odd integer. If $n > 1$, then $b^2 \equiv 1 \pmod 4$ and therefore $10^n - 1 \equiv 1 \pmod 4$, which is impossible. Hence, $n = 1$ and $b = 3$, and it follows that the only positive rational numbers having the required property are $1/3$ and $2/3$.

(348) First assume that $10^h \equiv 1 \pmod n$, that is that there exists an integer k such that $10^h = 1 + kn$. Then, for each fraction m/n, we have

(1) $$10^h \frac{m}{n} = km + \frac{m}{n}.$$

Assume that $m/n = 0.a_1 a_2 a_3 \ldots$; then equation (1) allows us to write

$$km + \frac{m}{n} = a_1 a_2 \ldots a_h . a_{h+1} a_{h+2} \ldots .$$

Equating integer parts and equating fractional parts shows that

(2) $$km = a_1 a_2 \ldots a_h$$

and that

(3) $$\frac{m}{n} = 0.a_{h+1} a_{h+2} \ldots .$$

But equation (3) confirms that the digits a_{h+1}, a_{h+2}, ... are precisely the digits a_1, a_2, ... This means that the expansion of m/n repeats itself after h digits and therefore that the period of m/n is h.

Conversely, if m/n is of period h, that is

$$\frac{m}{n} = 0.a_1 a_2 \ldots a_h a_1 \ldots a_h \ldots,$$

then

$$10^h \frac{m}{n} - a_1 a_2 \ldots a_h = 0.a_1 a_2 \ldots a_h \ldots = \frac{m}{n}.$$

Consequently

$$\frac{(10^h - 1)m}{n} = a_1 a_2 \ldots a_h$$

is an integer. Since m and n are relatively prime, then we have $n|(10^h - 1)$.

Finally, assume that the period of m/n is h and that $10^{h_0} \equiv 1 \pmod n$. Then, m/n also has h_0 digits which repeat themselves and $h_0 \geq h$. In particular, h is the smallest positive integer satisfying $10^h \equiv 1 \pmod n$.

(349) In the solution of Problem 348, it is proved that $km = a_1 a_2 \ldots a_h$, which yields the result.

(350) This follows from the fact that $10^r(m/n) - (m/n) = a_1 a_2 \ldots a_r$.

(351) Let $N = 2^{n-1} + 2^{d-1} - 1$. We will show that $2^d - 1 \geq 3$ is a proper divisor of N, thereby showing that N is a composite number. Since $2^d - 1$ is an odd number, it is enough to show that $2^d - 1 | 2N$. But

$$2N = 2^n + 2^d - 2 = 2^n - 1 + 2^d - 1 = (2^d)^{n/d} - 1 + 2^d - 1$$
$$= (2^d - 1)(2^{d(\frac{n}{d}-1)} + 2^{d(\frac{n}{d}-2)} + \cdots + 2^d + 1) + (2^d - 1),$$

which proves the result.

(352) Let $n = 2^q - 1$, where q is a prime, be such a number. Since q is odd and $\mu^2(n) = 0$, there exists an odd prime number p such that $p^2 | n$. We then have

(1) $2^q \equiv 1 \pmod{p^2}$.

On the other hand, using Euler's Theorem, we have $2^{\phi(p^2)} \equiv 1 \pmod{p^2}$, so that

(2) $2^{p(p-1)} \equiv 1 \pmod{p^2}$.

It follows from (1) and (2) that $q | p(p-1)$, which implies that $q | (p-1)$ (since if $q = p$, then $2^q \equiv 1 \pmod q$, contradicting the fact that $2^{q-1} \equiv 1 \pmod q$). Hence, there exists a positive integer a such that $p - 1 = aq$, which in light of (1) gives

$$2^{p-1} = (2^q)^a \equiv 1^a = 1 \pmod{p^2},$$

thus establishing that p is a Wieferich prime.

REMARK: Only two Wieferich primes have been found so far, namely 1093 and 3511; it is known that there are no other such primes smaller than 1.25×10^{15}.

(353) We will show that the three smallest prime factors of n are 2, 3 and 11. First of all, it is clear that $2 | n$. To see that $3 | n$, it is sufficient to observe that

$$5^{96} - 7^{112} \equiv 2^{96} - 1^{112} \equiv (-1)^{96} - 1 = 1 - 1 = 0 \pmod 3.$$

Clearly, 5 and 7 are not prime factors of n. Let us check if 11 divides n. By Fermat's Little Theorem, we have $5^{10} \equiv 1 \pmod{11}$ and $7^{10} \equiv 1 \pmod{11}$, so that

$$5^{96} \equiv 5^{90} \cdot 5^6 \equiv 1 \cdot 125^2 \equiv 4^2 = 16 \equiv 5 \pmod{11},$$
$$7^{112} \equiv 7^{110} \cdot 7^2 \equiv 1 \cdot 49 = 49 \equiv 5 \pmod{11}.$$

Combining these two congruences, we easily conclude that $11 | n$.

(354) Let $N = \dfrac{m^a}{2} + \dfrac{m}{2} - 1$. We will show that $m - 1 | N$. To do so, since $m - 1$ is odd, it is clear that we shall reach our goal if we can manage to show that $m - 1 | 2N$. But

$$2N = m^a + m - 2 = m^a - 1 + m - 1 = (m-1)(m^{a-1} + m^{a-2} + \cdots$$
$$+ m + 1) + (m - 1),$$

which proves the result.

(355) Let $u_n = 2^{2^n} + 3$. As soon as $2^n = 3a - 1$, then u_n can be written as

$$u_n = 2^{3a-1} + 2^2 - 1 = \frac{(2^3)^a}{2} + \frac{2^3}{2} - 1,$$

a number which is composite by Problem 354 if $a \geq 2$. Now, there exist infinitely many positive integers n such that $2^n = 3a - 1$ for a certain positive integer a. Indeed, if n is odd, then $2^n + 1 \equiv (-1)^n + 1 = 0$ (mod 3), and this is why all numbers of the form $2^{2^n} + 3$, with $n \geq 3$ odd, are composite.

(356) We first write

$$N = 2^{2^6} + 15 = 2^{64} + 2^4 - 1 = \frac{(2^5)^{13}}{2} + \frac{2^5}{2} - 1.$$

Applying the result of Problem 354 with $m = 2^5$ and $a = 13$, we obtain that $m - 1 = 2^5 - 1 = 31$ divides N, and the result follows.

(357) The answer is NO. Indeed, although $2^{2^n} + 15$ is prime for $n = 0, 1, 2, 3, 4, 5$, when $n = 6$, we have

$$2^{2^6} + 15 = 18\,446\,744\,073\,709\,551\,631 = 31 \cdot 107 \cdot 5\,561\,273\,462\,077\,043.$$

To avoid using a computer in order to obtain this last factorization, one can consult the solution of Problem 356 to learn that $31 | (2^{2^6} + 15)$ and therefore that this last number is composite.

(358) This follows from the fact that $973 = 1000 - 27 = 10^3 - 3^3$ is a divisor of $10^9 - 3^9$ and from the fact that $139 | 973$.

(359) We have $n = 10^6 - 7^2 = (10^3 - 7)(10^3 + 7) = 993 \cdot 1007 = 3 \cdot 331 \cdot 1007$. But since n has a prime factor p such that $300 < p < 400$ and since 1007 is not divisible by 3, we conclude that $p = 331$.

(360) We have

$$2^{21} - 1 = (2^7)^3 - 1 = (2^7 - 1)\left((2^7)^2 + (2^7) + 1\right) = 127 \cdot \left((2^7)^2 + (2^7) + 1\right)$$

and the result follows.

(361) We have

$$
\begin{aligned}
2^{2^6} - 1 &= 2^{64} - 1 = \left(2^{32} + 1\right)\left(2^{32} - 1\right) = \left(2^{32} + 1\right) \\
&\qquad\qquad\qquad\qquad\qquad \cdot \left(2^{16} + 1\right)\left(2^{16} - 1\right) \\
&= \left(2^{32} + 1\right)\left(2^{16} + 1\right)\left(2^8 + 1\right)\left(2^8 - 1\right) \\
&= \left(2^{32} + 1\right)\left(2^{16} + 1\right)\left(2^8 + 1\right)\left(2^4 + 1\right)\left(2^4 - 1\right) \\
&= \left(2^{32} + 1\right)\left(2^{16} + 1\right)\left(2^8 + 1\right)\left(2^4 + 1\right)\left(2^2 + 1\right)\left(2^2 - 1\right) \\
&= \left(2^{32} + 1\right)\left(2^{16} + 1\right) \cdot 257 \cdot 17 \cdot 5 \cdot 3.
\end{aligned}
$$

The numbers 3, 5, 17 and 257 are therefore each a prime factor of $2^{2^6} - 1$. REMARK: The complete factorization of $2^{2^6} - 1$ is:

$$2^{2^6} - 1 = 3 \cdot 5 \cdot 7 \cdot 257 \cdot 641 \cdot 65537 \cdot 6700417.$$

(362) In fact, one can show slightly more. Indeed, if $S_+(N)$ stands for the number of positive integers $n \leq N$ such that $r_n := n10^n + 1$ is prime, we have that

$$(*) \qquad\qquad S_+(N) \geq \left[\frac{N}{3}\right] + \pi(N) \qquad (N \geq 11).$$

In order to prove $(*)$, we first observe that if $n \equiv 2 \pmod 3$, then $3|r_n$; this follows from the fact that, in this case, $n = 3k+2$ for a certain integer $k \geq 0$, so that

$$n10^n + 1 = (3k+2)10^{3k+2} + 1 \equiv 200 \cdot 10^{3k} + 1 \equiv 2 \cdot 1^{3k} + 1 \equiv 0 \pmod 3.$$

On the other hand, it follows from Fermat's Little Theorem that if $n+1$ is a prime number $p > 5$, then r_n is a multiple of p, since in this case we have $n = p - 1$, so that

$$n10^n + 1 = (p-1)10^{p-1} + 1 \equiv (p-1) + 1 \equiv 0 \pmod p.$$

From these two observations and the fact that $S_+(11) = 8$, inequality $(*)$ follows.

REMARK: Using a computer, we obtain that the smallest seven positive integers n such that $n10^n + 1$ is prime are 1, 3, 9, 21, 363, 2161 and 4839.

(363) In light of Problem 75, we have that $2^2 - 1 = 3$, $2^3 - 1 = 7$ and $2^5 - 1 = 31$ are divisors of $2^{30} - 1$. Similarly, we have that $2^5 - 1 = 31$ and $2^7 - 1 = 127$ are prime factors of $2^{35} - 1$.

(364) Let $n = 2^{30} - 1$. Since $n = 2^{30} - 1 = 2^{3(11-1)} - 1 = 8^{11-1} - 1$ and since $(8, 11) = 1$, Fermat's Little Theorem then yields the result.

(365) Using Problem 75, we have that $3^4 - 1 = 80$ and that $3^6 - 1 = (3^3 - 1)(3^3 + 1) = 26 \cdot 28$ divides $3^{12} - 1$. Then it follows easily that 2, 5, 7 and 13 are prime factors of $3^{12} - 1$. Similarly, we have that $3^8 - 1 = (3^4 - 1)(3^4 + 1) = 80 \cdot 82$ and that $3^{12} - 1 = (3^6 - 1)(3^6 + 1) = (3^3 - 1)(3^3 + 1)(3^6 + 1) = 26 \cdot 28 \cdot 730$ divides $3^{24} - 1$. From this, it follows that 2, 5, 7, 13, 41 and 73 are prime factors of $3^{24} - 1$.

(366) Since m is odd, we have

$$a^m + 1 = (a+1)(a^{m-1} - a^{m-2} + a^{m-3} - \cdots - a + 1),$$

and the result follows.

This shows that

$$1001 = 10^3 + 1 = (10+1)(10^2 - 10 + 1) = 11 \cdot 91 = 11 \cdot 7 \cdot 13.$$

(367) The result is immediate if we write

$$a^m + 1 = (a^{m/d})^d + 1$$

and we then apply the result of Problem 366.

This shows that

$$1000001 = 10^6 + 1 = (10^2 + 1)(10^4 - 10^2 + 1) = 101 \cdot 9901.$$

(368) It follows from Problem 367 that

$$10^{15} + 1 = (10^3 + 1)((10^3)^4 - (10^3)^3 + (10^3)^2 - 10^3 + 1).$$

The result then follows from the fact that 7, 11 and 13 are factors of 1001.

(369) It is enough to observe that $n^4 + 4 = (n^2 - 2n + 2)(n^2 + 2n + 2)$. For the general case, we only need to observe that

$$n^4 + a^2 = (n^2 - \sqrt{2a}\, n + a)(n^2 + \sqrt{2a}\, n + a).$$

Let us mention that the condition "$n \geq \sqrt{2a}$" is sufficient but not necessary.

(370) We will show that if $k \equiv 2 \pmod 6$, then $k10^k + 1$ is a multiple of 3. Indeed, if $k = 6j + 2$ for a certain nonnegative integer j, then

$$k10^k + 1 = (6j + 2)10^{6j+2} + 1 \equiv 6j + 2 + 1 \equiv 0 \pmod 3,$$

which proves the result.

REMARK: It is easy to see that, given a prime number $p \neq 2, 5$, then for each positive integer k of the form $k = (jp + 1)(p - 1)$, where j is a nonnegative integer, we have $p | k10^k + 1$. Indeed, for such a prime number p, it follows from Fermat's Little Theorem that $10^{p-1} \equiv 1 \pmod p$, so that

$$k10^k + 1 = (jp + 1)(p - 1)(10^{(p-1)})^{jp+1} + 1 \equiv (jp + 1)(p - 1) + 1$$
$$= jp^2 + p - jp - 1 + 1 \equiv 0 \pmod p.$$

(371) Indeed, let $p = k + 2$ be a prime number. Then, from Fermat's Little Theorem, we have $k^{p-1} \equiv 1 \pmod p$. Since $p - 1 = k + 1$, it follows that $k^{k+1} \equiv 1 \pmod p$. Finally, since $k = p - 2 \equiv -2 \pmod p$, we then obtain successively

$$k \cdot k^k \equiv 1 \pmod p,$$
$$-2 \cdot k^k \equiv 1 \pmod p,$$
$$2k^k \equiv -1 \pmod p,$$

which establishes that $p | (2k^k + 1)$, as required.

(372) The result follows from the identity

$$2^{4r-2} + 1 = (2^{2r-1} - 2^r + 1)(2^{2r-1} + 2^r + 1),$$

which is easily proved by a simple multiplication of the two factors of the right-hand side. Thus, we obtain

$$2^{58} + 1 = (2^{29} - 2^{15} + 1)(2^{29} + 2^{15} + 1).$$

Observing that

$$2^{29} - 2^{15} + 1 = 4^{14} \cdot 2 - 4^6 \cdot 2^3 + 1 \equiv (-1)^{14} \cdot 2 - (-1)^6 \cdot 8 + 1$$
$$\equiv 2 - 3 + 1 = 0 \pmod 5,$$

we quickly obtain 5 as a third factor. In fact, without any difficulty, we obtain that the factorization of $2^{58} + 1$ is

$$2^{58} + 1 = 5 \cdot 107\,367\,629 \cdot 536\,903\,681.$$

(373) First let $q | M_p$, q prime. By definition, we have $2^p \equiv 1 \pmod q$, while by Fermat's Little Theorem we have $2^{q-1} \equiv 1 \pmod q$. This means that $p | q - 1$, and this is why $q - 1 = \ell p$ for a certain positive integer ℓ. Since ℓ is necessarily even, the result follows. Finally, if $r | M_q$, r composite, then $r = q_1 q_2 \cdots q_s$ for certain prime numbers $q_1 \leq q_2 \leq \ldots \leq q_s$. But it follows from the first part that $q_i \equiv 1 \pmod{2p}$, $i = 1, 2, \ldots, s$, in which case $r = q_1 q_2 \cdots q_s \equiv 1 \pmod{2p}$, as required.

(374) In the case $p = 61$, the Lucas-Lehmer Test can be programmed as follows with MATHEMATICA:

```
p=61; s=4; j=1; mp=2^p-1;
While [j<p-1, {r=Mod[s^2-2,mp]);s=r;j++}];
If [Mod[r,mp]==0,Print["2^",p,"-1 is prime"],
```

```
Print["2^",p,"-1 is composite"]]
```

(375) Using several times the identities $a^2 - b^2 = (a+b)(a-b)$ and $a^3 + b^3 = (a+b)(a^2 - ab + b^2)$, we obtain

$$
\begin{aligned}
10^{48} - 1 &= (10^{24} + 1)(10^{24} - 1) \\
&= (10^{24} + 1)(10^{12} + 1)(10^{12} - 1) \\
&= (10^{24} + 1)(10^{12} + 1)(10^6 + 1)(10^6 - 1) \\
&= (10^{24} + 1)(10^{12} + 1)(10^6 + 1)(10^3 + 1)(10^3 - 1) \\
&= (10^{24} + 1)(10^{12} + 1)(10^6 + 1)(10 + 1)(10^2 - 10 + 1) \\
&\qquad\qquad\qquad\qquad\qquad\qquad\cdot(10 - 1)(10^2 + 10 + 1) \\
&= (10^8 + 1)(10^{16} - 10^8 + 1)(10^4 + 1)(10^8 - 10^4 + 1) \\
&\qquad\qquad\cdot(10^2 + 1)(10^4 - 10^2 + 1) \cdot 11 \cdot 91 \cdot 9 \cdot 111.
\end{aligned}
$$

On the one hand, $10^8 + 1 = 17 \cdot 5\,882\,353$ and $10^4 + 1 = 73 \cdot 137$. On the other hand, using the Lucas-Lehmer Test, we obtain, taking successively $a = 7$, $a = 13$ and $a = 6$, that $10^{16} - 10^8 + 1$, $10^8 - 10^4 + 1$ and $10^4 - 10^2 + 1$ are prime. Gathering these results, we obtain that

$$
10^{48} - 1 = 3^3 \cdot 7 \cdot 11 \cdot 13 \cdot 17 \cdot 37 \cdot 73 \cdot 101 \cdot 137 \cdot 9\,901 \cdot 5\,882\,353
$$
$$
\cdot 99\,990\,001 \cdot 9\,999\,999\,900\,000\,001.
$$

(376) We first look for positive integers n for which the corresponding number N is divisible by 7. Since $78\,557 \equiv 3 \pmod 7$, then we are looking for integers n such that

$$
N = 78\,557 \cdot 2^n + 1 \equiv 3 \cdot 2 \cdot 2^{n-1} + 1 \equiv 6 \cdot 2^{n-1} + 1 \equiv 0 \pmod 7,
$$

that is integers n such that $2^{n-1} \equiv 1 \pmod 7$. Since the order of 2 modulo 7 is 3, it follows that $3|n - 1$; that is $n \equiv 1 \pmod 3$. Then, we look for positive integers n for which the corresponding number N is divisible by 13. Since $78\,557 \equiv -2 \pmod{13}$, we need to solve $2^{n+1} \equiv 1 \pmod{13}$. Since the order of 2 modulo 13 is 12, we obtain that $12|n + 1$; that is $n \equiv -1 \pmod{12}$. Similarly, we obtain that $3|N$ if and only if $n \equiv 0 \pmod 2$, that $5|N$ if and only if $n \equiv 1 \pmod 4$, that $19|N$ if and only if $n \equiv -3 \pmod{18}$, that $37|N$ if and only if $n \equiv -9 \pmod{36}$ and finally that $73|N$ if and only if $n \equiv 3 \pmod 9$. To show that each positive integer n satisfies at the very least one of the seven congruences, we only need to observe that the seven congruences obtained are respectively equivalent to the seven families of congruences enumerated below:

$$
\begin{aligned}
n &\equiv 1, 4, 7, 10, 13, 16, 19, 22, 25, 28, 31, 34 \pmod{36}, \\
n &\equiv 11, 23, 35 \pmod{36}, \\
n &\equiv 0, 2, 4, 6, 8, 10, 12, 14, 16, 17, 18, 20, 22, 24, 26, \\
&\qquad\qquad\qquad\quad 28, 30, 32, 34 \pmod{36}, \\
n &\equiv 1, 5, 9, 13, 17, 21, 25, 29, 33 \pmod{36}, \\
n &\equiv 15, 33 \pmod{36}, \\
n &\equiv 27 \pmod{36}, \\
n &\equiv 3, 12, 21, 30 \pmod{36}.
\end{aligned}
$$

Since these congruences cover the set of all equivalence classes modulo 36, the result follows.

(377) Using the identity $n^3 + 1 = (n+1)(n^2 - n + 1)$ several times, we can write

$$
\begin{aligned}
10^{27} + 1 &= (10^9 + 1)(10^{18} - 10^9 + 1) \\
&= (10^3 + 1)(10^6 - 10^3 + 1)(10^{18} - 10^9 + 1) \\
&= (10 + 1)(10^2 - 10 + 1)(10^6 - 10^3 + 1)(10^{18} - 10^9 + 1) \\
&= 11 \cdot 91 \cdot (10^6 - 10^3 + 1)(10^{18} - 10^9 + 1) \\
&= 7 \cdot 11 \cdot 13 \cdot (10^6 - 10^3 + 1)(10^{18} - 10^9 + 1).
\end{aligned}
$$

We have thus found the three prime factors 7, 11 and 13.
REMARK: The complete factorization of $10^{27} + 1$ is:

$$7 \cdot 11 \cdot 13 \cdot 19 \cdot 52\,579 \cdot 70\,541\,929 \cdot 14\,175\,966\,169.$$

(378) Starting with $x = 45$, we obtain that $x^2 - n = 45^2 - 2009 = 16$, a perfect square. It follows that

$$2009 = 2025 - 16 = 45^2 - 4^2 = (45 + 4)(45 - 4) = 49 \cdot 41.$$

We first compute $\sqrt{n} = 538.285\ldots$. We therefore choose $x = 539$, and we compute $x^2 - n = 290521 - 289751 = 770$, which is not a perfect square. We then take $x = 540$, and we obtain $x^2 - n = 291600 - 289751 = 1849 = 43^2$, the desired perfect square. We thus have

$$289751 = 540^2 - 43^2 = (540 + 43)(540 - 43) = 583 \cdot 497.$$

Observe that $289751 = 7 \cdot 11 \cdot 53 \cdot 71$.

(379) We first try with $p_0 = 3$. We then have

$$m = 3n = 3 \cdot 1254713 = 3764139.$$

Then, set $a = [\sqrt{m}] + 1 = 1940 + 1 = 1941$. Since $a^2 - m = 1941^2 - 3764139 = 3342$ is not a perfect square, we set $a = 1942$, in which case we obtain $a^2 - m = 1942^2 - 3764139 = 7225 = 85^2$. We conclude that $m = 1942^2 - 85^2 = (1942 - 85)(1942 + 85) = 1857 \cdot 2027 = 3 \cdot 619 \cdot 2027$, from which we obtain that the factorization of n is

$$n = 619 \cdot 2027.$$

(380) Fermat's Factorization Method consists of finding integers a and b such that $n = a^2 - b^2$. We then have $n = (a - b)(a + b) = pq$, so that

$$a = \frac{p + q}{2}, \qquad b = \frac{q - p}{2}.$$

The first choice of a is $a = [\sqrt{n}] + 1$. As long as $\sqrt{a^2 - n}$ is not an integer, we set $a = a + 1$, until eventually $\sqrt{([\sqrt{n}] + k)^2 - n}$ is an integer. If k is the required number, it is clear that $k \approx a - \sqrt{n}$. But

$$(1) \qquad a - \sqrt{n} = \frac{p + q}{2} - \sqrt{pq} = p\left(1 + \frac{\delta}{2} - \sqrt{1 + \delta}\right).$$

The series expansion of $1 + \frac{\delta}{2} - \sqrt{1 + \delta}$ gives

$$(2) \qquad 1 + \frac{\delta}{2} - \sqrt{1 + \delta} = \frac{\delta^2}{8} + R(\delta^3),$$

where $|R(\delta^3)| < \delta^3$. Combining (1) and (2) gives the result.

(381) By hypothesis, we have $p \approx 3q$. We shall therefore apply Fermat's Factorization Method to the number $m = 3n = 3pq = 566058039$. We first observe that $[\sqrt{m}] = 23792$. Choosing $a = 23792$, we quickly realize that $a^2 - m = 566059264 - m = 1225 = 35^2 = b^2$, so that

$$566058039 = 23792^2 - 35^2 = 23757 \cdot 23827.$$

Since it is easy to see that it is 23757 which is divisible by 3, we conclude that the complete factorization of 188686013 is given by

$$188686013 = 7919 \cdot 23827.$$

(382) Since $\ell := \frac{k+1}{2}$ is an integer ≥ 3, it is clear that

$$(*) \qquad\qquad n(r - 1) = r^{k+1} - 1 = \left(r^\ell + 1\right)\left(r^\ell - 1\right).$$

If ℓ is even,

$$r^\ell - 1 = \left(r^{\ell/2} + 1\right)\left(r^{\ell/2} - 1\right),$$

in which case $(*)$ becomes

$$n(r - 1) = r^{k+1} - 1 = \left(r^\ell + 1\right)\left(r^\ell - 1\right)$$
$$= \left(r^\ell + 1\right)\left(r^{\ell/2} + 1\right)\left(r^{\ell/2} - 1\right),$$

so that

$$n = \begin{cases} \left(r^\ell + 1\right)\left(r^{\ell/2} + 1\right)\left(r^{\ell/4} + 1\right)\frac{r^{\ell/4}-1}{r-1} & \text{if } \ell/2 \text{ is even,} \\ \left(r^\ell + 1\right)\left(r^{\ell/2} + 1\right)\left(r^{m-1} + r^{m-2} + \cdots + r + 1\right) & \text{if } \ell/2 = m \text{ is odd,} \end{cases}$$

that is, in both cases, the product of at least three distinct integers larger than 1.

On the other hand, if ℓ is odd, then

$$r^\ell + 1 = (r + 1)\left(r^{\ell-1} - r^{\ell-2} + r^{\ell-3} - \cdots - r + 1\right),$$

in which case $(*)$ becomes

$$n(r-1) = r^{k+1} - 1 = (r + 1)\left(r^{\ell-1} - r^{\ell-2} + r^{\ell-3} - \cdots - r + 1\right)\left(r^\ell - 1\right),$$

so that

$$n = (r + 1)(r^{\ell-1} - r^{\ell-2} + \cdots - r + 1)(r^{\ell-1} + r^{\ell-2} + \cdots + r + 1),$$

that is again a product of three integers larger than 1, which are in fact distinct if $r \geq 3$ or if $r = 2$ and $k \geq 7$.

(383) First of all, it is clear that each of the elements of the set represents a positive odd number and that this set contains at most 2^k integers. To prove that two numbers of the type $2^k \pm 2^{k-1} \pm 2^{k-2} \pm \cdots \pm 2^1 \pm 1$ cannot be equal, assume that there exist $\alpha_0, \alpha_1, \ldots, \alpha_{k-1}, \beta_0, \beta_1, \ldots, \beta_{k-1} \in \{-1, 1\}$ such that

$$(1) \quad 2^k + \alpha_{k-1}2^{k-1} + \alpha_{k-2}2^{k-2} + \cdots + \alpha_1 2^1 + \alpha_0 = 2^k + \beta_{k-1}2^{k-1}$$
$$+ \beta_{k-2}2^{k-2} + \cdots + \beta_1 2^1 + \beta_0$$

without having each $\alpha_i = \beta_i$. If such is the case, then let j be the first subscript such that $\alpha_j \neq \beta_j$. We may assume that $\alpha_j > \beta_j$, in which case $\alpha_j - \beta_j = 2$ and therefore (1) becomes

$$(2) \qquad 2^{j+1} + (\alpha_{j-1} - \beta_{j-1})2^{j-1} + \cdots + (\alpha_1 - \beta_1)2 + (\alpha_0 - \beta_0) = 0.$$

But it is easy to establish that

$$\left|(\alpha_{j-1} - \beta_{j-1})2^{j-1} + \cdots + (\alpha_1 - \beta_1)2 + (\alpha_0 - \beta_0)\right|$$
$$\leq 2^{j-1} + 2^{j-2} + \cdots + 2 + 1 = 2^j - 1 < 2^{j+1},$$

and this is why (2) is impossible, in which case we must have

$$\alpha_i = \beta_i \qquad \text{for } i = 0, 1, 2, \ldots, k-1.$$

Thus the result.

(384) (i) Such a number k must be prime, because if it is not, then $k = ab$ with $1 < a \leq b < n$, in which case $\frac{10^{ab}-1}{9} = \frac{10^{ab}-1}{10^b-1} \cdot \frac{10^b-1}{9}$, the product of two integers > 1.

(ii) So let p be a prime number, and let q be a prime factor of $(10^p - 1)/9$. Since it is clear that $q \neq 2, 5$, we have from Fermat's Little Theorem that $10^{q-1} \equiv 1 \pmod{q}$. On the other hand, we have by our hypothesis that $10^p \equiv 1 \pmod{q}$. Combining these two relations, we may conclude that $p | q - 1$ and therefore that there exists a positive integer i such that $q - 1 = ip$. Since $q - 1$ is even, it follows that $i = 2j$ for a certain positive integer j, and the result follows.

(iii) We obtain that the five smallest prime numbers p such that the corresponding number $(10^p - 1)/9$ is prime are 2, 19, 23, 317 and 1031.

(iv) The following is a table displaying the factorization of the numbers $(10^p - 1)/9$, for each prime number $3 \leq p \leq 67$, $p \neq 19, 23$:

p	factorization of $(10^p - 1)/9$
3	$3 \cdot 37$
5	$41 \cdot 271$
7	$239 \cdot 4649$
11	$21649 \cdot 513239$
13	$53 \cdot 79 \cdot 265371653$
17	$2071723 \cdot 5363222357$
29	$3191 \cdot 16763 \cdot 43037 \cdot 62003 \cdot 77843839397$
31	$2791 \cdot 6943319 \cdot 57336415063790604359$
37	$2028119 \cdot 247629013 \cdot 2212394296770203368013$
41	$83 \cdot 1231 \cdot 538987 \cdot 201763709900322803748657942361$
43	$173 \cdot 1527791 \cdot 1963506722254397 \cdot 2140992015395526641$
47	$35121409 \cdot 316362908763458525001406154038726382279$
53	$107 \cdot 1659431 \cdot 1325815267337711173$ $\cdot 4719885879949142566020071$
59	2559647034361 $\cdot 4340876285657460212144534289928559826755746751$
61	$733 \cdot 4637 \cdot 329401 \cdot 974293 \cdot 1360682471$ $\cdot 106007173861643 \cdot 70617099990156159479$
67	$493121 \cdot 79863595778924342083$ $\cdot 28213380943176667001263153660999177245677$

REMARK: It is interesting to observe that each of the above numbers $(10^p - 1)/9$ is squarefree (as is the case for Mersenne numbers $2^p - 1$ for which the factorization is known).

(385) From (∗), we obtain that

$$n = x^{2k+1} + y^{2k+1}$$
$$= (x+y)(x^{2k} - x^{2k-1}y + x^{2k-2}y^2 - \cdots - xy^{2k-1} + y^{2k}).$$

Since $2 \le x + y < x^{2k+1} + y^{2k+1}$, it follows that $x + y$ is a proper divisor of n and therefore that n is composite.

(386) By hypothesis, there exist positive integers x and y such that

$$p_0 \cdot n = x^3 + y^3 = (x+y)(x^2 - xy + y^2).$$

The result then follows if we can prove that (i) $x + y \ne p_0$ and that (ii) $x + y$ is a proper divisor of n.

First assume that (i) is false; that is $x + y = p_0$. We will then have

$$p_0^3 = (x+y)^3 > x^3 + y^3 = p_0 n,$$

which implies that $p_0^2 > n$ and therefore that $p_0 > \sqrt{n}$, which contradicts the hypothesis $p_0 < \sqrt{n}$.

Since $x + y \ge 2$, in order to prove (ii), it is enough to prove that $x + y < n$. Assume the contrary, that is that $x + y = n$. Without any loss in generality, we may assume that $x > y \ge 1$, in which case $2x > x + y = n$, so that $x > n/2$. But, in this case,

$$p_0 n = x^3 + y^3 = (x+y)(x^2 - xy + y^2) = n(x^2 - xy + y^2),$$

which implies that

$$p_0 = x^2 - xy + y^2 > x^2 - xy = x(x - y) > x > \frac{n}{2},$$

contradicting the fact that $p_0 < \sqrt{n}$.

(387) We easily observe that

$$7n = 2^3 + 717^3 = (2 + 717)(2^2 - 2 \cdot 717 + 717^2),$$

so that $719 | 7n$. Since $(7, 719) = 1$, it follows that n is composite and that $719 | n$.

(388) We easily observe that

$$11n = 1^5 + 1212^5$$
$$= (1 + 1212)(1^4 - 1^3 \cdot 1212 + 1^2 \cdot 1212^2 - 1 \cdot 1212^3 + 1212^4),$$

so that $1213 | 11n$. Since $(11, 1213) = 1$, it follows that n is composite and that $1213 | n$. In fact,

$$237\,749\,938\,896\,803 = 41 \cdot 1213 \cdot 4780526791.$$

(389) Let b be an integer such that $(b, n) = 1$, and let p be a prime number such that $p | n$. Then, $(b, p) = 1$ so that

$$b^{p-1} \equiv 1 \pmod{p}.$$

It follows that

$$b^{n-1} = \left(b^{p-1}\right)^{(n-1)/(p-1)} \equiv 1^{(n-1)/(p-1)} \equiv 1 \pmod{p},$$

which implies that n is a Carmichael number.

REMARK: The reciprocal of this result is true (see Giblin [14], p. 156) so that an odd composite number $n \ge 3$ is a Carmichael number if and only if it is squarefree and $p|n \Rightarrow p-1|n-1$. This result is called *"Korselt's Criterion"*.

(390) First of all, we show that there exists a positive integer m such that $p = 6m + 1$. Indeed, if this is not the case, then $p = 6m + 5$ for a certain integer m, in which case $2p - 1 = 12m + 9 = 3(4m + 3)$ would not be prime. We can therefore write

$$n = (6m + 1)(12m + 1)(18m + 1) = p \cdot q \cdot r,$$

say. Since

$$n - 1 = 6 \cdot 12 \cdot 18 \cdot m^3 + (6 \cdot 12 + 6 \cdot 18 + 12 \cdot 18)m^2 + (6 + 12 + 18)m,$$

it is easy to see that

$$p - 1 = 6m|n - 1, \qquad q - 1 = 12m|n - 1 \quad \text{and} \quad r - 1 = 18m|n - 1,$$

which guarantees that n is a Carmichael number, from Problem 389.

REMARK: Unfortunately, no one has ever proved that there exist infinitely many triples of prime numbers of the form $(p, 2p - 1, 3p - 2)$. Let us also mention that $(7, 13, 19)$, $(37, 73, 109)$ and $(211, 421, 631)$ are such triples. Using a computer, we obtain that there exist 228 triples of prime numbers $(p, 2p - 1, 3p - 2)$ with $p < 100\,000$.

(391) Let n be a Carmichael number. By definition, n cannot be a prime number. In light of Korselt's Criterion, n cannot be a prime power (since it must be squarefree). In fact, it is enough to show that $n \neq pq$, where $p < q$ are two odd prime numbers. Assume that $n = pq$. Then, we should have $p - 1|n - 1$ and $q - 1|n - 1$, so that $q - 1|n - 1 = pq - 1 = pq - p + p - 1 = p(q - 1) + (p - 1)$, which implies that $q - 1|p - 1$ and that $p - 1|n - 1 = pq - 1 = pq - q + q - 1 = q(p - 1) + (q - 1)$, and therefore $p - 1|q - 1$. Combining these relations, it follows that $p - 1 = q - 1$, that is $p = q$, a contradiction.

(392) First consider the case $j = 1$. Since

$$q_2 q_3 \cdots q_k - 1 = q_1 q_2 q_3 \cdots q_k - q_1 q_2 q_3 \cdots q_k + q_2 q_3 \cdots q_k - 1$$
$$= q_1 q_2 q_3 \cdots q_k - q_2 q_3 \cdots q_k(q_1 - 1),$$

it follows that $q_1 - 1|q_2 q_3 \cdots q_k - 1$ if and only if $q_1 - 1|q_1 q_2 q_3 \cdots q_k - 1$, and the result follows.

(393) Since $x^3 + y^3 = (x + y)(x^2 - xy + y^2)$, we immediately derive from $(*)$ that 97 and 109 are (prime) divisors of 327 763. Since

$$\frac{327\,763}{97 \cdot 109} = 31,$$

the complete factorization of 327 763 is then

$$327\,763 = 31 \cdot 97 \cdot 109.$$

(394) Since $x^3 + y^3 = (x + y)(x^2 - xy + y^2)$, we have that

$$7 \cdot n = 341\,532\,611 = 699^3 + 8^3 = (699 + 8)(699^2 - 699 \cdot 8 + 8^2),$$

so that $7 \cdot n = 707 \cdot a$ for a certain integer a, which means that $n = 101 \cdot a$. Hence, 101 is a prime factor of n.

REMARK: The complete factorization of 48 790 373 is

$$48\,790\,373 = 31 \cdot 101 \cdot 15\,583.$$

(395) First of all, if $n > 3$ is even, say $n = 2k$, $k \geq 2$, then

$$2^n - 7 = 2^{2k} - 7 = 4^k - 7 \equiv 1^k - 1 = 0 \pmod 3.$$

Hence, if n is even, the number $2^n - 7$ is divisible by 3 and is therefore composite.

If $n \equiv 1 \pmod 4$, $n > 3$, then there exists a positive integer k such that $n = 4k + 1$, so that

$$2^n - 7 = 2^{4k+1} - 7 = 16^k \cdot 2 - 7 \equiv 1^k \cdot 2 - 2 = 0 \pmod 5.$$

Hence, if $n = 4k + 1$, the number $2^n - 7$ is divisible by 5 and is therefore composite.

If $n \equiv 7 \pmod{10}$, $n > 3$, then there exists an integer $k \geq 0$ such that $n = 10k + 7$, so that

$$2^n - 7 = 2^{10k+7} - 7 = 1024^k \cdot 2^7 - 7 \equiv 1^k \cdot 7 - 7 = 0 \pmod{11}.$$

Hence, if $n = 10k + 7$, the number $2^n - 7$ is divisible by 11 and is therefore composite.

If $n \equiv 11 \pmod{12}$, $n > 3$, then there exists an integer $k \geq 0$ such that $n = 12k + 11$, so that

$$2^n - 7 = 2^{12k+11} - 7 = 4096^k \cdot 2^{11} - 7 \equiv 1^k \cdot 7 - 7 = 0 \pmod{13}.$$

Hence, if $n = 12k + 11$, the number $2^n - 7$ is divisible by 13 and is therefore composite.

Therefore, we only need to consider the numbers $n = 15$, 19, 31, 39, ... Using the instruction PrimeQ[2^n - 7] of MATHEMATICA, we find that $2^{39} - 7$ is prime.

Continuing this process, we obtain that the next five values of n with the above property are 715, 1983, 2319, 2499 and 3775.

To find these values, using the MAPLE software, we type in the program

```
> for n to 4000 do if isprime(2^n-7)
> then print(n) else fi; od;
```

(396) Let

$$n = n_1 \cdot n_2 = \frac{a^p - 1}{a - 1} \cdot \frac{a^p + 1}{a + 1} = \frac{a^{2p} - 1}{a^2 - 1}.$$

It is clear that n_1 and n_2 are odd. On the other hand, since $n_1 \equiv 1 \pmod{2p}$ and $n_2 \equiv 1 \pmod{2p}$, we have that $n \equiv 1 \pmod{2p}$. Since $n | a^{2p} - 1$, we have $a^{2p} \equiv 1 \pmod n$ and therefore $a^{n-1} \equiv 1 \pmod n$, which means that n is pseudoprime in basis a. Applying this method, we easily find that the numbers 341 and 7381 are pseudoprime numbers in basis 2 and 3 respectively.

REMARK: This method was imagined in 1904 by Cipolla to generate pseudoprime numbers.

(397) (Malo, 1903; see Williams [41]). Let n be a pseudoprime number in basis 2. If we show that the number $N = 2^n - 1$ is also pseudoprime, we shall be done. First of all, N is composite, since $2^n - 1$ is divisible by $2^a - 1$ for each divisor a of n. Since n is pseudoprime, $n | 2^{n-1} - 1$, and this is why $n | 2^n - 2 = N - 1$. It follows that $N = 2^n - 1 | 2^{N-1} - 1$ and therefore that $2^{N-1} \equiv 1 \pmod N$ as required.

(398) If $n = ks + r$, $0 \leq r < s$, then

$$a^n = a^{ks+r} = (a^s)^k \, a^r$$

and therefore, since $a^n \equiv 1 \pmod{m}$, we have $a^r \equiv 1 \pmod{m}$. But since $r < s$, this contradicts the minimal choice of s, unless $r = 0$, in which case $s|n$.

(399) (Brillhart, Lehmer and Selfridge, 1975). It is clear that we only need to prove that $\phi(n) = n - 1$. Since $\phi(n) \leq n - 1$, it is enough to show that $n - 1|\phi(n)$. But if $n - 1$ does not divide $\phi(n)$, then there exists a prime power q^α such that q^α divides $n - 1$ but does not divide $\phi(n)$. If e is the smallest exponent such that $a^e \equiv 1 \pmod{n}$ (see Problem 398), then $e|n-1$, but e does not divide $(n-1)/q$, in which case $q^\alpha|e$. Since $a^{\phi(n)} \equiv 1 \pmod{n}$, we have that $e|\phi(n)$, and this is why $q^\alpha|\phi(n)$, which contradicts the above statement.

(400) With MATHEMATICA, we write

```
n=10^12+61;w={2,5,3947,12667849};
Do[Print[q=w[[i]]," ",PowerMod[7,(n-1)/q,n]],{i,1,4}]
```

We then obtain the following table:

q	$7^{(n-1)/q} \pmod{n}$
2	1000000000060
5	49990566449
3947	818653818766
12667849	991362440375

so that indeed $7^{(n-1)/q} \not\equiv 1 \pmod{n}$ for each prime divisor q of $n - 1$. It then follows from Lucas' Test that n is prime.

(401) Since $n - 1 = r^4$, it is enough to find, for each r, a number a such that $a^{r^4} \equiv 1 \pmod{n}$ and such that $a^{r^4/q} \not\equiv 1 \pmod{n}$ for each prime divisor q of r. In the case $r = 1910 = 2 \cdot 5 \cdot 191$, choosing $a = 13$ is convenient. In the case $r = 1916 = 4 \cdot 479$, the choice $a = 3$ is appropriate. Finally, in the case $r = 1926 = 2 \cdot 3^2 \cdot 107$, choosing $a = 5$ is adequate. Thus the result.

(402) With MATHEMATICA, we write

```
n=10^12+63;w={2,3,7,47,168861871};
Do[Print[q=w[[i]]," ",PowerMod[5,(n-1)/q,n]],{i,1,5}]
```

We then obtain the following table:

q	$5^{(n-1)/q} \pmod{n}$
2	1000000000062
3	144114448610
7	385212254787
47	465337403973
168861871	578084528999

Hence by Lucas' Test, we conclude that n is prime.

(403) By Fermat's Little Theorem, we have that $2^{p-1} \equiv 1 \pmod{p}$. Therefore, since $p - 1|k!$, we have $2^{k!} \equiv 1 \pmod{p}$ and therefore $m \equiv 1 \pmod{p}$. We

have thus shown that $p|m-1$. This implies that $p|g$ and therefore that $g > 1$.

(404) Using the following MATHEMATICA program

```
n=2^(2^9)+1;k=40;i=2;s=3;
While[i<=k,{s=PowerMod[s,i,n]; i++}];s=s-1; b=GCD[s,n];
Print[b]
```

we then find the prime factor 2424833.

(405) Let $n = 252123019542987435093029$. Using MATHEMATICA,

```
n=252123019542987435093029;k=500;i=2;s=2; While[i<=k,
{s=PowerMod[s,i,n];i++}]; s=s-1;b=GCD[s,n];Print[b]
```

we obtain the prime factor $p = 252097807237$. Since $n/p = 1000100010017$ is a prime number, the complete factorization of n is

$$252123019542987435093029 = 252097807237 \cdot 1000100010017.$$

(406) The following MATHEMATICA program

```
n=2^71-1;k=100000;i=2;s=3;
While[i<=k,{s=PowerMod[s,i,n];
If[IntegerQ[i/100]&&
((p=GCD[s-1,n])>1),
Print[i," ",p," ",n=n/p] ];i++}]
```

gives the following table:

i	p	n/p
1700	228479	10334355636337793
7400	212885833	48544121
17100	48544121	1

which establishes the factorization

$$2^{71} - 1 = 228479 \cdot 48544121 \cdot 212885833.$$

(407) The factorization is $136258390321 = 104831 \cdot 1299791$.

(408) We first find $m = 269\,146\,942$ and $g = (m - 1, n) = 17\,389$. Applying the Pollard $p - 1$ Test, we obtain the result.

(409) By hypothesis, we have

$$2^{2^n} \equiv -1 \pmod{p} \quad \text{and therefore} \quad 2^{2^{n+1}} \equiv 1 \pmod{p}.$$

Therefore, 2^{n+1} is the smallest exponent such that $2^{2^{n+1}} \equiv 1 \pmod{p}$. But by Fermat's Little Theorem, $2^{p-1} \equiv 1 \pmod{p}$, which means that $2^{n+1}|p-1$ and in particular that $8|p-1$ since $n \geq 2$. Therefore, using the Euler's Criterion, we have

$$2^{\frac{p-1}{2}} \equiv \left(\frac{2}{p}\right) \equiv 1 \pmod{p}.$$

It follows that $2^{n+1} \left| \dfrac{p-1}{2} \right.$ and therefore that $2^{n+2}|p-1$. We have thus proved that there exists a positive integer k such that $p - 1 = k \cdot 2^{n+2}$, thus the result.

(410) In light of Problem 409, each prime factor p of F_5 is of the form $p = k \cdot 2^7 + 1$. The first value of k for which the corresponding number $p = k \cdot 2^7 + 1$

divides F_5 is $k = 5$, which gives $p = 641$, and we conclude that 641 is a prime divisor of F_5.

(411) In light of Problem 409, each prime factor p of F_6 is of the form $p = k \cdot 2^8 + 1$. Using a computer, we verify that the first value of k such that $F_6/(k \cdot 2^8 + 1)$ is an integer is $k = 1\,071$. The number p corresponding to $k = 1\,071$ is $p = 274\,177$, which is indeed a prime number. Thus the result.

(412) First assume that F_n is prime and that $\left(\dfrac{k}{F_n}\right) = -1$. Then, by Euler's Criterion, we have

$$k^{\frac{F_n-1}{2}} \equiv \left(\frac{k}{F_n}\right) \equiv -1 \pmod{F_n}.$$

Reciprocally, if $k^{\frac{F_n-1}{2}} \equiv -1 \pmod{F_n}$, let r be the residue of k modulo F_n. Since $r^{\frac{F_n-1}{2}} \equiv -1 \pmod{F_n}$, we have that $r^{F_n-1} \equiv 1 \pmod{F_n}$. Using the result of Problem 399, it follows that F_n is prime. Therefore,

$$\left(\frac{k}{F_n}\right) \equiv k^{\frac{F_n-1}{2}} \equiv -1 \pmod{F_n}.$$

(413) Let $\alpha = n + a$, with $n \in \mathbb{Z}$ and $0 \le a < 1$, and let $\beta = m + b$, $0 \le b < 1$.

(a) Proving this inequality amounts to proving that $[a+b] \le [2a]+[2b]$. If $0 \le a + b < 1$, the result is immediate. If $1 \le a + b$, then $2a + 2b \ge 2$ and therefore $[2a] \ge 1$ or $[2b] \ge 1$, and the result follows.

(b) It is enough to show that $2[a + b] \le [3a] + [3b]$. If $0 \le a + b < 1$, the result follows. On the other hand, if $a + b \ge 1$, then we must show that $2 \le [3a] + [3b]$. Clearly, $3a + 3b \ge 3$. Now, since $[x] + [y] \ge [x+y] - 1$ for all $x, y \in \mathbb{R}$, it follows that $[3a] + [3b] \ge [3(a+b)] - 1 \ge 2$, which gives the result.

(c) It is enough to show that $3[a + b] \le [4a] + [4b]$. If $0 \le a + b < 1$, the result is immediate. On the other hand, if $a + b \ge 1$, then $4a + 4b \ge 4$ and since $[4a] + [4b] \ge [4a + 4b] - 1 \ge 3$, we obtain the result.

Inequalities (d) and (e) are obtained in a similar manner.

(414) This follows by observing that

$$\frac{(2n)!}{(n!)^2} = \binom{2n}{n} = 2\binom{2n-1}{n-1},$$

for each integer $n \ge 1$.

(415) For part (a), in light of Theorem 27, it is enough to show that

$$\left[\frac{2m}{p^i}\right] + \left[\frac{2n}{p^i}\right] \ge \left[\frac{m}{p^i}\right] + \left[\frac{n}{p^i}\right] + \left[\frac{m+n}{p^i}\right],$$

an inequality which is a consequence of Problem 413(a).

For part (b), again in light of Theorem 27, it is enough to show that

$$\left[\frac{4m}{p^i}\right] + \left[\frac{4n}{p^i}\right] \ge \left[\frac{m}{p^i}\right] + \left[\frac{n}{p^i}\right] + 3\left[\frac{m+n}{p^i}\right],$$

which itself follows from Problem 413(c).

(416) Because of Theorem 27, it is enough to show that

$$\sum_{i=1}^{\infty} \left[\frac{n}{p^i} \right] \geq \sum_{i=1}^{\infty} \left[\frac{a_1}{p^i} \right] + \cdots + \sum_{i=1}^{\infty} \left[\frac{a_r}{p^i} \right].$$

Since

$$\left[\frac{a_1}{p^i} \right] + \cdots + \left[\frac{a_r}{p^i} \right] \leq \left[\frac{a_1 + \cdots + a_r}{p^i} \right] = \left[\frac{n}{p^i} \right],$$

we obtain the result after summing on i.

(417) To obtain the number of zeros placed at the end of the number 23!, we must find the largest number α such that $10^{\alpha} \| 23!$. Since 23! contains more 2's than 5's, it is enough to compute the largest power of 5 which divides 23!. We are therefore looking for the largest integer α such that $5^{\alpha} \| 23!$. This number α is given by

$$\alpha = \left[\frac{23}{5} \right] = 4.$$

There are therefore four zeros at the end of 23!.

(418) We may write $n! = 2^a 5^b m$ where $(m, 10) = 1$. The largest power of 10 which divides $n!$ is b and since $n!/10^b = 2^{a-b} m$ is an even integer, the result follows.

(419) The integers n whose number of zeros appearing at the end of the decimal expansion of $n!$ is 57 are those whose largest power of 5 which divides $n!$ is 57. We are therefore looking for an integer n such that $S_n = 57$ where $S_n = \left[\frac{n}{5} \right] + \left[\frac{n}{5^2} \right] + \left[\frac{n}{5^3} \right] + \cdots$. If $n = 200$, then $S_n = 49$; if $n = 250$, then $S_n = 62$. Hence, if $S_n = 57$, we must search amongst integers n such that $200 < n < 250$. We write $n = 125 + 25a + 5b + c$, where $a = 3$ or 4, $0 \leq b \leq 4$ and $0 \leq c \leq 4$. Then, $S_n = 1 + (5+a) + (25+5a+b) = 31 + 6a + b$; thus $S_n = 57$ if and only if $6a + b = 26$. Since $0 \leq b \leq 4$, we must have $a = 4$ and $b = 2$. Hence, $n = 235 + c$ where $c = 0, 1, 2, 3, 4$. We conclude that the integers n whose number of zeros appearing at the end of the decimal expansion of $n!$ is 57 are 235, 236, 237, 238 and 239.

For the second part, we proceed as with the first. We only need to consider the largest power of 5 which divides $n!$. For $n = 249$,

$$\left[\frac{249}{5} \right] + \left[\frac{249}{25} \right] + \left[\frac{249}{125} \right] = 59,$$

and for $n = 250$ the corresponding sum is 62. This shows that there does not exist any integer whose number of zeros appearing at the end of the decimal expansion of $n!$ is 60 or 61.

(420) (a) The number α is given by

$$\alpha = \left[\frac{5^n - 3}{5} \right] + \left[\frac{5^n - 3}{5^2} \right] + \left[\frac{5^n - 3}{5^3} \right] + \cdots + \left[\frac{5^n - 3}{5^{n-1}} \right].$$

Since $[x + m] = [x] + m$ for $m \in \mathbb{N}$ and since $[-x] = -[x] - 1$ for $x \notin \mathbb{Z}$, we have

$$\alpha = 5^{n-1} + 5^{n-2} + \cdots + 5 + \left[\frac{-3}{5} \right] + \left[\frac{-3}{5^2} \right] + \cdots + \left[\frac{-3}{5^{n-1}} \right]$$

$$= \frac{5}{4}(5^{n-1} - 1) - (n - 1) = \frac{5^n - 4n - 1}{4}.$$

(b) We have

$$\alpha = \left[\frac{p^n - i}{p}\right] + \left[\frac{p^n - i}{p^2}\right] + \left[\frac{p^n - i}{p^3}\right] + \cdots + \left[\frac{p^n - i}{p^{n-1}}\right],$$

and since $[-x + m] = [-x] + m = -[x] - 1 + m$ if $x \notin \mathbb{Z}$, then

$$\alpha = p + p^2 + \cdots + p^{n-1} + \left[\frac{-i}{p}\right] + \left[\frac{-i}{p^2}\right] + \cdots + \left[\frac{-i}{p^{n-1}}\right]$$

$$= \frac{p^n - p}{p - 1} - (n - 1) = \frac{p^n - (p-1)n - 1}{p - 1}.$$

(421) We must consider two cases. If $p = 2$, we have $\alpha = n + \sum_{j=1}^{\infty}\left[\frac{n}{2^j}\right]$. If $p > 2$,

we have $\alpha = \sum_{j=1}^{\infty}\left[\frac{n}{p^j}\right]$.

(422) It is clear that if $p = 2$, then $\alpha = 0$. We may therefore limit our search to the case $p > 2$. Since

$$\prod_{i=0}^{n}(2i + 1) = 1 \cdot 3 \cdot 5 \cdots (2n - 1) \cdot (2n + 1) = \frac{(2n + 1)!}{2^n n!},$$

it is then easy to see that

$$\alpha = \sum_{j=1}^{\infty}\left[\frac{2n + 1}{p^j}\right] - \sum_{j=1}^{\infty}\left[\frac{n}{p^j}\right].$$

Moreover, since the largest power of 2 which divides $\dfrac{(2n + 1)!}{2^n n!}$ is 0, we have

$$0 = \sum_{k=1}^{\infty}\left[\frac{2n + 1}{2^k}\right] - \sum_{k=1}^{\infty}\left[\frac{n}{2^k}\right] - n,$$

which proves the given relation.

(423) (*AMM, Vol. 82, 1975, p. 854*). In fact, we must identify each integer n which is divisible by an integer m such that $m^2 \le n < (m + 1)^2$. These numbers n are obviously those of the form m^2, $m^2 + m$ or $m^2 + 2m$.

(424) First of all, it is clear that

$$n = \sqrt{n \cdot n} < \sqrt{n(n + 1)} < \sqrt{(n + 1)^2} = n + 1,$$

so that

(1) $$2n < 2\sqrt{n(n + 1)} < 2n + 2.$$

Moreover,

$$2\sqrt{n(n + 1)} = (\sqrt{n} + \sqrt{n + 1})^2 - (2n + 1).$$

Therefore, we derive from (1) that

$$2n < (\sqrt{n} + \sqrt{n + 1})^2 - (2n + 1) < 2n + 2,$$

which implies that

$$4n + 1 < (\sqrt{n} + \sqrt{n + 1})^2 < 4n + 3$$

and therefore that

(2) $$\sqrt{4n+1} < \sqrt{n} + \sqrt{n+1} < \sqrt{4n+3}.$$

Let k be the unique integer such that

$$k^2 \leq 4n+1 < (k+1)^2.$$

Hence, between the integers k^2 and $(k+1)^2$, we find the integers $4n+1$, $4n+2$ and $4n+3$. How can we confirm that these last two numbers are not "at the right of $(k+1)^2$"? The reason is that no perfect square is of the form $4n+2$ or $4n+3$. Therefore, we must have

$$k^2 \leq 4n+1 < 4n+2 < 4n+3 < (k+1)^2.$$

Hence, it follows that

$$k \leq \sqrt{4n+1} < \sqrt{4n+2} < \sqrt{4n+3} < k+1.$$

Finally, since, in light of (2), the quantity $\sqrt{n} + \sqrt{n+1}$ shows up between $\sqrt{4n+1}$ and $\sqrt{4n+3}$, the result follows.

(425) (*AMM, Vol. 95, 1988, p. 133*). Let $x = \sqrt{n} + \sqrt{n+1} + \sqrt{n+2}$. Then,

$$x^2 = 3n + 3 + 2\left(\sqrt{n(n+1)} + \sqrt{n(n+2)} + \sqrt{(n+1)(n+2)}\right).$$

For $n \geq 1$, the inequalities $(n+2/5)^2 < n(n+1) < (n+1/2)^2$, $(n+7/10)^2 < n(n+2) < (n+1)^2$ and $(n+7/5)^2 \leq (n+1)(n+2) < (n+3/2)^2$ lead to $9n+8 < x^2 < 9n+9$, in which case $[x] = [\sqrt{9n+8}]$. The case $n = 0$ is verified directly.

(426) (*MMAG, Vol. 48, 1975, p. 292*). Let $m = ak+b$, with $a, b \in \mathbb{N}$, $0 \leq b < k$. We then have

$$\left[\frac{m}{k} - 1\right] + \left[-\frac{(m+1)}{k}\right] = \left[\frac{m}{k}\right] - 1 + \left[-\frac{(m+1)}{k}\right]$$

$$= a + \left[-\frac{(ak+b+1)}{k}\right] - 1 = a + \left[-\frac{(b+1)}{k}\right] - a - 1 = -2.$$

(427) First let $x \in \mathbb{Q}$; that is $x = \frac{a}{b}$, say. Then, $b!x \in \mathbb{N}$ and $m!x \in \mathbb{N}$ for each $m \geq b$. Hence, as soon as $m \geq b$, we have $\cos^2(m!\pi x) = 1$ and this is why $\lim_{m\to\infty}[\cos^2(m!\pi x)] = 1$. On the other hand, if $x \in \mathbb{R} \setminus \mathbb{Q}$, then $m!x \in \mathbb{R} \setminus \mathbb{Q}$, which implies that $0 \leq \cos^2(m!\pi x) < 1$ for all $m \geq 1$ and that $[\cos^2(m!\pi x)] = 0$ for all $m \geq 1$, which of course implies that $\lim_{m\to\infty}[\cos^2(m!\pi x)] = 0$.

(428) We first write n in basis 2:

$$n = e_0 + e_1 2 + e_2 2^2 + e_3 2^3 + e_4 2^4 + e_5 2^5 + \cdots,$$

where each $e_i \in \{0, 1\}$. Then, it is easy to establish that

$$\left[\frac{n+1}{2}\right] = e_0 + e_1 + e_2 2 + e_3 2^2 + e_4 2^3 + e_5 2^4 + e_6 2^5 + \cdots,$$

$$\left[\frac{n+2}{4}\right] = e_1 + e_2 + e_3 2 + e_4 2^2 + e_5 2^3 + e_6 2^4 + \cdots,$$

$$\left[\frac{n+4}{8}\right] = e_2 + e_3 + e_4 2 + e_5 2^2 + e_6 2^3 + \cdots,$$

$$\left[\frac{n+8}{16}\right] = e_3 + e_4 + e_5 2 + e_6 2^2 + \cdots,$$

and so on. By adding the respective columns, we easily obtain the result.

(429) If m is an integer satisfying this equality, it is easy to verify that $m \pm 25$ also satisfy this equality. Consequently, it is enough to show that this equation is true for all integers m such that $17 \le m \le 41$. But for such integers m, we have $[(m-17)/25] = 0$, and therefore it is enough to show that, for these integers m,

$$(*) \qquad \left[\frac{m}{3}\right] = \left[\frac{8m+13}{25}\right].$$

If $17 \le m \le 39$, then

$$\frac{m}{3} \le \frac{8m+13}{25} < \frac{m+0.9}{3}$$

and therefore $(*)$ is verified. For $m = 40$ and $m = 41$, the result is immediate.

(430) For each $m \in \mathbb{Z}$, the value of this quantity is 7. Indeed, since $[a+m] = [a] + m$ for $m \in \mathbb{Z}$, the expression of the statement can be written as

$$7 + \left[\frac{3m+4}{13}\right] - \left[\frac{m - \left[\frac{m-7}{13}\right]}{4}\right].$$

We will show that, for all $m \in \mathbb{Z}$,

$$\left[\frac{3m+4}{13}\right] = \left[\frac{m - \left[\frac{m-7}{13}\right]}{4}\right].$$

If m is an integer satisfying this equality, it is easy to verify that $m \pm 13$ also satisfy this equality. Consequently, it is enough to show that this equation is true for all integers m such that $7 \le m \le 19$. But for such integers m, we have $[(m-7)/13] = 0$, and therefore it is sufficient to show that, for these integers m,

$$(*) \qquad \left[\frac{m}{4}\right] = \left[\frac{3m+4}{13}\right].$$

If $7 \le m \le 16$, then

$$\frac{m}{4} \le \frac{3m+4}{13} < \frac{m+0.9}{4},$$

in which case $(*)$ is verified; for $m = 17, 18$ and 19, $(*)$ is immediate. The expression of the statement is therefore always equal to 7 and thus does not depend on m.

(431) We consider the two possible cases: $k \geq n/2$ and $k < n/2$. In the first case, for $1 \leq i \leq k$, we have

$$\frac{n-i}{k} < \frac{n}{k} \leq n \cdot \frac{2}{n} = 2,$$

so that

$$\left[\frac{n-j}{k} \right] \leq 1.$$

On the other hand, still with this case, we have $(n-i)/k \geq 1$ if and only if $i \leq n - k$. It follows that

$$\sum_{j=1}^{k} \left[\frac{n-j}{k} \right] = \sum_{j=1}^{n-k} \left[\frac{n-j}{k} \right] = \sum_{j=1}^{n-k} 1 = n - k,$$

which establishes $(*)$ in this first case. To study the second case, we have that there exist positive integers a and $r < k$ such that $n = ak + r$, in which case

$$\sum_{j=1}^{k} \left[\frac{n-j}{k} \right] = \underbrace{a + a + \cdots + a}_{r} + \underbrace{(a-1) + \cdots + (a-1)}_{k-r}$$

$$= ar + (a-1)(k-r) = ak - k + r = n - k,$$

which completes the proof of the second case.

(432) Let $\alpha = [\alpha] + \frac{\beta}{n}$, with $0 \leq \beta < n$, then

(1) $[n\alpha] = [n[\alpha] + \beta] = n[\alpha] + [\beta],$

while

(2) $[\alpha] + \left[\alpha + \frac{1}{n} \right] + \cdots + \left[\alpha + \frac{n-1}{n} \right] = \sum_{j=0}^{n-1} \left[[\alpha] + \frac{\beta + j}{n} \right]$

$$= n[\alpha] + \sum_{j=0}^{n-1} \left[\frac{\beta + j}{n} \right] = n[\alpha] + \sum_{j=1}^{n} \left[\frac{\beta + n - j}{n} \right]$$

$$= n[\alpha] + \sum_{j=1}^{[\beta]} \left[\frac{\beta + n - j}{n} \right] + \sum_{j=[\beta]+1}^{n} \left[\frac{\beta + n - j}{n} \right]$$

$$= n[\alpha] + \sum_{j=1}^{[\beta]} 1 = n[\alpha] + [\beta].$$

The result then follows by comparing (1) and (2).

(433) It is enough to replace α by α/n in Problem 432.

(434) Let

$$S = \{(x, y) \in \mathbb{N} \times \mathbb{N} \mid 1 \leq x \leq n - 1, 1 \leq y \leq m - 1\}.$$

It is clear that $\#S = (n-1)(m-1)$. Now let $S_1 = \{(x, y) \in S \mid mx > ny\}$ and $S_2 = \{(x, y) \in S \mid mx < ny\}$. Since $(m, n) = 1$, there is no point with integer coordinates on the line $y = mx/n$, and this is why

$$S_1 \cup S_2 = S \quad \text{and} \quad S_1 \cap S_2 = \emptyset,$$

and by symmetry we have

$$\#S_1 = \#S_2 = \frac{\#S}{2} = \frac{(n-1)(m-1)}{2}.$$

For a fixed integer x chosen arbitrarily in the interval $[1, n-1]$, there is exactly $\left[\dfrac{mx}{n}\right]$ points with integer coordinates of abscissa x located on the line $mx = ny$. Therefore,

$$\#S_1 = \sum_{x=1}^{n-1} \sum_{1 \le y < mx/n} 1 = \sum_{x=1}^{n-1} \left[\frac{mx}{n}\right] = \frac{(n-1)(m-1)}{2},$$

as required.

(435) Let

$$S = \{(x,y) \in \mathbb{N} \times \mathbb{N} \mid 1 \le x \le n-1, 1 \le y \le m-1\}.$$

It is clear that $\#S = (n-1)(m-1)$. Let

$$S_1 = \{(x,y) \in S \mid mx \ge ny\} \quad \text{and} \quad S_2 = \{(x,y) \in S \mid mx \le ny\}.$$

Since $y = mx/n = m_1 x/n_1$, where $m = m_1 d$, $n = n_1 d$, it follows that for $x < n$, there are $d-1$ points with integer coordinates on the diagonal. For a fixed integer x chosen arbitrarily in the interval $[1, n-1]$, there are exactly $\left[\dfrac{mx}{n}\right]$ points with integer coordinates of abscissa x located on the line $mx = ny$ and $d-1$ on the diagonal. Therefore,

$$S_1 = \sum_{x=1}^{n-1} \sum_{1 \le y \le mx/n} 1 = \sum_{x=1}^{n-1} \left[\frac{mx}{n}\right],$$

and similarly

$$S_2 = \sum_{y=1}^{m-1} \sum_{1 \le x \le ny/m} 1 = \sum_{y=1}^{m-1} \left[\frac{ny}{m}\right].$$

Hence,

$$\#S_1 + \#S_2 = (m-1)(n-1) + (d-1),$$

and since the number of points with integer coordinates is the same for both sets, we obtain the result.

(436) The number of points with integer coordinates inside or on the contour of the rectangle formed by $1 \le x \le n$ and $1 \le y \le m$ is mn. Let $y = mx/n$ where $(m, n) = d$. Then, on the diagonal, there are $d = (m, n)$ points with integer coordinates. Since both sums

$$\sum_{j=1}^{m} \left[\frac{jn}{m}\right] \quad \text{and} \quad \sum_{j=1}^{n} \left[\frac{jm}{n}\right]$$

contain (m, n) integer coordinates on the diagonal, it follows that

$$(*) \qquad mn = \sum_{j=1}^{m} \left[\frac{jn}{m}\right] + \sum_{j=1}^{n} \left[\frac{jm}{n}\right] - (m, n).$$

But using the result of Problem 435,

$$\sum_{j=1}^{m-1}\left[\frac{jn}{m}\right]=\sum_{j=1}^{n-1}\left[\frac{jm}{n}\right].$$

Combining this last relation with $(*)$, we easily obtain the result.

(437) (*CRUX, 1988*). Observe that

$$\sqrt{(n+1)^2+(n+1)}=\sqrt{(n+1)(n+2)}.$$

But

$$(n+1)^2<(n+1)(n+2)<(n+2)^2.$$

Hence,

$$n+1<\sqrt{(n+1)(n+2)}<n+2,$$

and therefore

$$\left[\sqrt{(n+1)(n+2)}\right]=n+1.$$

Hence,

$$\left[\sqrt{(n+1)^2+n+1}\right]^2=(n+1)^2,$$

and it follows that

$$f(n)=(n+1)^2+n-(n+1)^2=n,$$

which proves that $f(n)-n=0$ for each positive integer n.

(438) From Theorem 27, for each prime number $p\le n$, we have

$$(1)\qquad\qquad\alpha_p=\sum_{i=1}^{\infty}\left[\frac{n}{p^i}\right].$$

So let $p\le n$, p fixed. Writing $n=[d_1,d_2,\ldots,d_k]_p$, where $d_1,d_2,\ldots,d_k$ are the digits of n in basis p, we have successively

$$
\begin{aligned}
(2)\qquad n&=d_1p^{k-1}+d_2p^{k-2}+\cdots+d_{k-1}p+d_k,\\
\left[\frac{n}{p}\right]&=d_1p^{k-2}+d_2p^{k-3}+\cdots+d_{k-1},\\
\left[\frac{n}{p^2}\right]&=d_1p^{k-3}+d_2p^{k-4}+\cdots+d_{k-2},\\
\left[\frac{n}{p^3}\right]&=d_1p^{k-4}+d_2p^{k-5}+\cdots+d_{k-3},\\
&\vdots\qquad\vdots\\
\left[\frac{n}{p^{k-2}}\right]&=d_1p+d_2,\\
\left[\frac{n}{p^{k-1}}\right]&=d_1.
\end{aligned}
$$

Combining (1) and (2), we obtain

$$\alpha_p=d_1\left(1+p+p^2+\cdots+p^{k-2}\right)+d_2\left(1+p+p^2+\cdots+p^{k-3}\right)$$
$$+\cdots+d_{k-2}(1+p)+d_{k-1}.$$

Multiplying each side of this last equation by $(p-1)$, we then have

$$
\begin{aligned}
(p-1)\alpha_p \\
&= d_1(p^{k-1}-1) + d_2(p^{k-2}-1) + \cdots + d_{k-2}(p^2-1) + d_{k-1}(p-1) \\
&= d_1 p^{k-1} + d_2 p^{k-2} + \cdots + d_{k-1}p - (d_1 + d_2 + \cdots + d_{k-1}) \\
&= d_1 p^{k-1} + d_2 p^{k-2} + \cdots + d_{k-1}p + d_k - (d_1 + d_2 + \cdots + d_{k-1} + d_k) \\
&= n - s_p(n),
\end{aligned}
$$

as was to be shown.

(439) (*Putnam, December 2001*). Since $\|x\| = [x+1/2]$, we first want to evaluate $[\sqrt{n} + 1/2]$. But we know that for each $x \in \mathbb{R}^+$, there exists a positive integer m such that $m - \frac{1}{2} \le x < m + \frac{1}{2}$, that is $[x + \frac{1}{2}] = m$. Setting $x = \sqrt{n}$, we obtain

$$
\sqrt{n} < \frac{2m+1}{2} \iff n \le \frac{4m^2 + 4m + 1}{4} = m^2 + m + \frac{1}{4},
$$

that is $n \le m(m+1)$, and

$$
\sqrt{n} \ge m - \frac{1}{2} \iff n \ge m^2 - m + \frac{1}{4},
$$

that is $n \ge m^2 - m + 1$. Using this observation and letting S be the series to evaluate, we obtain successively that

$$
\begin{aligned}
S &= \sum_{m=1}^{\infty} \left(2^m + \frac{1}{2^m}\right) \sum_{n=m^2-m+1}^{m^2+m} \frac{1}{2^n} \\
&= 2\sum_{m=1}^{\infty} \left(2^m + \frac{1}{2^m}\right) \frac{1}{2^{m^2-m+1}} \left(1 - \frac{1}{2^{2m}}\right) \\
&= 2\left\{ \sum_{m=1}^{\infty} \frac{1}{2^{m^2-2m+1}} \left(1 - \frac{1}{2^{2m}}\right) + \sum_{m=1}^{\infty} \frac{1}{2^{m^2+1}} \left(1 - \frac{1}{2^{2m}}\right) \right\} \\
&= 2\left\{ \sum_{m=1}^{\infty} \frac{1}{2^{(m-1)^2}} \left(1 - \frac{1}{2^{2m}}\right) + \frac{1}{2}\sum_{m=1}^{\infty} \frac{1}{2^{m^2}} \left(1 - \frac{1}{2^{2m}}\right) \right\} \\
&= 2\left\{ \sum_{m=1}^{\infty} \frac{1}{2^{(m-1)^2}} - \frac{1}{2}\sum_{m=1}^{\infty} \frac{1}{2^{m^2}} + \frac{1}{2}\sum_{m=1}^{\infty} \frac{1}{2^{m^2}} - \sum_{m=1}^{\infty} \frac{1}{2^{(m+1)^2}} \right\} \\
&= 2\left\{ 1 + \frac{1}{2} + \sum_{m=3}^{\infty} \frac{1}{2^{(m-1)^2}} - \sum_{m=1}^{\infty} \frac{1}{2^{(m+1)^2}} \right\} \\
&= 2 \cdot \frac{3}{2} = 3.
\end{aligned}
$$

(440) Let $\{x\} = x - [x]$ be the fractional part of x. Since

$$
2\left\{\frac{n}{2}\right\} + 1 = \begin{cases} 1 & \text{if } n \text{ is even,} \\ 2 & \text{if } n \text{ is odd,} \end{cases}
$$

it is easy to see that $f(n) = 2\{\frac{n+1}{2}\} + 1$ and therefore that

$$f(n) = 2\left\{\frac{n+1}{2}\right\} + 1 = 2\left(\frac{n+1}{2} - \left[\frac{n+1}{2}\right]\right) + 1$$

$$= n + 1 - 2\left[\frac{n+1}{2}\right] + 1.$$

The required formulation is therefore

$$f(n) = n + 2 - 2\left[\frac{n+1}{2}\right].$$

(441) Let us start with the computation of A_n and examine the first terms of this sum. We have

$$A_n = [1^{1/2}] + [2^{1/2}] + [3^{1/2}] + [4^{1/2}] + [5^{1/2}] + [6^{1/2}] + [7^{1/2}] + [8^{1/2}]$$
$$+ [9^{1/2}] + \cdots = 1 + 1 + 1 + 2 + 2 + 2 + 2 + 2 + 3 + \cdots .$$

We quickly notice that the number of times that the integer i ($1 \leq i \leq n-1$) appears in A_n is equal to $(i+1)^2 - i^2$, so that

$$A_n = 1(2^2 - 1^2) + 2(3^2 - 2^2) + 3(4^2 - 3^2) + \cdots + (n-1)(n^2 - (n-1)^2),$$

an expression which can be simplified as follows:

$$A_n = 2^2 - 1^2 + 2 \cdot 3^2 - 2 \cdot 2^2 + 3 \cdot 4^2 - 3 \cdot 3^3 + \cdots$$
$$+ (n-1)n^2 - (n-1)(n-1)^2$$
$$= -1^2 - 2^2 - 3^2 - \cdots - (n-1)^2 + (n-1)n^2$$
$$= (n-1)n^2 - \frac{(n-1)n(2n-1)}{6} = \frac{n(n-1)(4n+1)}{6},$$

where we have called upon the first formula given in the statement of the problem. We have thus established that

$$A_n = \frac{n(n-1)(4n+1)}{6} \qquad (n = 2, 3, 4, \ldots).$$

Proceeding in the same manner, this time by calling upon the second formula given in the statement of the problem, we prove that

$$B_n = \frac{(n-1)n^2(3n+1)}{4} \qquad (n = 2, 3, 4, \ldots).$$

(442) Let $y = an - [an]$. Since $a^2 = a + 1$, we have

$$a^2 n - [a^2 n] = (a+1)n - [(a+1)n] = an - [an] = y.$$

Hence,

$(*) \quad -y/a = y(1-a) = (a^2 n - [a^2 n]) - (a^2 n - a[an]) = a[an] - [a^2 n],$

and since $0 < y < 1 < a = (1 + \sqrt{5})/2$, by taking integer parts on each side of $(*)$, we obtain the result.

(443) It is enough to observe that the solution a also verifies equation $x^3 - 2x - 1 = 0$. Hence, setting $y = 2an - [2an]$, we obtain $a^3 n - [a^3 n] = (2a+1)n - [(2a+1)n] = y$, and this is why

$(*) \qquad\qquad -\dfrac{y}{a} = y(2 - a^2) = a^2[2an] - 2[a^3 n].$

Since $0 < y < 1 < a$, taking integer parts on each side of $(*)$, we obtain the result.

(444) Under the conditions of the system, the inequality $xy \leq n$ is equivalent to $x \leq n/y$ and, for y fixed, the number of such x's is $[n/y]$. As y varies from 1 to n, we therefore obtain that

$$N = \left[\frac{n}{1}\right] + \left[\frac{n}{2}\right] + \cdots + \left[\frac{n}{n}\right].$$

To obtain the other expression to which N must be equal, we only need to observe that the number of points with integer coordinates located on the vertical $x = k$ and below the curve $xy = n$ is equal to $[n/k]$.

(445) The number of integers amongst the numbers $1, 2, \ldots, n$ which are divisible by 2^k and not by 2^{k+1} is $[n/2^k] - [n/2^{k+1}]$. Since $[x] + [x + \frac{1}{2}] = [2x]$, it follows that setting $x = n/2^{k+1}$, we have that $[n/2^k] - [n/2^{k+1}] = [n/2^{k+1} + 1/2]$. If we sum this expression for $k = 0, 1, 2, \ldots$, we have then counted all integers from 1 to n.

(446) This follows from the fact that $\dfrac{[n\alpha]}{n} = \dfrac{n\alpha - \{n\alpha\}}{n} = \alpha - \dfrac{\{n\alpha\}}{n} = \alpha + \dfrac{\theta}{n}$, where $\{y\}$ stands for the fractional part of y and $0 \leq \theta < 1$.

(447) Let m be the unique positive integer such that $m^k \leq [\alpha] \leq \alpha < (m+1)^k$. Then, $[\sqrt[k]{\alpha}] = m = \left[\sqrt[k]{[\alpha]}\right]$.

(448) The following is a program written with MAPLE
```
> with(numtheory):
> F:=proc(fonct,n)
> local r:
> r:=divisors(n);
> sum(fonct(r[i]), i=1..tau(n));
> end:
```
and we write F(phi,n);

(449) The following is a program written with MAPLE
```
> for i from 1 to 1000 do
> if irem(F(tau,i),3) <> 0 then print(ifactor(i))
> else fi; od:
```

(450) The following program written with MAPLE generates perfect numbers.
```
> for i to 89 do
> Mersenne := 2^i - 1;
> if isprime(Mersenne)
> then print(sigma(i) = 2^(i - 1)*Mersenne)
> else fi; od;
```
In both cases, the answer is NO.

(451) This follows from the definition. For the second part, kf is not a multiplicative function, except when $k = 1$, because $kf(1) = kf(1 \cdot 1) = kf(1) \cdot kf(1)$, in which case $k^2 = k$. When $k = 0$, the function $0f = 0$ is not multiplicative because $0(1) = 0 \neq 1$.

Finally, $f + g$ is not necessarily multiplicative: for instance, consider $f = g = 1$.

(452) The answer is NO. Indeed, $f(2 \cdot 3 \cdot 5) = f(2) \cdot f(3) \cdot f(5) = 0$ and $f(3 \cdot 5 \cdot 7) = f(3) \cdot f(5) \cdot f(7) = 1$ implies that $f(2) = 0$, which contradicts the fact that $f(2 \cdot 5 \cdot 7) = f(2) \cdot f(5) \cdot f(7) = 1$.

(453) By definition, we have

$$\frac{(k-1)k}{2} < n \le \frac{k(k+1)}{2},$$
$$(k-1)k < 2n \le k(k+1).$$

Since $(k-1)k$ and $2n$ are two even integers, we may write $(k-1)k \le 2n - 2 < k^2 + k$, so that, by setting $N = n - 1$, we have successively

$$(k-1)k \le 2N < k^2 + k,$$
$$4k^2 - 4k \le 8N < 4k^2 + 4k,$$
$$(2k-1)^2 \le 8N + 1 < (2k+1)^2,$$
$$2k - 1 \le \sqrt{8N+1} < 2k + 1,$$
$$2k \le 1 + \sqrt{8N+1} < 2k + 2,$$
$$k \le \frac{1 + \sqrt{8N+1}}{2} < k + 1,$$
$$k \le \frac{1 + \sqrt{8n-7}}{2} < k + 1.$$

It follows from this that

$$k = \left[\frac{1 + \sqrt{8n-7}}{2}\right] = \left\|\frac{1}{2}\sqrt{8n-7}\right\|,$$

since $\|x\| = [x + \frac{1}{2}]$. Because $f(n) = 1/k$, the result follows.

(454) Since

$$[\sqrt{n} - 1] = \sum_{\substack{a^2 \le n \\ a \ge 2}} 1, \quad [\sqrt[3]{n} - 1] = \sum_{\substack{a^3 \le n \\ a \ge 2}} 1, \quad [\sqrt[4]{n} - 1] = \sum_{\substack{a^4 \le n \\ a \ge 2}} 1,$$

and so on, we have that

$$f(n) = \sum_{k \ge 2} \sum_{\substack{a^k \le n \\ a \ge 2}} 1.$$

Let α be an arbitrary positive integer. If we can find an integer n such that

$$(*) \qquad\qquad f(n) - f(n-1) = \alpha,$$

then the result will be proved. In fact, it is enough to choose

$$n = 2^{2^\alpha}.$$

Indeed, since

$$n = \left(2^{2^\alpha}\right)^{2^0} = \left(2^{2^{\alpha-1}}\right)^{2^1} = \left(2^{2^{\alpha-2}}\right)^{2^2} = \ldots = \left(2^{2^{\alpha-(\alpha-1)}}\right)^{2^{\alpha-1}},$$

thereby displaying α representations of the form a^k of the integer n, we obtain $(*)$.

(455) The only totally multiplicative function is the function $\rho(n)$.

(456) If $a < \sqrt{n} < a + 1$, then $a \le \sqrt{n-1} < a + 1$, while if $\sqrt{n} = a$, we have $[\sqrt{n-1}] = a - 1$. Therefore,

$$[\sqrt{n}] - [\sqrt{n-1}] = \begin{cases} 1 & \text{if } n = a^2, a \in \mathbb{N}, \\ 0 & \text{otherwise.} \end{cases}$$

It follows that f is multiplicative. However, f is not completely multiplicative since $f(4) = 1 \neq f(2)f(2)$.

(457) The functions $\gamma(n)$, $g(n)$ and $h(n)$ are strongly multiplicative.

(458) The answer is YES. First of all, it is clear that g is multiplicative. Now, for $k \geq 2$ and p prime, we have

$$
\begin{aligned}
g(p^k) &= \sum_{d|p^k} \mu^2(d)f(d) \\
&= \mu^2(1)f(1) + \mu^2(p)f(p) + \mu^2(p^2)f(p^2) + \cdots + \mu^2(p^k)f(p^k) \\
&= \mu^2(1)f(1) + \mu^2(p)f(p) + 0 + \cdots + 0 = 1 + \mu^2(p)f(p) = g(p),
\end{aligned}
$$

which proves that f is strongly multiplicative.

(459) The answer is YES. Indeed, by hypothesis, if p is an arbitrary prime number and k a positive integer, we have

$$
f(p) = f(p^k) = (f(p))^k.
$$

Hence, if $f(p) \neq 0$ and $k = 2$, it follows that $f(p) = 1$. We then have established that the only possible values of $f(p)$ are 0 and 1. But since f is entirely determined by the set of values of $f(p)$, it follows that $\{f(n) : n = 1, 2, 3, \ldots\} \subseteq \{0, 1\}$.

(460) This function g is not multiplicative. Indeed, if g were multiplicative, we would have $g(20) = g(4)g(5)$. But this last equality is not verified since $g(20) = 3$ while $g(4) = 2$ and $g(5) = 3$.

(461) Let $(*)$ be the equation which is to be proved. Set $m = q_1^{\alpha_1} \cdots q_r^{\alpha_r}$ and $n = q_1^{\beta_1} \cdots q_r^{\beta_r}$. Then,

$$
(m, n) = q_1^{\min\{\alpha_1, \beta_1\}} \cdots q_r^{\min\{\alpha_r, \beta_r\}}
$$

and

$$
[m, n] = q_1^{\max\{\alpha_1, \beta_1\}} \cdots q_r^{\max\{\alpha_r, \beta_r\}}.
$$

If f is multiplicative, we have

$$
\begin{aligned}
f((m, n))f([m, n]) = f(q_1^{\min\{\alpha_1, \beta_1\}}) \cdots f(q_r^{\min\{\alpha_r, \beta_r\}}) \cdot \\
\cdot f(q_1^{\max\{\alpha_1, \beta_1\}}) \cdots f(q_r^{\max\{\alpha_r, \beta_r\}}).
\end{aligned}
$$

Let k be an integer such that $1 \leq k \leq r$. We examine what happens with the factor q_k and its exponent. On the left-hand side of $(*)$, the contribution of the factor q_k is

$$
f(q_k^{\min\{\alpha_k, \beta_k\}})f(q_k^{\max\{\alpha_k, \beta_k\}}),
$$

while in the right-hand side of $(*)$, the contribution is

$$
f(q_k^{\alpha_k})f(q_k^{\beta_k}).
$$

Since these last two quantities are equal, the result follows.

The reciprocal is immediate because for $(m, n) = 1$, we have $[m, n] = mn$ and the equation gives $f(mn) = f(m)f(n)$, which implies that f is a multiplicative function.

(462) Part (a) follows from the fact that if $(m, n) = 1$, then

$$
\gamma(mn) = \prod_{p|mn} p = \prod_{p|m} p \cdot \prod_{p|n} p = \gamma(m)\gamma(n).
$$

Part (b) can be obtained by observing that

$$\sum_{d|n} |\mu(d)|\phi(d) = \prod_{p|n}(1 + \phi(p)) = \prod_{p|n}(1 + p - 1) = \prod_{p|n} p = \gamma(n).$$

(463) An immediate application of the abc conjecture to the numbers $n^2 - 1$, 1, n^2 (since $(n^2 - 1) + 1 = n^2$) yields

$$n^2 < M(\varepsilon) \cdot \gamma(n^2 - 1)^{1+\varepsilon} \cdot n^{1+\varepsilon},$$

which implies

$$n^{1-\varepsilon} < M(\varepsilon) \cdot \gamma(n^2 - 1)^{1+\varepsilon}.$$

We then have

$$n < M(\varepsilon)^{1/(1-\varepsilon)} \cdot \gamma(n^2 - 1)^{\frac{1+\varepsilon}{1-\varepsilon}} = M(\varepsilon)^{1/(1-\varepsilon)} \cdot \gamma(n^2 - 1)^{1+\frac{2\varepsilon}{1-\varepsilon}}.$$

Setting $\varepsilon' = \frac{2\varepsilon}{1-\varepsilon}$, we thus have that for each $\varepsilon' > 0$, there exists a positive constant $M_0(\varepsilon')$ such that for each $n \geq 2$, we have

$$n < M_0(\varepsilon') \cdot \gamma(n^2 - 1)^{1+\varepsilon'},$$

as was to be shown.

(464) From Problem 461, we have for $(m, n) = 1$,

$$f(km)f(kn) = f((km, kn))f([km, kn]) = f(k)f(kmn)$$

and therefore

$$\frac{f(km)}{f(k)}\frac{f(kn)}{f(k)} = \frac{f(kmn)}{f(k)},$$

which means that $\dfrac{f(kn)}{f(k)}$ is a multiplicative function.

(465) Let $k \in \mathbb{N} \cup \{0\}$ be defined implicitly by $f(3) = 3 + k$. Then we have successively, using the fact that f is strictly increasing and multiplicative,

$$\begin{aligned}
f(6) &= f(2)f(3) = 6 + 2k, \\
f(5) &\leq 5 + 2k, \\
f(10) &= f(2)f(5) \leq 10 + 4k, \\
f(9) &\leq 9 + 4k, \\
f(18) &= f(2)f(9) \leq 18 + 8k, \\
f(15) &\leq 15 + 8k.
\end{aligned}$$

(1)

On the other hand, since $f(3) = 3 + k$, we have that $f(5) \geq 5 + k$ and therefore that

(2) $$f(15) = f(3)f(5) \geq 15 + 8k + k^2.$$

From (1) and (2), it follows that $k = 0$ and therefore that $f(3) = 3$. We have then proved that $f(2^1 + 1) = 2^1 + 1$. Let us show that, more generally, we have

(3) $$f(2^\nu + 1) = 2^\nu + 1 \quad (\nu = 1, 2, \ldots).$$

For this, we use induction. Assume that the relation (3) is true for $\nu = r$ and show that it is then true for $\nu = r + 1$. But

$$f(2^{r+1} + 2) = f(2)f(2^r + 1) = 2(2^r + 1) = 2^{r+1} + 2.$$

Since f is strictly increasing, this means that $f(2^{r+1}+1) = 2^{r+1}+1$. Relation (3) is thus proved. The fact that f is strictly increasing then implies that $f(m) = m$ for all $m \in \mathbb{N}$.

(466) We only need to prove that $g(p^a) = ag(p)$ for each positive integer a and each prime number p. But since

$$g(p^a) = \lim_{k \to \infty} \frac{f(p^{ka})}{k} = \lim_{r \to \infty} \frac{f(p^r)}{r/a} = a \lim_{r \to \infty} \frac{f(p^r)}{r} = ag(p),$$

the claim is proved.

(467) We have $h(1) = 1$ if $f(1) = g(1) = 1$. Let $n = n_1 n_2$, where $(n_1, n_2) = 1$. If $n = dr$, then d and r have unique factorizations $d = d_1 d_2$ and $r = r_1 r_2$ such that $n_1 = d_1 r_1$ and $n_2 = d_2 r_2$. Moreover, $(d, r) = 1$ if and only if $(d_1, r_1) = 1 = (d_2, r_2)$. This shows that $h(n_1 n_2) = h(n_1)h(n_2)$, that is that the function h is multiplicative.

(468) Assume that $(m, n) = 1$ and that $[d, r] = mn$. Then d can be written in a unique way as $d = d_1 d_2$ with $d_1 | m$ and $d_2 | n$ and also $r = r_1 r_2$ with $r_1 | m$ and $r_2 | n$. In this case, $[d_1, r_1] = m$ and $[d_2, r_2] = n$, so that

$$h(mn) = \sum_{[d,r]=mn} f(d)g(r) = \sum_{\substack{[d_1,r_1]=m \\ [d_2,r_2]=n}} f(d_1)f(d_2)g(r_1)g(r_2)$$

$$= \sum_{[d_1,r_1]=m} f(d_1)g(r_1) \cdot \sum_{[d_2 r_2]=n} f(d_2)g(r_2) = h(m)h(n),$$

and the result follows.

(469) The answer is NO. Indeed, if $f(p) = 1$ for each prime number p, and $f(p^a) = 0$ for each integer $a \geq 2$ and each prime number p, then $\lim_{n \to \infty} f(n)$ does not exist.

(470) The answer is NO. Indeed, if $f(2^k) = 1$ for each positive integer k, and $f(p^k) = 0$ for each prime number $p \geq 3$ and each positive integer k, then $\lim_{n \to \infty} f(n)$ does not exist.

(471) (*This problem is a result due to Paul Erdős*). Let p and q be two arbitrary distinct prime numbers. Let k and $\ell = \ell(k)$ be two integers such that $p^k < q^\ell < p^{k+1}$. It is clear that one can find two sequences $k_1, k_2, \ldots$ and $\ell_1, \ell_2, \ldots$ (with $\ell_i = \ell_i(k_i)$ for each i) such that

(1) $p^{k_i} < q^{\ell_i} < p^{k_i+1}$ $(i = 1, 2, \ldots)$,

and this is why

(2) $k_i \log p < \ell_i \log q < (k_i + 1) \log p$ $(i = 1, 2, \ldots)$.

Dividing both sides of (2) by $\ell_i \log p$, we obtain

(3) $\frac{k_i}{\ell_i} < \frac{\log q}{\log p} < \frac{k_i}{\ell_i} + \frac{1}{\ell_i}.$

We have thus proved that the sequence $\dfrac{k_1}{\ell_1}, \dfrac{k_2}{\ell_2}, \ldots$ converges and that

(4) $\lim_{i \to \infty} \frac{k_i}{\ell_i} = \frac{\log q}{\log p}.$

Since f is monotone and totally additive, it follows from (1) that

(5) $k_i f(p) < \ell_i f(q) < (k_i + 1) f(p)$ $(i = 1, 2, \ldots)$.

Dividing both sides of (5) by $\ell_i f(p)$, we obtain

$$(6) \qquad \frac{k_i}{\ell_i} < \frac{f(q)}{f(p)} < \frac{k_i}{\ell_i} + \frac{1}{\ell_i}.$$

These equalities also confirm that the sequence $\dfrac{k_1}{\ell_1}, \dfrac{k_2}{\ell_2}, \dots$ converges and moreover that

$$(7) \qquad \lim_{i\to\infty} \frac{k_i}{\ell_i} = \frac{f(q)}{f(p)}.$$

Since the limit of a sequence is unique, it follows from (4) and (7) that

$$(8) \qquad \frac{f(q)}{f(p)} = \frac{\log q}{\log p}, \quad \text{that is that} \quad \frac{f(p)}{\log p} = \frac{f(q)}{\log q}.$$

Since the prime numbers p and q have been chosen arbitrarily, taking $p = 2$ and choosing any q, (8) implies

$$\frac{f(q)}{\log q} = \frac{f(2)}{\log 2} \quad \text{for each prime number } q.$$

Setting $c = f(2)/\log 2$, the result follows.

(472) (a) We obtain

$$f^j(2^\alpha) = \begin{cases} 2^{\alpha-j} & \text{if } j \le \alpha, \\ 1 & \text{if } j \ge \alpha. \end{cases}$$

(b) This is a particular case of the preceding question. We have indeed that $f^j(2^\alpha) = 1$ if $j \ge \alpha$.
(c) It is for $n = 7$ (with $k = 2$).
(d) If $n = 4k + 1$ (here $a \to b$ means that $f(a) = b$),

$$n = 4k + 1 \to 3n + 1 = 12k + 4 \to 6k + 2 \to 3k + 1,$$

which implies that

$$f^3(n) = 3k + 1 = 3\left(\frac{n-1}{4}\right) + 1 = \frac{3n}{4} + \frac{1}{4} < \frac{3n}{4} + 1.$$

On the other hand, if $n = 4k + 3$, we have

$$n = 4k + 3 \to 3n + 1 = 12k + 10 \to 6k + 5 \to 3(6k + 5) + 1 = 18k + 16,$$

which means that

$$f^3(n) = 18k + 16 > 4(4k + 3) = 4n.$$

(e) The integer $n = 62$ will do. We have indeed

$$62 \to 31 \to 94 \to 47 \to 142 \to 71 \to 214 \to 107 \to 322 \to 161 \to 484.$$

(f) The answer is YES.
(g) Since n is odd, we have $f(n) = 3n + 1 = 2^\alpha r$, with $\alpha \ge 1$ and r odd. It follows that $f^2(n) = 2^{\alpha-1}r$ and therefore that

$$f^3(n) = \begin{cases} 2^{\alpha-2}r & \text{if } \alpha \ge 2, \\ 3r + 1 & \text{if } \alpha = 1. \end{cases}$$

It follows from this that $f^3(n) > n$ if and only if $\alpha = 1$, which occurs if $2\|3n + 1$, and it happens with a probability of $\frac{1}{2}$.

(h) Choosing $n = 5$, we find that the iteration $f(5)$, $f^2(5)$, ... enters an endless loop, namely

$$5 \to 26 \to 13 \to 66 \to 33 \to 166 \to 83 \to 416 \to 208$$
$$\to 104 \to 52 \to 26 \to 13,$$

of which 13 is the turning point. Hence, the process never reaches 1.

(i) Let k be the integer defined by the relation $2^k < n \leq 2^{k+1}$. It is clear that $\mathsf{Syr}(n) > k$. Since $2^{k+1} \geq n$, we have $k + 1 \geq \log_2 n$ and therefore $k \geq \log_2 n - 1$. It follows that

$$\mathsf{Syr}(n) > k \geq \log_2 n - 1$$

and therefore that $\mathsf{Syr}(n) \geq \log_2 n$, as required.

(j) We proceed as with the preceding problem. However, since $f(n) = 3n + 1$, we are thereby adding a stage to the process (by passing from n to $3n + 1$) and in fact at least two more, since we more than double the number n.

(k) The result can easily be proved by successive iterations.

(l) We simply use (k).

(m) The result is essentially a generalization of parts (k) and (l).

(n) With the choice $n = 2^{\alpha+1} - 1$, and calling upon (k), we obtain

$$\frac{f^{2\alpha+1}(n)}{n} = \frac{2(3^{\alpha+1} - 1)}{2 \cdot 2^\alpha - 1} > \frac{2(3^{\alpha+1} - 1)}{2 \cdot 2^\alpha} = \frac{3^{\alpha+1} - 1}{2^\alpha} > \frac{3^{\alpha+1}}{2^{\alpha+1}}$$
$$= \left(\frac{3}{2}\right)^{\alpha+1} > M,$$

since we can choose α as large as we want.

(o) This function is obviously almost identical to the function f defined above.

(473) The answer is NO. It is enough to choose for example $a = 4$, $b = 6$ and $c = 7$.

(474) (a) It is clear that we must examine the different ways of writing 9 as a product of integers ≥ 2. The only possible choices are 9 and $3 \cdot 3$, so that $n = 2^8$ or $n = 2^2 \cdot 3^2$. It is clear that $n = 36$ is the smallest.

(b) Reasoning as above, we obtain $n = 2^4 \cdot 3 = 48$.

(c) The smallest integer is 144.

(475) Since $n > 1$, we have that $n = q_1^{a_1} \cdots q_r^{a_r}$ and $\tau(n) = (a_1 + 1) \cdots (a_r + 1) = 14 = 2 \cdot 7$. Then, either $r = 2$ with $a_1 = 1$ and $a_2 = 6$ or $r = 1$ with $a_1 = 13$. It follows that the positive numbers with exactly 14 divisors are of two kinds: the numbers pq^6, where p and q are distinct prime numbers, and the numbers p^{13}, where p is an arbitrary prime number.

(476) (a) Since $20! = 2^{18} \cdot 3^8 \cdot 5^4 \cdot 7^2 \cdot 11 \cdot 13 \cdot 17 \cdot 19$, we find

$$\tau(20!) = 2^4 \cdot 3^3 \cdot 5 \cdot 19,$$

and this is why the largest prime number dividing $\tau(20!)$ is 19.

(b) Similarly, we find

$$\sigma(20!) = (2^{19} - 1)(3^9 - 1)(5^5 - 1) \cdot 2^3 \cdot 3^4 \cdot 5 \cdot 7 \cdot 19.$$

Since $2^{19} - 1$ is a Mersenne prime larger than all the other prime factors, we conclude that $2^{19} - 1$ is the largest prime number dividing $\sigma(20!)$.

(c) 3.

(d) 61.

(477) (*Problem A–1 of Putnam, 1983*). Let $\tau(m)$ be the number of divisors of m. Then the number of positive integers n such that $n|a$ or $n|b$ is equal to

$$\tau(a) + \tau(b) - \tau((a,b)).$$

On the other hand, since the function τ is multiplicative and since $\tau(p^r) = r + 1$ for each prime number p and each integer $r \geq 1$, the number of required positive integers n is

$$\tau(10^{40}) + \tau(20^{30}) - \tau(2^{40} \cdot 5^{30}) = \tau(2^{40} \cdot 5^{40}) + \tau(2^{60} \cdot 5^{30})$$
$$-\tau(2^{40} \cdot 5^{30}) = 41^2 + 61 \cdot 31 - 41 \cdot 31 = 2301.$$

(478) Expanding the right-hand side, we obtain

$$\sum_{n=1}^{\infty} \frac{1}{2^n - 1} \;=\; \sum_{n=1}^{\infty} \left(\frac{1}{2^n} \cdot \frac{1}{1 - 1/2^n} \right) = \sum_{n=1}^{\infty} \frac{1}{2^n} \left(1 + \frac{1}{2^n} + \frac{1}{2^{2n}} + \cdots \right)$$
$$= \sum_{n=1}^{\infty} \left(\frac{1}{2^n} + \frac{1}{2^{2n}} + \frac{1}{2^{3n}} + \cdots \right) = \sum_{d_2=1}^{\infty} \sum_{d_1=1}^{\infty} \frac{1}{2^{d_1 d_2}}$$
$$= \sum_{m=1}^{\infty} \frac{1}{2^m} \sum_{d_1 d_2 = m} 1 = \sum_{m=1}^{\infty} \frac{\tau(m)}{2^m}.$$

REMARK: The value of the series is $1.6006695152\ldots$, a number which Paul Erdős [**10**] has proved to be irrational.

(479) Part (i) can easily be obtained using induction. To prove part (ii), we first observe that for each integer $k \geq 2$,

$$(*) \qquad\qquad \frac{1}{b_k} = \frac{1}{b_1 b_2 \cdots b_{k-1}} - \frac{1}{b_1 b_2 \cdots b_k},$$

an equality that follows from the fact that

$$\frac{1}{b_k} \;=\; \frac{1}{b_k - 1} - \left(\frac{1}{b_k - 1} - \frac{1}{b_k} \right) = \frac{1}{b_1 b_2 \cdots b_{k-1}} - \frac{1}{b_k(b_k - 1)}$$
$$=\; \frac{1}{b_1 b_2 \cdots b_{k-1}} - \frac{1}{b_k^2 - b_k} = \frac{1}{b_1 b_2 \cdots b_{k-1}} - \frac{1}{b_{k+1} - 1}$$
$$=\; \frac{1}{b_1 b_2 \cdots b_{k-1}} - \frac{1}{b_1 b_2 \cdots b_k},$$

where we have used (i) and the definition of the sequence (b_k).

Now using $(*)$, we obtain that for each integer $k \geq 4$,

$$\frac{1}{b_1} + \frac{1}{b_2} + \cdots + \frac{1}{b_k} \;=\; \frac{1}{2} + \left(\frac{1}{2} - \frac{1}{2 \cdot 3} \right) + \left(\frac{1}{2 \cdot 3} - \frac{1}{2 \cdot 3 \cdot 7} \right)$$
$$+ \left(\frac{1}{2 \cdot 3 \cdot 7} - \frac{1}{2 \cdot 3 \cdot 7 \cdot 43} \right)$$
$$+ \cdots + \left(\frac{1}{b_1 b_2 \cdots b_{k-1}} - \frac{1}{b_1 b_2 \cdots b_k} \right),$$
$$=\; \frac{1}{2} + \frac{1}{2} - \frac{1}{b_1 b_2 \cdots b_k},$$

and the result (ii) follows by letting k tend to $+\infty$.

To obtain the last relation, we write, using (ii),

$$g(n) = \sum_{j=1}^{\infty} \frac{n-1}{b_j} - \sum_{j=1}^{\infty} \left[\frac{n-1}{b_j} \right] = \sum_{j=1}^{\infty} \left\{ \frac{n-1}{b_j} \right\},$$

as required.

REMARK: The reader interested by this sequence will appreciate the recent papers of J.W. Sander and G. Myerson; see in particular J.W. Sander [34].

(480) Let $(m, n) = 1$. If m and n are odd, then, since τ is multiplicative,

$$\tau_1(mn) = \tau(mn) = \tau(m)\tau(n) = \tau_1(m)\tau_1(n),$$

and the result is proved in this case. On the other hand, if one of these two integers is even, say m, then there exists a positive integer α such that $m = 2^{\alpha}r$, with r odd (and $(r, n) = 1$). We then have

$$\tau_1(mn) = \tau_1(rn) = \tau(rn) = \tau(r)\tau(n) = \tau_1(2^{\alpha}r)\tau_1(n) = \tau_1(m)\tau_1(n).$$

All cases are thus covered, and the result is proved.

(481) For each prime number p, we cannot have both $p|a$ and $p|b$, and therefore either a contains the largest power of p which divides n or it does not contain any factor of n. This leaves two choices for each prime factor p, and we therefore have a total of $2^{\omega(n)}$ choices.

(482) Let $f(n) = \#\{(a, b) \mid a \geq 1, b \geq 1 \text{ and } [a, b] = n\}$. It is immediate that f is a multiplicative function. Indeed, assume that $(m, n) = 1$ and that $[a, b] = mn$. Then, a can be written in a unique way as $a = a_1 a_2$, with $a_1 | m$ and $a_2 | n$. Similarly, $b = b_1 b_2$, with $b_1 | m$ and $b_2 | n$. In this case, $[a_1, b_1] = m$ and $[a_2, b_2] = n$. We therefore only need to consider $n = p^r$, in which case we have $a = p^{\alpha}$, $b = p^{\beta}$, with $r = \max\{\alpha, \beta\}$. If $r = \alpha$, then $\beta = 0, 1, \ldots, r$ ($r + 1$ possibilities), or else $r = \beta$ and $\alpha = 0, 1, \ldots, r - 1$ (r possibilities). We therefore have a total of $2r + 1$ possibilities, and this is why $2r + 1 = \tau(p^{2r}) = \tau(n^2)$, as was to be shown.

(483) If $(a, b) = d$ and $ab = n$, we set $A = a/d$ and $B = b/d$. The integers A and B are relatively prime and satisfy $AB = n/d^2$. Conversely, if there exist relatively prime integers A and B such that $AB = n/d^2$, then setting $a = dA$ and $b = dB$, we find that $(a, b) = d$ and $ab = n$. We only need to find the number of ordered pairs (A, B) such that $(A, B) = 1$ and $AB = n/d^2$. In light of Problem 481, the number of such pairs is $2^{\omega(n/d^2)}$.

Finally, since $2^{\omega(n/d^2)}$ stands for the number of ordered pairs a, b such that $(a, b) = d$ and $ab = n$, then summing over all the d's such that $d^2 | n$, we obtain the total number of ordered pairs a, b such that $ab = n$, that is the number of divisors of n.

(484) By hypothesis, we have $n = 2^{\alpha}m$, with m odd. Then,

$$(1) \qquad \tau(2n) = \tau(2^{\alpha+1}m) = \tau(2^{\alpha+1})\tau(m) = (\alpha + 2)\tau(m).$$

On the other hand,

$$(2) \qquad \tau(n) = \tau(2^{\alpha}m) = \tau(2^{\alpha})\tau(m) = (\alpha + 1)\tau(m).$$

Combining (1) and (2), we obtain

$$\frac{\tau(2n)}{\tau(n)} = \frac{\alpha+2}{\alpha+1}.$$

(485) Let $n = \prod_{i=1}^{r} q_i^{a_i}$; then $\tau(n) = \prod_{i=1}^{r}(a_i+1)$, in which case it is clear that $\tau(n)$ is odd if and only if a_i is even. Hence, we only need to show that each a_i is even if and only if n is a perfect square. It is immediate that if each a_i is even, then n is a perfect square. Conversely, if n is a perfect square, then $n = m^2$, $m \in \mathbb{N}$. If $m = \prod_{i=1}^{r} q_i^{e_i}$, we therefore obtain that $n = \prod_{i=1}^{r} q_i^{2e_i}$, and the uniqueness of the canonical representation of n then implies that $a_i = 2e_i$ for $i = 1, 2, \ldots, r$.

(486) Let $n = \prod_{i=1}^{r} q_i^{\alpha_i}$ (with $q_1 < q_2 < \ldots < q_r$ primes, $\alpha_1, \alpha_2, \ldots, \alpha_r$ positive integers) be a number such that $\sigma(n)$ is prime. Using the formula

$$\sigma(n) = \prod_{i=1}^{r} \frac{q_i^{\alpha_i+1} - 1}{q_i - 1}$$

given in Theorem 31, it is clear that we must have $n = q^\alpha$ for a certain prime number q and a certain positive integer α, in which case $\tau(n) = \tau(q^\alpha) = \alpha + 1$. Therefore we only need to prove that $\alpha + 1$ is a prime number. Now, if it is not the case, that is if $\alpha + 1 = ab$ with $2 \leq a \leq b < \alpha + 1$, we will have

$$\sigma(n) = \sigma(q^\alpha) = \frac{q^{\alpha+1} - 1}{q - 1} = \frac{q^{ab} - 1}{q - 1} = \frac{q^a - 1}{q - 1} \cdot$$
$$\cdot (q^{a(b-1)} + q^{a(b-2)} + \cdots + q^a + 1),$$

that is the product of two integers ≥ 2, which contradicts the fact that $\sigma(n)$ is prime. Thus, the result follows.

(487) If $n = q_1^{a_1} \cdots q_r^{a_r}$, then $\sigma(n) = \prod_{i=1}^{r}(1 + q_i + \cdots + q_i^{a_i})$. If $q_i = 2$, then $Q_i = 1 + q_i + \cdots + q_i^{a_i}$ is odd. If q_i is an odd prime number, then Q_i is odd if and only if there exists an odd number of terms in Q_i. Hence, in order to have Q_i odd, a_i must be even. Since $\sigma(n)$ is odd if and only if each Q_i is odd, it follows that n must be the product of 2^k ($k \geq 0$) and of a perfect square.

(488) If d runs through the set of divisors of n, then n/d does also. Therefore, we have

$$\left(\prod_{d|n} d\right)^2 = \prod_{d|n} d \cdot \prod_{d|n} \frac{n}{d} = \prod_{d|n} n = n^{\tau(n)},$$

thus the result. When $\tau(n)$ is an odd number, the formula still holds because, as we have shown in Problem 485, n is then a perfect square.

(489) We have proved in Problem 488 that $\prod_{d|n} d = n^{\tau(n)/2}$. Therefore, calling upon the inequality comparing the geometric mean and the arithmetic

mean (see Theorem 5), we obtain

$$n = \left(\prod_{d|n} d^2\right)^{1/\tau(n)} \leq \frac{1}{\tau(n)}\sum_{d|n}d^2 = \frac{\sigma_2(n)}{\tau(n)},$$

and the result follows.

(490) The minimal value is 6. To prove this, consider separately the cases "n prime" and "n composite". If n is prime and larger than 2, then $n+1$ is composite, in which case $\tau(n+1) \geq 3$. It follows that

$$\tau(n(n+1)) = \tau(n)\cdot\tau(n+1) = 2\cdot\tau(n+1) \geq 2\cdot 3 = 6.$$

On the other hand, if n is not prime and larger than 2, then $\tau(n) \geq 3$, so that

$$\tau(n(n+1)) = \tau(n)\cdot\tau(n+1) \geq 3\cdot\tau(n+1) \geq 3\cdot 2 = 6.$$

We have thus established that $\tau(n(n+1)) \geq 6$, for all $n \geq 3$. The minimum is therefore attained when $n=3$, since in this case $\tau(n(n+1)) = 2\cdot 3 = 6$.

(491) We have proved in Problem 488 that $\prod_{d|n} d = n^{\tau(n)/2}$. Defining m and α by $n = 2^\alpha m$, with m odd, the relation

(1) $$f_1(n) = m^{\tau(m)/2}$$

is immediate. To establish the relation

(2) $$f_2(n) = \left(2^{\alpha(\alpha+1)}m^\alpha\right)^{\tau(m)/2} = (2n)^{\alpha\tau(m)/2},$$

we first observe that $f_2(n)\cdot f_1(n) = n^{\tau(n)/2}$, so that $f_2(n) = n^{\tau(n)/2}/f_1(n)$. Substituting (1) in this last equation, we easily obtain (2).

(492) (*MMAG, Vol. 48, 1975, p. 185*). This is equivalent to showing that

$$\sum_{m=1}^n \left(2[n/m]-\tau(m)\right)\log m = 0.$$

But this follows from

$$\sum_{m=1}^n 2[n/m]\log m = \sum_{i=1}^n [n/i]\log i + \sum_{j=1}^n [n/j]\log j$$

$$= \sum_{i=1}^n \log i \sum_{j=1}^{[n/i]} 1 + \sum_{j=1}^n \log j \sum_{i=1}^{[n/j]} 1$$

$$= \sum_{m=1}^n \sum_{ij=m} (\log i + \log j) = \sum_{m=1}^n \log m \sum_{d|m} 1$$

$$= \sum_{m=1}^n \tau(m)\log m.$$

(493) To each divisor $d \leq \sqrt{n}$ we can associate the divisor n/d, which is therefore $\geq \sqrt{n}$. Since these two categories of divisors end up covering all the divisors of n, it is clear that $\tau(n) \leq 2[\sqrt{n}]$. If n is a perfect square, then $\sqrt{n}$ is a divisor of n and the quantity $n/\sqrt{n}$ does not introduce any new divisor of n; this is why in this case we have $\tau(n) \leq 2[\sqrt{n}]-1 < 2\sqrt{n}$. On

the other hand, if n is not a perfect square, then $\sqrt{n}$ is not an integer, so that $[\sqrt{n}] < \sqrt{n}$ and therefore $\tau(n) \leq 2[\sqrt{n}] < 2\sqrt{n}$. Hence, in all cases, we have

$$\tau(n) < 2\sqrt{n}.$$

Finally, for $n \geq 5$, we have $2\sqrt{n} < n$, which implies that $\tau(n) < n$. Since $\tau(3) = 2 < 3$ and $\tau(4) = 3 < 4$, we have that

$$\tau(n) < n \qquad (n \geq 3).$$

It follows that the sequence

$$n, \ \tau(n), \ \tau(\tau(n)), \ \tau(\tau(\tau(n))), \ldots$$

is strictly decreasing as long as its terms are larger than 2. On the other hand, it is clear that the number 1 does not show up in any sequence, except of course in the constant sequence 1,1,1,.... It is therefore easy to see that each sequence decreases until it reaches the number 2, after which the number 2 repeats itself indefinitely. We know that $\tau(n) = 2$ if and only if n is prime. Therefore, the sequences whose first element is a prime number p do not generate any perfect square, since they are as follows:

$$p, 2, 2, 2, 2, 2, \ldots .$$

In fact, these are the only sequences which do no generate any perfect square. Indeed, let us examine such a sequence starting with the first prime number "2" appearing in the sequence. This prime number "2" will necessarily be preceded by a prime number p. But this prime p must itself be preceded by a perfect square, since as we proved in Problem 485, $\tau(n)$ is odd if and only if n is a perfect square. This is why the only sequence which does not produce any perfect square is the one which starts with a prime number.

(494) Since $f(n) = n^a$ is a multiplicative function, we easily establish the formula by finding the value of $\sigma_a(p^\alpha)$ for each $p^\alpha \| n$.

(495) (*AMM, Vol. 80, 1973, p. 948*). Since

$$\sigma(p^a) = 1 + p + \cdots + p^a \equiv 1 \pmod{p},$$

then $(p^a, \sigma(p^a)) = 1$ and hence if $p^a | \sigma(p^a)\sigma(q^b)$, we obtain $p^a | \sigma(q^b)$. But

$$\sigma(q^b) = 1 + q + \cdots + q^{b-1} + q^b = \frac{q^b - 1}{q - 1} + q^b < 2q^b$$

and since $q^b < p^a$, it follows that $1 < \sigma(q^b) < 2p^a$. Hence, $p^a | \sigma(q^b)$ implies that $\sigma(q^b) = p^a$.

(496) Let $(m, n) = 1$. If m and n are odd, then, since σ is multiplicative,

$$\sigma^*(mn) = \sigma(mn) = \sigma(m)\sigma(n) = \sigma^*(m)\sigma^*(n),$$

and the result is proved in this particular case. On the other hand, if one of these two integers is even, say m, then there exists a positive integer α such that $m = 2^\alpha r$, with r odd (and $(r, n) = 1$). We then have

$$\sigma^*(mn) = \sigma^*(rn) = \sigma(rn) = \sigma(r)\sigma(n) = \sigma^*(2^\alpha r)\sigma^*(n) = \sigma^*(m)\sigma^*(n).$$

All cases having been covered, the result is proved.

(497) We first show that, for each integer $n \geq 1$, we have $3|\sigma(3n-1)$. Let $N = 3n - 1$. If N was a a perfect square, ir would be of the form $N = 3n$ or $N = 3n+1$. Hence, N is not a perfect square, so that by Problem 485, the number r of divisors is even. Letting $d_1 < d_2 < \ldots < d_r$ be the divisors of N, we get

$$\sigma(N) = d_1 + d_2 + \cdots + d_{r/2} + \frac{N}{d_{r/2}} + \cdots + \frac{N}{d_1}.$$

It is clear that in order to prove that $3|\sigma(N)$, it is enough to show that $d_i + \dfrac{N}{d_i}$ is divisible by 3, for $i = 1, 2, \ldots, r/2$. First of all, in the case $d = 1$, we have

$$d + \frac{N}{d} = 1 + N = 1 + (3n - 1) = 3n,$$

a multiple of 3. If $d|n$ and $d > 1$, then $d = 3k + a$ for certain integers k and a, while $\dfrac{N}{d} = 3\ell + b$ for certain integers ℓ and b. Of course,

$$d \cdot \frac{N}{d} = N,$$

and this is why

$$(3k + a)(3\ell + b) = 3n - 1,$$

which means that $ab \equiv -1 \pmod 3$. But this can occur only if one of the integers a or b is congruent to 1 modulo 3 and the other to -1 modulo 3, so that in all cases,

$$a + b \equiv 0 \pmod 3.$$

But then

$$d + \frac{N}{d} = 3k + a + 3\ell + b \equiv 0 \pmod 3,$$

as required.

To prove that $4|\sigma(4n-1)$ for each integer $n \geq 1$, we proceed essentially in the same manner, but this time by working with congruences modulo 4.

By the same method, we easily prove that $12|\sigma(12n - 1)$ for each integer $n \geq 1$.

(498) Since $\sigma(p^r) = (p^{r+1} - 1)/(p - 1)$, then $\sigma(p^a)|\sigma(p^b)$ if and only if $(p^{a+1} - 1)|(p^{b+1} - 1)$, and this is true if and only if $(a + 1)|(b + 1)$, by way of Problem 75.

(499) We easily establish that

$$\sigma_{-a}(n) = \sum_{d|n} d^{-a} = n^{-a} \sum_{d|n} \left(\frac{n}{d}\right)^a = n^{-a} \sum_{d|n} d^a = n^{-a}\sigma_a(n).$$

For the second part, it is enough to set $a = 1$ in the result of the first part.

(500) If $2^k - 1$ is a prime number, then $n = 2^{k-1}(2^k - 1)$ satisfies $\sigma(n) = 2n$.

For the second part, assume that n is an even perfect number; that is assume that $n = 2^t m$, where m is an odd number and $t > 0$. Then,

$$2n = \sigma(n) = \sigma(2^t)\sigma(m) = (2^{t+1} - 1)\sigma(m),$$

in which case

$$(*) \qquad 2^{t+1}m = (2^{t+1} - 1)\sigma(m).$$

It follows that $(2^{t+1} - 1)|2^{t+1}m$, and since $(2^{t+1} - 1, 2^{t+1}) = 1$, then $2^{t+1} - 1|m$; that is $m = (2^{t+1} - 1)M$ for a certain positive integer M. Replacing this value of m in $(*)$, we obtain $\sigma(m) = 2^{t+1}M$; and since m and M $(M < m)$ are divisors of m, then

$$2^{t+1}M = \sigma(m) \geq m + M = 2^{t+1}M.$$

Therefore, $\sigma(m) = m + M$. But this equality implies that m has only the divisors m and M, which in turn implies that m is prime and $M = 1$.
REMARK: In Euclid's *Elements*, the result follows from the fact that if $2^k - 1$ is prime, then the number $2^{k-1}(2^k - 1)$ is a perfect number. In the eighteenth century, Euler proved that if n is an even perfect number, then there exists a prime number of the form $2^k - 1$ such that $n = 2^{k-1}(2^k - 1)$.

(501) The answer is YES. Indeed, we know that if n is a even perfect number, then $n = 2^{q-1}(2^q - 1)$, with $2^q - 1$ prime, in which case

$$S := \sum_{p|n} \frac{1}{p} = \frac{1}{2} + \frac{1}{2^q - 1} > \frac{1}{2}$$

and

$$S = \frac{1}{2} + \frac{1}{2^q - 1} \leq \frac{1}{2} + \frac{1}{2^2 - 1} = \frac{1}{2} + \frac{1}{3} = \frac{5}{6} < 2\log\frac{\pi}{2},$$

since $2\log\frac{\pi}{2} \approx 0.903$.

(502) Since n is an even perfect number, it follows that $n = 2^{k-1}(2^k - 1)$ for a certain integer $k \geq 2$, in which case $8n+1 = 2^{2k+2} - 2^{k+2} + 1 = (2^{k+1} - 1)^2$.

(503) If p^a is a perfect number, then we must have

$$\sigma(p^a) = 1 + p + p^2 + \cdots + p^a = 2p^a.$$

This is equivalent to

$$1 + p + p^2 + \cdots + p^{a-1} = \frac{p^a - 1}{p - 1} = p^a,$$

that is to

$$p^a(p - 2) = -1,$$

which is impossible.

(504) Each even perfect number n is of the form $2^{p-1}(2^p - 1)$, where $2^p - 1$ is a prime number and p a prime number. If $p = 2$, then $n = 6$. Assume that $p > 2$. But each prime number > 2 is of the form $4m + 1$ or $4m + 3$. If p is of the form $4m + 1$, then

$$n = 2^{4m}(2^{4m+1} - 1) = 16^m(2 \cdot 16^m - 1), \quad \text{where } m \geq 1.$$

Since 16^m ends with the digit 6, we have that $2 \cdot 16^m - 1$ ends with the digit 1, and the number n ends with a 6.
   In the same manner, if $p = 4m + 3$, then

$$n = 2^{4m+2}(2^{4m+3} - 1) = 4 \cdot 16^m(8 \cdot 16^m - 1), \quad \text{where } m \geq 0.$$

Since 16^m ends with a 6, it follows that $4 \cdot 16^m$ ends with a 4; moreover, $8 \cdot 16^m - 1$ ends with a 7, and this is why n ends with the digit 8.

(505) If $n > 6$ is an even perfect number, then, by Problem 500, there exists a positive integer α such that

$$n = 2^\alpha \left(2^{\alpha+1} - 1\right), \qquad \text{with } 2^{\alpha+1} - 1 \text{ prime}.$$

Since $2^{\alpha+1} - 1$ is prime, it is clear that $\alpha + 1$ is odd, in which case there exists a positive integer β such that $\alpha = 2\beta$. We therefore have

$$n = 2^\alpha \left(2^{\alpha+1} - 1\right) = 2^{2\beta} \left(2^{2\beta+1} - 1\right) = (2^\beta)^2 \left(2 \cdot (2^\beta)^2 - 1\right) = k^2(2k^2 - 1).$$

The result then follows from Problem 22.

(506) First of all, if $n = p^\alpha$ for $p \geq 3$ prime and $\alpha \geq 1$ an integer, then

$$\sigma(n) = 1 + p + \cdots + p^\alpha < 2p^\alpha = 2n,$$

which contradicts the fact that n is perfect.

Hence, we only need to prove that n cannot be written as

$$n = p^\alpha q^\beta,$$

for certain prime numbers $3 \leq p < q$ and positive integers α and β.

But, if such a representation of n was possible, we would have

$$
\begin{aligned}
\sigma(n) &= \sigma(p^\alpha)\sigma(q^\beta) = (1 + p + \cdots + p^\alpha)(1 + q + \cdots + q^\beta) \\
&= p^\alpha \left(1 + \frac{1}{p} + \cdots + \frac{1}{p^\alpha}\right) q^\beta \left(1 + \frac{1}{q} + \cdots + \frac{1}{q^\beta}\right) \\
&< 2p^\alpha q^\beta = 2n,
\end{aligned}
$$

since

$$
\left(1 + \frac{1}{p} + \cdots + \frac{1}{p^\alpha}\right)\left(1 + \frac{1}{q} + \cdots + \frac{1}{q^\beta}\right) < \frac{1}{1 - \frac{1}{p^\alpha}} \cdot \frac{1}{1 - \frac{1}{q^\beta}}
$$

$$
\leq \frac{1}{1 - \frac{1}{3}}\frac{1}{1 - \frac{1}{5}} = \frac{15}{8} < 2,
$$

again a contradiction.

(507) (*AMM, Vol. 82, 1975, p. 1015*). The only positive integer with this property is 6. To prove this result, we consider separately two cases.

First case. If n is an even perfect number, then $n = 2^{p-1}(2^p - 1)$ where $2^p - 1$ and p are prime. Since σ is a multiplicative function, it follows that

$$\sigma(\sigma(n)) = \sigma(2^p(2^p - 1)) = \sigma(2^p)\sigma(2^p - 1) = (2^{p+1} - 1)2^p,$$

an even number. If this number is also perfect, then $2^{p+1} - 1$ and $p + 1$ are prime, which is possible only if $p = 2$ and $n = 6$.

Second case. If n is an odd perfect number, then $\sigma(n) = 2n$ implies

$$\sigma(\sigma(n)) = \sigma(2n) = \sigma(2)\sigma(n) = 6n.$$

If this last number is perfect, then, since it is even, it can be written as $6n = 2^{p-1}(2^p - 1)$ with $2^p - 1$ prime, which can occur only if $n = 1$, which is not perfect.

Therefore, there is only one solution, namely $n = 6$.

(508) Assume that n is odd and tri-perfect. Then, $n = q_1^{\alpha_1} \cdots q_r^{\alpha_r}$ with each prime number $q_i \geq 3$ and

(1) $\sigma(n) = \left(1 + q_1 + q_1^2 + \cdots + q_1^{\alpha_1}\right) \cdots \left(1 + q_r + q_r^2 + \cdots + q_r^{\alpha_r}\right).$

For $\sigma(n) = 3n$ to be verified, $\sigma(n)$ must be odd, and therefore each factor $1 + q_i + q_i^2 + \cdots + q_i^{\alpha_i}$ on the right of (1) is odd, which implies that each α_i must be even, meaning that n must be a perfect square.

(509) First of all, it is clear that if n is of the form 2^α, then n is not tri-perfect. Assume that $n = 2^\alpha q_1 \cdots q_r$ for a certain positive integer r, each q_i being an odd prime number. The equation

(1) $\sigma(n) = 3n$

can be written as

$$(2^{\alpha+1} - 1)(q_1 + 1) \cdots (q_r + 1) = 3 \cdot 2^\alpha \cdot q_1 \cdots q_r.$$

Each number $q_i + 1$ being even, we must necessarily have $r \leq \alpha$.

If $\alpha = 1$, we have $r = 1$, and therefore (1) can be written as $3(q_1+1) = 3 \cdot 2 \cdot q_1$, in which case $q_1 = 1$, which is nonsense.

If $\alpha = 2$, we have $r \leq 2$, and therefore (1) can be written as $7(q_1+1) = 12q_1$ in the case $r = 1$ and $7(q_1 + 1)(q_2 + 1) = 12q_1q_2$ in the case $r = 2$. The first case is reduced to $5q_1 = 7$, which makes no sense. The equation for the second case is impossible since the two q_i's are ≥ 3.

If $\alpha = 3$, we have $r \leq 3$, and therefore (1) can be written as

(2) $3 \cdot 5 \cdot \prod(q_i + 1) = 3 \cdot 2^3 \cdot \prod q_i,$

which means that one of the q_i's is 5, say $q_1 = 5$, in which case $q_1 + 1 = 6$, so that (2) becomes

$$2 \cdot 3 \cdot \prod_{i>1}(q_i + 1) = 2^3 \cdot \prod_{i>1} q_i,$$

which means that one of the q_i's is 3, say $q_2 = 3$, in which case $q_2 + 1 = 4$, which completes the process in the sense that $r = 2$ serves our purpose. We have thus obtained as a solution of (1): $n = 2^3 \cdot 3 \cdot 5 = 120$.

If $\alpha = 4$, we have $r \leq 4$, and therefore (1) can be written as

(3) $31 \cdot \prod(q_i + 1) = 3 \cdot 2^4 \cdot \prod q_i,$

which means that one of the q_i's is 31, say $q_1 = 31$, in which case $q_1 + 1 = 32 = 2^5$, which makes no sense since the largest power of 2 on the right of (3) is 2^4. Therefore, we have no solutions in the case $\alpha = 4$.

If $\alpha = 5$, we have $r \leq 5$, and therefore (1) can be written as

(4) $7 \cdot 9 \cdot \prod(q_i + 1) = 3 \cdot 2^5 \cdot \prod q_i,$

which means that one of the q_i's is 3 and the other 7, say $q_1 = 3$ and $q_2 = 7$, in which case $q_1 + 1 = 4$ and $q_2 + 1 = 8$, so that the equation (4) is complete in the sense that $r = 2$ serves our purpose. We have thus obtained as a solution of (1): $n = 2^5 \cdot 3 \cdot 7 = 672$.

If $\alpha = 6$, we have $r \leq 6$, and therefore (1) can be written as

(5) $127 \cdot \prod(q_i + 1) = 3 \cdot 2^6 \cdot \prod q_i,$

which means that one of the q_i's is 127, say $q_1 = 127$, in which case $q_1 + 1 = 128 = 2^7$, which makes no sense since the largest power of 2 on the right of (5) is 2^6. Hence, there are no solutions in the case $\alpha = 6$.

If $\alpha = 7$, we have $r \le 7$, and therefore (1) can be written as

$$(6) \qquad 3 \cdot 5 \cdot 17 \cdot \prod (q_i + 1) = 3 \cdot 2^7 \cdot \prod q_i,$$

which means that three of the q_i's are 3, 5 and 17, say $q_1 = 3$, $q_2 = 5$ and $q_3 = 17$, in which case $q_1 + 1 = 4$, $q_2 + 1 = 6$ and $q_3 + 1 = 18 = 2 \cdot 3^2$, which makes no sense since the largest power of 3 on the right of (6) (after simplification) is 3. Hence, there are no solutions in the case $\alpha = 7$.

If $\alpha = 8$, we have $r \le 8$, and therefore (1) can be written as

$$(7) \qquad 7 \cdot 73 \cdot \prod (q_i + 1) = 3 \cdot 2^8 \cdot \prod q_i,$$

which means that two of the q_i's are 7 and 73, say $q_1 = 7$ and $q_2 = 73$, in which case $q_1 + 1 = 8$ and $q_2 + 1 = 74$, so that equation (7) becomes (after simplification)

$$8 \cdot 74 \cdot \prod_{i \ge 3} (q_i + 1) = 3 \cdot 2^8 \cdot \prod_{i \ge 3} q_i,$$

that is

$$(8) \qquad 37 \cdot \prod_{i \ge 3} (q_i + 1) = 3 \cdot 2^4 \cdot \prod_{i \ge 3} q_i,$$

which means that one of the q_i's is 37, say $q_3 = 37$, in which case $q_3 + 1 = 38 = 2 \cdot 19$, which means that (8) becomes

$$(9) \qquad 2 \cdot 19 \cdot \prod_{i \ge 4} (q_i + 1) = 3 \cdot 2^4 \cdot \prod_{i \ge 4} q_i,$$

which means that one of the q_i's is 19, say $q_4 = 19$, in which case $q_4 + 1 = 20 = 2^2 \cdot 5$, which means that (9) becomes

$$(10) \qquad 2^2 \cdot 5 \cdot \prod_{i \ge 5} (q_i + 1) = 3 \cdot 2^3 \cdot \prod_{i \ge 5} q_i,$$

which leads to $q_5 = 5$, so that $q_5 + 1 = 6$, which reduces (10) to

$$2^2 \cdot 5 \cdot 6 \cdot \prod_{i \ge 6} (q_i + 1) = 3 \cdot 2^3 \cdot 5 \cdot \prod_{i \ge 6} q_i.$$

But the only possibility is $r = 6$, hence the solution $n = 2^8 \cdot 5 \cdot 7 \cdot 19 \cdot 37 \cdot 73 = 459\,818\,240$.

Continuing this method, the case $\alpha = 9$ yields the solution $n = 2^9 \cdot 3 \cdot 11 \cdot 31 = 523\,776$, while the case $\alpha = 10$ yields no solutions at all.

REMARK: There are reasons to believe (see R.K. Guy [16], p. 48) that there exist only 6 tri-perfect numbers, the other two being

$$1\,476\,304\,896 = 2^{13} \cdot 3 \cdot 11 \cdot 43 \cdot 127,$$

$$51\,001\,180\,160 = 2^{14} \cdot 5 \cdot 7 \cdot 19 \cdot 31 \cdot 151.$$

(510) (*AMM, Vol. 93, 1986, p. 813*).

(a) Let $d = (a, b)$, so that $\sigma(d) \leq 2d - 1$. Let $a = d\alpha$, $b = d\beta$, with $(\alpha, \beta) = 1$. From Dirichlet's Theorem, there exist infinitely many prime numbers of the form $\alpha + \beta m$. For each prime number $p > 2d$ of this form, let $n = dp$ and let x be defined by $n = a + bx$. We then have $n = dp = a + bx \equiv a \pmod{b}$. It follows that

$$
\begin{aligned}
\frac{\sigma(n)}{2n} &= \frac{\sigma(dp)}{2dp} = \frac{\sigma(d)}{2d} \cdot \frac{\sigma(p)}{p} \leq \frac{2d-1}{2d} \cdot \frac{p+1}{p} \\
&= \left(1 - \frac{1}{2d}\right)\left(1 + \frac{1}{p}\right) < \left(1 - \frac{1}{2d}\right)\left(1 + \frac{1}{2d}\right) \\
&= 1 - \frac{1}{4d^2} < 1,
\end{aligned}
$$

and therefore that n is deficient.

(b) Let $q_1, q_2, \ldots$ be the prime numbers $> b$. Since the sum of the reciprocals of the prime numbers diverges, there exists an integer k such that $1/q_1 + 1/q_2 + \cdots + 1/q_k > 1$. Given positive integers $a_1, a_2, \ldots, a_k$, one can solve the congruence

$$
n = a + bx \equiv 0 \pmod{q_1^{a_1} \cdots q_k^{a_k}}.
$$

For such an integer n, we obtain

$$
\frac{\sigma(n)}{n} \geq \left(1 + \frac{1}{q_1}\right)\left(1 + \frac{1}{q_2}\right) \cdots \left(1 + \frac{1}{q_k}\right) \geq 1 + \sum_{j=1}^{k} \frac{1}{q_j} > 2.
$$

It follows that n is abundant.

(511) Since n is an even perfect number, there exists a positive integer k such that

$$(*) \qquad n = 2^{k-1}\left(2^k - 1\right) = 2^{k-1} \cdot p,$$

where $p = 2^k - 1$ is prime. Then,

$$\tau(n) = \tau(2^{k-1})\tau(p) = 2k.$$

On the other hand, taking the logarithm in basis 2 on both sides of the first equation of $(*)$, we obtain successively

$$
\begin{aligned}
\log_2 n &= (k-1)\log_2 2 + \log_2(2^k - 1) \\
&= (k-1) + \log_2\left(2^k \cdot \left(1 - \frac{1}{2^k}\right)\right) \\
&= k - 1 + k\log_2 2 + \log_2\left(1 - \frac{1}{2^k}\right) \\
&= 2k - 1 + \log_2\left(1 - \frac{1}{2^k}\right) = 2k - 1 + \alpha_k,
\end{aligned}
$$

where it is easy to see that the expression α_k satisfies $-1 < \alpha_k < 0$. But since $\tau(n) = 2k$, this last equation can be written as

$$\tau(n) = \log_2 n + 1 - \alpha_k.$$

Since $0 < -\alpha_k < 1$, we may therefore write

$$\log_2 n + 1 + 0 < \tau(n) < \log_2 n + 1 + 1,$$

that is

$$\log_2 n + 1 < \tau(n) < \log_2 n + 2,$$

which means of course that

$$\tau(n) = [\log_2 n + 2] = [\log_2 n] + 2.$$

(512) A given number n has r digits if and only if $10^{r-1} \leq n < 10^r$, so that $(r-1) \leq \log_{10} n < r$, which means that

$$(*) \qquad\qquad r = [\log_{10} n] + 1,$$

where $[x]$ stands for the largest integer not exceeding x, is the required general formula. It then follows from $(*)$ that the number r of digits of the Spence prime number is given by $r = \left[\log_{10}\left(2^{2976221} - 1\right)\right] + 1$. But obviously

$$895931.4 < \log_{10} 2^{2976220} < \log_{10}\left(2^{2976221} - 1\right)$$
$$< \log_{10} 2^{2976221} < 895931.5,$$

so that $r = 895932$, allowing us to conclude that the Spence number has exactly 895932 digits.

REMARK: The reader is invited to consult the Web site www.mersenne.org to obtain the most recent results concerning Mersenne primes.

(513) Consider the sequence $\{(\sigma(n!)/n!)\}$, $n = 1, 2, \ldots$, for which it is clear that

$$\frac{\sigma(n!)}{n!} \geq \frac{1}{1} + \frac{1}{2} + \frac{1}{3} + \cdots + \frac{1}{n}.$$

The result then follows from the divergence of the harmonic series $\sum_{n=1}^{\infty} \frac{1}{n}$.

(514) Assume that n is an even perfect number. Then $n = 2^{k-1}(2^k - 1)$ for a certain positive integer k. Since each triangular number is of the form $m(m + 1)/2$, it is enough to choose $m = 2^k - 1$ in order to obtain the result.

(515) (*AMM, Vol. 94, 1987, p. 794*). We proceed by induction. Let $m = q_1^{\alpha_1} q_2^{\alpha_2} \cdots q_k^{\alpha_k}$ be a perfect number such that $q_1 < q_2 < \ldots < q_k$, $k \geq 2$, and assume that $q_1 \geq k + 1$, in which case $q_i \geq k + i$ for each i, $2 \leq i \leq k$. We then have

$$\begin{aligned}
\frac{\sigma(m)}{m} &= \left(1 + \frac{1}{q_1} + \cdots + \frac{1}{q_1^{\alpha_1}}\right) \cdots \left(1 + \frac{1}{q_k} + \cdots + \frac{1}{q_k^{\alpha_k}}\right) \\
&< \left(\sum_{i=0}^{\infty} \frac{1}{q_1^i}\right) \cdots \left(\sum_{i=0}^{\infty} \frac{1}{q_k^i}\right) = \prod_{j=1}^{k}\left(1 + \frac{1}{q_j - 1}\right) \\
&< \prod_{j=1}^{k}\left(1 + \frac{1}{k + j - 1}\right) = \prod_{j=1}^{k} \frac{k + j}{k + j - 1} = \frac{2k}{k} = 2.
\end{aligned}$$

But the inequality $\sigma(m)/m < 2$ contradicts the fact that m is a perfect number. Therefore, the hypothesis to the effect that "$q_1 \geq k + 1$" must be false, and this is why $q_1 \leq k$, as required.

(516) (*AMM, Vol. 75, 1969, p. 1149*). This double inequality follows from a more general result, namely: *If $n = \prod_{i=1}^{k} q_i^{\alpha_i}$ is an arbitrary integer, then*

$$\frac{\sigma(q_1 q_2 \cdots q_k)}{q_1 q_2 \cdots q_k} \leq \frac{\sigma(n)}{n} < \frac{q_1 q_2 \cdots q_k}{\phi(q_1 q_2 \cdots q_k)}.$$

Indeed, since

$$\frac{1 + q_i}{q_i} = 1 + \frac{1}{q_i} \leq 1 + \frac{1}{q_i} + \frac{1}{q_i^2} + \cdots + \frac{1}{q_i^{\alpha_i}} < 1 + \sum_{j=1}^{\infty} \frac{1}{q_i^j}$$

$$= \frac{1}{1 - \frac{1}{q_i}} = \frac{q_i}{\phi(q_i)},$$

we obtain

$$\frac{\sigma(q_i)}{q_i} \leq \frac{\sigma(q_i^{\alpha_i})}{q_i^{\alpha_i}} < \frac{q_i}{\phi(q_i)}.$$

The fact that both σ and ϕ are multiplicative finally yields the result. When n is a perfect number, we have $\sigma(n) = 2n$, and the result follows.

(517) Since n is an even perfect number, there exists a positive integer k such that $n = 2^{k-1}(2^k - 1)$, with $2^k - 1$ prime. We then have

$$\phi(n) = n \prod_{p|n} \left(1 - \frac{1}{p}\right) = 2^{k-1}(2^k - 1)\left(1 - \frac{1}{2}\right)\left(1 - \frac{1}{2^k - 1}\right)$$

$$= 2^{k-1}(2^{k-1} - 1),$$

as required.

(518) The answer is YES. It is enough to consider the sequence of integers $n = 11^k$, $k = 1, 2, \ldots$. Indeed, we have

$$\phi(n) = \phi(11^k) = 11^k \left(1 - \frac{1}{11}\right) = 11^k \cdot \frac{10}{11} = 10 \cdot 11^{k-1}.$$

(519) The required number is $600 - \phi(600) = 440$. In the second case, the required number is $2\phi(600) = 320$.

(520) The required number is $7\phi(600) = 1120$.

(521) For each $i = 0, 1, \ldots, k - 1$, set $E_i = \{n \mid im < n \leq (i+1)m\}$. Then, all the positive integers $\leq mk$ belong to $E = \cup_{i=0}^{k-1} E_i$. If $(n, m) = 1$ and $n \in E_i$, then $(n - im, m) = (n, m) = 1$ and $0 < n - im \leq m$, so that $n - im \in E_0$. The association $n \leftrightarrow n - im$ shows that the number of integers in E_i which are relatively prime with m is equal to the number of those which are in E_0, that is $\phi(m)$. Since each E_i contains $\phi(m)$ elements which are relatively prime with m, we have a total of $k\phi(m)$ integers, as was to be shown.

(522) We observe that each prime divisor of mn is either a prime divisor of m or of n, or else a prime divisor which divides both m and n. This is why

$$\frac{\phi(mn)}{mn} = \prod_{p|mn}\left(1-\frac{1}{p}\right) = \frac{\prod_{p|m}\left(1-\frac{1}{p}\right)\prod_{p|n}\left(1-\frac{1}{p}\right)}{\prod_{p|(m,n)}\left(1-\frac{1}{p}\right)}$$

$$= \frac{(\phi(m)/m)\cdot(\phi(n)/n)}{(\phi(d)/d)},$$

which gives after simplification the required equation.

(523) If $n = 2^r$, then, since $n > 2$, we have $r \geq 2$. Therefore, $\phi(n) = 2^{r-1}$ is an even number. If n has at least an odd prime factor, then

$$\phi(n) = n\prod_{p|n}\frac{p-1}{p} = \frac{n}{\prod_{p|n}p}\prod_{p|n}(p-1).$$

Since $n/\prod_{p|n}p$ is an integer and since $\prod_{p|n}(p-1)$ is an even number, the result follows.

(524) Since $1 \leq a < b$ and $(a, b) = 1$, the required number of fractions is

$$\#\{a \mid 1 \leq a \leq b, (a, b) = 1\},$$

that is $\phi(b)$.

(525) Since $m|n$, we have that $n = km$, $1 \leq k \leq n$. If $k = n$, we have that $m = 1$ and $\phi(m)|\phi(n)$. Assuming that $k < n$, we have, using the result of Problem 522,

$$(1) \qquad \phi(n) = \phi(km) = \phi(k)\phi(m)\frac{d}{\phi(d)} = d\phi(m)\frac{\phi(k)}{\phi(d)},$$

where $d = (m, k)$. We shall use induction on n. If $n = 1$, then $m = 1$ and we have the result. Assume that the result is true for all integers smaller than n. But, since $k < n$ and $d|k$, then $\phi(d)|\phi(k)$. Consequently, the right-hand side of (1) is a multiple of $\phi(m)$, which means that $\phi(m)|\phi(n)$.

    Another possible solution is the following. Let p be an arbitrary prime number such that $p^a\|m$ and $p^b\|n$. From the definition of the Euler function, p^a contributes to the factor $p^{a-1}(p-1)$ by $\phi(m)$, while p^b contributes to the factor $p^{b-1}(p-1)$ by $\phi(n)$. Since $a \leq b$, the result follows.

(526) (*AMM, Vol. 93, 1986, p. 656*). The case $n = 1$ is trivial. Assume that $n > 1$ and that $n = q_1^{\alpha_1} \cdots q_k^{\alpha_k}$. If $\phi(n)|n$, then $n = k \cdot \phi(n)$ for an integer k, so that

$$q_1 \cdots q_k = k(q_1 - 1) \cdots (q_k - 1).$$

Since $(q_1 - 1) \cdots (q_k - 1)$ is an even number, it is clear that, in order for the equality to be true, one of the q_j's must be equal to 2. We may therefore assume that $q_1 = 2$, in which case

$$2q_2 \cdots q_k = x(q_2 - 1) \cdots (q_k - 1).$$

Since $q_2, \ldots, q_k$ are odd primes, it follows from the above equality that n can contain at most one odd prime, say q_2. Let $q_2 - 1 = 2y$ for some positive integer y. Then $2q_2 = k(2y)$, which implies that $k = q_2$ and $y = 1$.

Hence, $q_2 - 1 = 2$; that is $q_2 = 3$. Therefore, $n = 2^a 3^b$, $a \geq 1$, $b \geq 0$. Reciprocally, we easily check that if $n = 2^a 3^b$, then $\phi(n)|n$.

(527) If $d|n$ and $k \in \mathbb{N}$, the set of prime numbers p such that $p|n$ coincides with the set of prime numbers p such that $p|nd^k$. Therefore,

$$\phi(nd^k) = nd^k \prod_{p|nd^k} \left(1 - \frac{1}{p}\right) = nd^k \prod_{p|n} \left(1 - \frac{1}{p}\right) = d^k \phi(n),$$

as was to be shown.

(528) In fact, we show that $5\phi(n)|2n$ if and only if $n = 2^a 5^b$. When $n = 2^a 5^b$, it is immediate that $5\phi(n)|2n$. Conversely, assume that $n = 2^a 5^b N$ where $(N, 10) = 1$ and $a \geq 0$, $b \geq 0$. The relation $5\phi(n)|2n$ implies that $\phi(N)|N$. But, since $(N, 10) = 1$, it follows from Problem 526 that $N = 3^c$, $c \geq 0$ and therefore that $n = 2^a 3^c 5^b$, $c \geq 0$. The fact that $5\phi(n)|2n$ then implies that $c = 0$. The given integers are therefore of the form $2^a 5^b$, with $a \geq 0$, $b \geq 0$.

(529) (a) These are the even integers. Indeed, using Problem 522, if $d = (n, 2)$,

$$\phi(2n) = \phi(n) \frac{d}{\phi(d)},$$

and therefore $\phi(2n) > \phi(n)$ implies that $\phi(d) < d$. But this occurs when $d \geq 2$ and therefore when $d = 2$; that is $2|n$.

(b) These are the odd integers, because if $d = (2, n)$, the relation of the statement implies that $d = 1$.

(c) Let $n = 2^a 3^b N$ with $a \geq 0$, $b \geq 0$ and $(6, N) = 1$. The relation $\phi(2n) = \phi(3n)$ implies that $2^a \phi(3^b)|2 3^b \phi(2^a)$. It follows that a cannot be zero, and the equality can therefore be written as $2^a \phi(3^b) = 2^a 3^b$. This implies that $b = 0$. The required numbers are therefore the even numbers which are not multiples of 3.

(530) Since p is an odd prime number, it follows that $\phi(4p) = \phi(4)\phi(p) = 2(p - 1)$, and since $2p + 1$ is a prime number, then

$$\phi(4p + 2) = \phi(2(2p + 1)) = \phi(2p + 1) = 2p,$$

and the result follows.

REMARK: Each prime number p such that $2p + 1$ is also prime is called a *Sophie Germain prime*. The Sophie Germain primes smaller than 100 are 2, 3, 5, 11, 23, 29, 41, 53, 83 and 89. Sophie Germain (1776-1831) proved that if p is such a number larger than 2, then any solution in positive integers x, y, z of the famous Fermat equation $x^p + y^p = z^p$ is such that $p|xyz$. Even though we still don't know how to prove that there exist infinitely many Sophie Germain primes, there is a heuristic argument suggesting that the number of such numbers not exceeding a given number x is approximately $cx/\log^2 x$ for some positive constant c (see R. Guy [**16**]).

(531) Let $r_1, \ldots, r_{\phi(n)}$ be the integers smaller than n which are relatively prime with n. Since $(a, n) = 1$ if and only if $(n - a, n) = 1$, we have

$$\sum_{i=1}^{\phi(n)} r_i = \sum_{i=1}^{\phi(n)} (n - r_i) = n\phi(n) - \sum_{i=1}^{\phi(n)} r_i,$$

and the result follows.

(532) For $n > 1$, let $n = q_1^{a_1} \cdots q_r^{a_r}$ be its representation as a product of distinct prime powers. If $q_1 = 2$, then n has $r - 1$ odd prime factors. Since $\phi(n)$ contains the factor $\prod\limits_{\substack{q_i | n \\ i \geq 2}} (q_i - 1)$, the factor 2 appears $r - 1$ times and therefore $2^{r-1} | \phi(n)$.

(533) It is clear that if n is of the given form, then $\phi(n) = 2^r$ for a certain positive integer r. So let us consider the reciprocal.

Since ϕ is a multiplicative function, it is enough to consider the numbers of the form p^k. But

$$\phi(p^k) = \begin{cases} p^{k-1}(p-1) = 2^\beta \Rightarrow p = 2 & \text{if } k > 1, \\ p - 1 = 2^\beta \Rightarrow p = 2^\beta + 1 & \text{if } k = 1. \end{cases}$$

Using Problem 186, $2^\beta + 1$ is a prime number only if $\beta = 2^k$, and the result follows.

(534) Combining the formula of Theorem 27 and the second formula of Theorem 30, we obtain the following results: (a) 47; (b) 41; (c) 29; (d) 23.

(535) The smallest positive integer divisible by six distinct prime numbers is $2 \cdot 3 \cdot 5 \cdot 7 \cdot 11 \cdot 13 = 30030$. Hence, every integer < 30030 has at most five distinct prime factors. Hence, for each integer $n \in [2, 30029]$, we have

$$\phi(n) \geq n \prod_{i=1}^{5} \left(1 - \frac{1}{p_i}\right) = \frac{16}{77} n,$$

so that

$$n \leq \frac{77}{16} \phi(n) < 5\phi(n).$$

It follows that for $1 < n < 30030$, $\phi(n) < n < 5\phi(n)$. Therefore, if $\phi(n) \leq 500$, we have $n \leq 2500$. This means that we only need to search amongst the numbers $500 \leq n \leq 2500$. With the help of a computer, we find that the largest integer n such that $\phi(n) \leq 500$ is $n = 2310$ (for which we have $\phi(n) = \phi(2310) = 480$).

(536) The answer is YES. Indeed, since $(2, 8m + 4) = 2$, using Problem 522, we have

$$\phi(8m + 4) = \phi(2)\phi(4m + 2)\frac{2}{\phi(2)} = 2\phi(4m + 2),$$

thus the result.

(537) We have, using Problem 522,

$$\phi(a^2 n + ab) = a\phi(an + b) \quad \text{and} \quad \phi(abn + a^2) = a\phi(bn + a),$$

which gives the result.

(538) Let $n = q_1^{\alpha_1} \cdots q_r^{\alpha_r}$, $r \leq 9$ be the representation of n as a product of distinct prime powers. Since $q_i \geq p_i$, for $i = 1, 2, \ldots, r$, we have

$$\frac{\phi(n)}{n} = \prod_{i=1}^{r} \left(1 - \frac{1}{q_i}\right) \geq \prod_{i=1}^{r} \left(1 - \frac{1}{p_i}\right) \geq \prod_{i=1}^{9} \left(1 - \frac{1}{p_i}\right) = \frac{110592}{676039} > \frac{1}{7},$$

and the result follows.

(539) (*AMM, Vol. 100, 1993, p. 404*). If n has at least two distinct odd prime factors and if p stands for one of them, then $(p-1)|(\phi(n)/2)$, and calling upon Fermat's Little Theorem, it follows that $p|(2^{\phi(n)/2}-1)$. Since

$$2^{\phi(n)} - 1 = (2^{\phi(n)/2} - 1)(2^{\phi(n)/2)} + 1)$$

and since these two factors are relatively prime, then $p \nmid 2^{\phi(n)/2} + 1$. We then obtain $(2^{\phi(n)/2} + 1, n) = 1$, and it follows that each prime divisor of $2^{\phi(n)/2} + 1$ is relatively prime with n and therefore divides $2^{\phi(n)} - 1$. It remains to consider the case $n = p^m$. If $n = 3^m$ with $m \geq 2$, then $3|\phi(n)$, and therefore 7 divides $2^{\phi(n)} - 1$ but not n. If $n = p^m$ with $p > 3$, then $2|\phi(n)$, and therefore 3 divides $2^{\phi(n)} - 1$ but not n.

(540) (*AMM, Vol. 85, 1978, p. 199*). If n is prime, then $\phi(n) = n - 1$ and $\sigma(n) = n + 1$.

Reciprocally, let $n > 2$ and assume that $\phi(n)|(n-1)$ and that $(n+1)|\sigma(n)$. Since $\phi(n)$ is even, it follows that n must be odd. Let p be an odd prime number such that $p^r|n$, $r \geq 2$. Then $p^{r-1}|\phi(n)$ so that $p^{r-1}|(n-1)$, a contradiction. It follows that n is a product $q_1 q_2 \cdots q_k$ of distinct odd prime numbers. It then follows that $\phi(n) = (q_1 - 1) \cdots (q_k - 1)$ and that $\sigma(n) = (q_1 + 1) \cdots (q_k + 1)$, which implies that $2^k|\phi(n)$ and $2^k|\sigma(n)$. If $k \geq 2$, then $4|\phi(n)$, so that $4|(n-1)$ and therefore $4 \nmid (n+1)$. Since $n+1$ is even, we have that $2|n+1$ and it follows that $2^{k-1} \mid \sigma(n)/(n+1)$, so that $2^{k-1} \leq \sigma(n)/(n+1) < \sigma(n)/n$. We derive from this that

$$2^{k-1} < \frac{\sigma(n)}{n} = \left(1 + \frac{1}{q_1}\right) \cdots \left(1 + \frac{1}{q_k}\right) < \left(\frac{4}{3}\right)^k.$$

Since this is impossible, we conclude that $k = 1$ and therefore that n is prime.

(541) (*AMM, Vol. 73, 1966, p. 1026*). Assume that $n = 2^a m$ where m is odd. Then, using the first formula of Theorem 30, we obtain

$$\sum_{\substack{d|n \\ d \text{ odd}}} \phi(n/d) = \sum_{d|m} \phi(2^a m/d) = \phi(2^a) \sum_{d|m} \phi(m/d)$$

$$= \phi(2^a) \sum_{d|m} \phi(d) = m\phi(2^a),$$

implying that the quotient which we need to evaluate is

$$\left(n - \sum_{\substack{d|n \\ d \text{ odd}}} \phi(n/d)\right) \Big/ \sum_{\substack{d|n \\ d \text{ odd}}} \phi(n/d) = \frac{m2^a - m\phi(2^a)}{m\phi(2^a)} = \frac{2^a - \phi(2^a)}{\phi(2^a)},$$

and the result follows.

(542) (*AMM, Vol. 75, 1968, p. 551*). Since $m > 2$, it follows by Bertrand's Postulate that there exists a prime number p such that $m/2 < p < m$. The given sum runs through the set

$$\{m_i \mid (m_i, m) = 1, \quad 1 \leq i \leq \phi(m)\}$$

and can therefore be written as

$$\sum_{i=1}^{\phi(m)} \frac{1}{m_i} = \left\{ \sum_{i=1}^{\phi(m)} \frac{1}{m_i} \prod_{k=1}^{\phi(m)} m_k \right\} \Big/ \prod_{k=1}^{\phi(m)} m_k.$$

But it is clear that p divides the denominator (being one of the m_k's) but cannot divide the numerator (since it divides all the terms except one). It follows that the denominator cannot divide the numerator.

(543) The answer is YES. Assume that $f(m) = f(n)$ and let us show that $m = n$. For each prime divisor p of m, let $p^a \| m$ and $p^b \| n$. We write $m = p^a M$. If $p^k \| f(m)$, we obtain that $k = a(a+1)\tau(M)/2$. Since $\tau(m) = (a+1)\tau(M)$, it follows that $k = a\tau(m)/2$. Since $f(m) = f(n)$, then $k = b\tau(n)/2$ and we have $a/b = \tau(n)/\tau(m)$. Therefore, the ratio a/b is the same for each prime divisor p of m. Since $a/b < 1$ implies that m is a proper divisor of n, then $f(m) < f(n)$. We draw a similar conclusion if $a/b > 1$. We therefore conclude that $a = b$ and also that $m = n$.

(544) (*AMM, Vol. 95, 1988, p. 962*). Let $f(n)$ be the sum to estimate. We can write

$$f(n) = \sum_{d|n} 2^{\omega(d)} \sum_{\substack{1 \le k \le n/d \\ (k,n/d)=1}} 1 = \sum_{d|n} 2^{\omega(d)} \phi(n/d).$$

Since $2^{\omega(n)}$ and $\phi(n)$ are multiplicative functions, the function f is also multiplicative. We easily obtain that $f(p^m) = p^m(1+1/p)$ for each positive integer m, and this is why

$$f(n) = n \prod_{p|n} \left(1 + \frac{1}{p}\right).$$

(545) We have

$$S(x) = \sum_{\substack{n \le x \\ n \text{ even}}} 1 + \sum_{\substack{n \le x \\ n \text{ odd}}} 2 = \left[\frac{x}{2}\right] \times 1 + \left[\frac{x+1}{2}\right] \times 2.$$

It follows that

$$\frac{x}{2} - 1 + 2\left(\frac{x+1}{2} - 1\right) \le S(x) \le \frac{x}{2} + (x+1),$$
$$\frac{3x}{2} - 2 \le S(x) \le \frac{3x}{2} + 1.$$

From this, it easily follows that

$$\lim_{x \to \infty} \frac{S(x)}{x} = \frac{3}{2}.$$

(546) This is an immediate consequence of the divergence of the series $\sum_p 1/p$ (see Theorem 16). Indeed, if we consider the sequence $n_1 < n_2 < \dots$ defined by $n_k = p_1 p_2 \cdots p_k$, $k = 1, 2, \dots$, then for each $k \ge 1$,

$$\sum_{p|n_k} \frac{1}{p} = \sum_{p \le p_k} \frac{1}{p},$$

an expression that tends to $+\infty$ as $k \to +\infty$.

(547) (*AMM, Vol. 78, 1971, p. 1140*). If f is totally multiplicative, the result is immediate. Reciprocally, assume that (∗) is satisfied. For $n = 1$, we have $f(1) = 1$. Assume that $n \geq 2$ and let $n = q_1^{\alpha_1} \cdots q_r^{\alpha_r}$ be the representation of n as a product of distinct prime powers, and set $\Omega(n) = \alpha_1 + \cdots + \alpha_r$. It is enough to show that $f(n) = f(1)f(q_1)^{\alpha_1} \cdots f(q_r)^{\alpha_r}$. We proceed by induction on the value of $\Omega(n)$. If $\Omega(n) = 1$, then n is prime (say $n = p$), and the proposition is true because

$$2f(p) = \tau(p)f(p) = f(1)f(p) + f(p)f(1) = 2f(1)f(p).$$

Assume that the proposition is true for all n such that $\Omega(n) \leq k$, $k \geq 1$. Choose n such that $\Omega(n) = k + 1$. Then,

$$\tau(n)f(n) = 2f(1)f(n) + \sum_{\substack{ab=n \\ 1<a,b<n}} f(a)f(b).$$

It follows that $\Omega(a) \leq k$ and $\Omega(b) \leq k$. The induction hypothesis then allows us to obtain

$$\tau(n)f(n) = 2f(1)f(n) + (\tau(n) - 2)f(1)^2 f(q_1)^{\alpha_1} \cdots f(q_r)^{\alpha_r}.$$

Since n is not prime, then $\tau(n) > 2$ and we obtain the result, regardless whether $f(1) = 0$ or $f(1) = 1$.

(548) Since the three expressions appearing in the chain of inequalities represent multiplicative functions, it is clear that it is enough to prove that the inequalities are true when n is a prime power, say $n = p^\alpha$. Therefore, it is enough to show that

$$2 \leq \alpha + 1 \leq 2^\alpha,$$

which is of course true for each positive integer $\alpha \geq 1$.

(549) (*AMM, Vol. 62, 1955, p. 348*). By definition, n is a perfect number if and only if $\sigma(n) = 2n$. Therefore n is a perfect number if and only if

$$H(n) = \frac{n\tau(n)}{\sum_{d|n}(n/d)} = \frac{n\tau(n)}{\sigma(n)} = \frac{n\tau(n)}{2n} = \frac{\tau(n)}{2}.$$

Since an even perfect number n is of the form $n = 2^{p-1}(2^p - 1)$, with p prime and $2^p - 1$ prime, then $\tau(n) = (p - 1 + 1)2 = 2p$. It follows that $H(n) = p$. Hence, if n is an even perfect number, then $n = 2^{H(n)-1}(2^{H(n)} - 1)$.

It remains to prove the implication

$$n = 2^{H(n)-1}(2^{H(n)} - 1) \implies n \text{ even perfect.}$$

First of all, it is clear that it is enough to prove that $2^{H(n)}-1$ is prime, since it will then follow that $H(n)$ is a prime p, in which case $n = 2^{p-1}(2^p - 1)$, that is a perfect number (see Problem 500). We have by hypothesis

(1) $$H(n) = \frac{n\tau(n)}{\sigma(n)}.$$

It follows from (1) that H is multiplicative. Thus, since $(2^{H(n)-1}, 2^{H(n)} - 1) = 1$, we have by hypothesis

$$H(n) = H\left(2^{H(n)-1}\left(2^{H(n)} - 1\right)\right) = H(2^{H(n)-1})H(2^{H(n)} - 1)$$
$$= \frac{2^{H(n)-1} \cdot H(n)}{2^{H(n)} - 1}\left(2^{H(n)} - 1\right)\frac{\tau(2^{H(n)} - 1)}{\sigma(2^{H(n)} - 1)}.$$

Simplifying this last relation shows that

$$\frac{\sigma(2^{H(n)} - 1)}{\tau(2^{H(n)} - 1)} = 2^{H(n)-1} = \frac{2^{H(n)}}{2}.$$

Set $m = 2^{H(n)} - 1$. We have thus established that

(2) $$\frac{\sigma(m)}{\tau(m)} = \frac{m+1}{2}, \quad \text{with } m \text{ odd.}$$

But this can occur only if m is prime. Indeed, if m is composite, then $\tau(m) \geq 3$ and we obtain, since the two smallest divisors of m are 1 and some divisor $d \geq 3$,

$$\frac{\sigma(m)}{\tau(m)} < \frac{m + (m/3)(\tau(m) - 2) + 1}{\tau(m)} = \frac{(m/3) + 1}{\tau(m)} + \frac{m}{3} \leq \frac{m}{9} + \frac{1}{3} + \frac{m}{3}$$
$$= \frac{4m}{9} + \frac{1}{3} < \frac{m+1}{2},$$

which contradicts (2). It follows that $2^{H(n)} - 1$ is prime, and the result follows.

(550) *(AMM, Vol. 78, 1971, p. 406)*. Let

$$f(b) = \sum_{a|b}\mu^2(a), \quad g(c) = \sum_{b|c}f(b) \quad \text{and} \quad h(n) = \sum_{c|n}g(c).$$

It is enough to show that $h(n) = \tau^2(n)$. Since the function μ is multiplicative, the functions f, g and h are also multiplicative; and since the function τ is multiplicative, it is enough to show that the result holds when n is a prime power, say $n = p^r$. In this case, we have

$$f(p^r) = \mu^2(1) + \mu^2(p) = 2, \quad f(1) = 1,$$
$$g(p^r) = f(1) + \sum_{s=1}^{r}f(p^s) = 1 + 2r, \quad g(1) = 1,$$
$$h(p^r) = g(1) + \sum_{s=1}^{r}g(p^s) = 1 + \sum_{s=1}^{r}(1 + 2s)$$
$$= 1 + r + r(r+1) = (r+1)^2 = \tau^2(p^r), \quad h(1) = 1.$$

Combining these relations, we find the result.

(551) *(AMM, Vol. 80, 1973, p. 76)*. Let m and n be positive integers such that $(m,n) = 1$, and let $1 \leq a \leq m$, $1 \leq b \leq n$. From the Chinese Remainder Theorem and the properties of f, we have that $m|f(a)$ and $n|f(b)$ if and only if $mn|f(x)$, where $x = x(a,b)$ is the unique integer such that $x \equiv a \pmod{m}$, $x \equiv b \pmod{n}$, and $1 \leq x \leq mn$. It follows that g

is multiplicative. For $d|n$, the number of values of $f(1),\ldots,f(n)$ divisible by d is equal to $(n/d)g(d)$; by the inclusion-exclusion principle, we have

$$h(n) = n - \sum (n/p)g(p) + \sum (n/pq)g(pq) - \cdots ,$$

where the first sum runs over all the prime numbers p such that $p|n$, the second runs over all distinct prime pairs $\{p,q\}$ such that $pq|n$, and so on. It follows that

$$h(n) = n \prod_{p|n} \left(1 - \frac{g(p)}{p}\right),$$

as required.

(552) First of all, it is clear that the function λ is multiplicative, and so is the function $\sum_{d|n} \lambda(d)$. Therefore, it is enough to check that

$$\sum_{d|p^\alpha} \lambda(d) = \begin{cases} 1 & \text{if } 2|\alpha, \\ 0 & \text{otherwise.} \end{cases}$$

But

$$\sum_{d|p^\alpha} \lambda(d) = 1 + \lambda(p) + \lambda(p^2) + \cdots + \lambda(p^\alpha) = 1 - 1 + 1 - 1 + \cdots + (-1)^\alpha,$$

and this last quantity is equal to 1 if α is even and 0 if α is odd, which proves our claim.

(553) Since f is multiplicative, we have

$$\sum_{d|n} \mu(d)f(d) = \prod_{p|n}(1 - f(p)) = (1 - f(2)) \prod_{\substack{p|n \\ p \neq 2}}(1 - f(p))$$

$$= (1 - 1) \prod_{\substack{p|n \\ p \neq 2}}(1 - f(p)) = 0.$$

(554) (T.M. Apostol [1], p. 48). We may reduce the fractions k/n, $k = 1, 2, \ldots, n$, by writing $k/n = a/d$, where $d|n$, $1 \le a \le d$, $(a,d) = 1$, a and d being uniquely determined by the integers k and n. Reciprocally, every fraction a/d such that $d|n$, $1 \le a \le d$, $(a,d) = 1$, can be written in the form k/n, $1 \le k \le n$. From this equivalence, we derive the result.

(555) We have

$$\sum_{d|n} \frac{\phi_m(d)}{d^m} = \sum_{d|n} \sum_{\substack{k=1 \\ (k,d)=1}}^{d} \left(\frac{k}{d}\right)^m .$$

Using Problem 554, we have

$$\sum_{d|n} \sum_{\substack{k=1 \\ (k,d)=1}}^{d} \left(\frac{k}{d}\right)^m = \sum_{k=1}^{n} \left(\frac{k}{n}\right)^m = \frac{1^m + \cdots + n^m}{n^m} .$$

Using the Moebius Inversion Formula, we obtain

$$\frac{\phi_m(n)}{n^m} = \sum_{d|n} \mu(d) F\left(\frac{n}{d}\right),$$

so that

$$\phi_m(n) = \sum_{d|n} d^m \mu(d) \left(1^m + 2^m + \cdots + \left(\frac{n}{d}\right)^m\right).$$

(556) This follows from Problem 555.

(557) This follows from Problem 555.

(558) This follows from Problem 555.

(559) Taking into account the multiplicative character of f, we obtain the formula

$$f(n) = \prod_{p^\alpha} f(p^\alpha) = \prod_{p^\alpha} \left(1 + \frac{1}{2}\right) = \left(\frac{3}{2}\right)^{\omega(n)}.$$

(560) The answer is YES. Indeed, since $\sigma(n) = \sum_{d|n} d$, the Moebius Inversion Formula easily yields the result.

(561) Indeed, we have

$$\sum_{d|n} \frac{1}{d^2} = \frac{1}{n^2} \sum_{d|n} \frac{n^2}{d^2} = \frac{1}{n^2} \sum_{d|n} d^2 = \frac{\sigma_2(n)}{n^2}.$$

(562) It follows from Problem 1 that

$$\sum_{r=1}^m r^3 = \left(\sum_{r=1}^m r\right)^2,$$

so that if $n = q_1^{\alpha_1} \cdots q_k^{\alpha_k}$ is the representation of n as a product of distinct prime powers, then

$$\sum_{d|n} (\tau(d))^3 = \sum_{d_1|q_1^{\alpha_1},\ldots,d_k|q_k^{\alpha_k}} \tau^3(d_1) \cdots \tau^3(d_k)$$

$$= \prod_{j=1}^k \sum_{d_j|q_j^{\alpha_j}} \tau^3(d_j) = \prod_{j=1}^k (1^3 + 2^3 + \cdots + (\alpha_j + 1)^3)$$

$$= \left(\prod_{j=1}^k \left(1 + 2 + \cdots + (\alpha_j + 1)\right)\right)^2 = \left(\sum_{d|n} \tau(d)\right)^2.$$

(563) These results are easily obtained from the relation

$$\sum_{d|n} \mu(d) f(d) = \prod_{p|n} (1 - f(p))$$

proved in the solution of Problem 553.

(564) Since μ and ϕ are multiplicative functions, μ^2/ϕ is clearly multiplicative and therefore $\sum_{d|n} \mu^2(d)/\phi(d) = F(n)$ is also multiplicative. But if p is prime and $a \in \mathbb{N}$, we have

$$F(p^a) = 1 + \frac{1}{p-1} = \frac{p}{p-1} = \frac{p^a}{p^{a-1}(p-1)} = \frac{p^a}{\phi(p^a)},$$

and we easily obtain that $F(n) = n/\phi(n)$.

(565) Since the function F is multiplicative, it is enough to find the value of $F(p^\alpha)$ for p prime and $\alpha \in \mathbb{N}$. Since $F(p^\alpha) = g(p^\alpha) - g(p^{\alpha-1})$, the result then follows immediately.

(566) Since the function f is multiplicative (see Problem 468), it is enough to find the value of

$$f(p^\alpha) = \sum_{[d,r]=p^\alpha} \phi(d)\phi(r),$$

with p prime and $\alpha \in \mathbb{N}$. Since $[d,r] = p^\alpha$ implies $d = p^a$, $r = p^b$ and $\max\{a,b\} = \alpha$, it follows that

$$f(p^\alpha) = \sum_{\substack{a=\alpha \\ b<\alpha}} \phi(p^a)\phi(p^b) + \sum_{\substack{a\le\alpha \\ b=\alpha}} \phi(p^a)\phi(p^b)$$

$$= \phi(p^\alpha) \sum_{d|p^{\alpha-1}} \phi(d) + \phi(p^\alpha) \sum_{d|p^\alpha} \phi(d) = p^{2\alpha}\left(1 - \frac{1}{p^2}\right).$$

(567) We have

$$\sum_{d|n} \Lambda(d) = \sum_{\substack{d|n \\ d=p^\alpha}} \Lambda(d) = \sum_{p^\alpha|n} \Lambda(p^\alpha) = \sum_{p^\alpha|n} \log p$$

$$= \sum_{p^\alpha\|n} \alpha \log p = \sum_{p^\alpha\|n} \log p^\alpha = \log \prod_{p^\alpha\|n} p^\alpha = \log n.$$

(568) For $n = 1$, the result is trivial. So let $n \ge 2$. Using Problem 567 and the Moebius Inversion Formula, we then obtain

$$\Lambda(n) = \sum_{d|n} \mu(d) \log(n/d) = \log n \sum_{d|n} \mu(d) - \sum_{d|n} \mu(d) \log d$$

$$= - \sum_{d|n} \mu(d) \log d.$$

(569) If n is odd, n/d is odd and therefore $(-1)^{n/d} = -1$, so that the result is established for n odd.

If n is even, set $n = 2^k m$, where $k \in \mathbb{N}$ and $(m,2) = 1$. Let $d_1, d_2, \ldots, d_r$ be the divisors of m. These being clearly all odd numbers, it follows that

$$d_1, d_2, \ldots, d_r,$$
$$2d_1, 2d_2, \ldots, 2d_r,$$
$$2^2 d_1, 2^2 d_2, \ldots, 2^2 d_r,$$
$$\vdots$$
$$2^k d_1, 2^k d_2, \ldots, 2^k d_r$$

represent all the divisors of n. We then have

$$\sum_{d|n}(-1)^{n/d}f(d) = \sum_{i=1}^{r}\left(\sum_{d_i|n}(-1)^{n/d_i}f(d_i) + \sum_{2d_i|n}(-1)^{n/2d_i}f(2d_i) + \cdots\right.$$

$$\left. + \sum_{2^{k-1}d_i|n}(-1)^{n/2^{k-1}d_i}f(2^{k-1}d_i) + \sum_{2^kd_i|n}(-1)^{n/2^kd_i}f(2^kd_i)\right)$$

$$= \sum_{i=1}^{r}\left(\sum_{d_i|n}f(d_i) + \sum_{2d_i|n}f(2d_i) + \cdots + \sum_{2^{k-1}d_i|n}f(2^{k-1}d_i) - \sum_{2^kd_i|n}f(2^kd_i)\right)$$

$$= \left(f(1) + f(2) + \cdots + f(2^{k-1}) - f(2^k)\right)\sum_{i=1}^{r}\sum_{d_i|m}f(d_i)$$

$$= \sum_{d|2^k}f(d)\sum_{i=1}^{r}\sum_{d_i|m}f(d_i) - 2f(2^k)\sum_{i=1}^{r}\sum_{d_i|m}f(d_i),$$

and the result follows.

(570) Set $n = 2^km$, where $k \in \mathbb{N}$ and $(m, 2) = 1$. Let $d_1, d_2, \ldots, d_r$ be the divisors of m. These being all clearly odd, it follows that

$$d_1, d_2, \ldots, d_r,$$
$$2d_1, 2d_2, \ldots, 2d_r,$$
$$2^2d_1, 2^2d_2, \ldots, 2^2d_r,$$
$$\vdots$$
$$2^kd_1, 2^kd_2, \ldots, 2^kd_r$$

represent all the divisors of n. We then have

$$\sum_{d|n}(-1)^{n/d}\phi(d) = \sum_{i=1}^{r}\left(\sum_{d_i|n}(-1)^{n/d_i}\phi(d_i) + \sum_{2d_i|n}(-1)^{n/2d_i}\phi(2d_i) + \cdots\right.$$

$$\left. + \sum_{2^{k-1}d_i|n}(-1)^{n/2^{k-1}d_i}\phi(2^{k-1}d_i) + \sum_{2^kd_i|n}(-1)^{n/2^kd_i}\phi(2^kd_i)\right)$$

$$= \sum_{i=1}^{r}\left(\sum_{d_i|n}\phi(d_i) + \sum_{2d_i|n}\phi(2d_i) + \cdots + \sum_{2^{k-1}d_i|n}\phi(2^{k-1}d_i) - \sum_{2^kd_i|n}\phi(2^kd_i)\right)$$

$$= \left(\phi(1) + \phi(2) + \cdots + \phi(2^{k-1}) - \phi(2^k)\right)\sum_{i=1}^{r}\left(\sum_{d_i|m}\phi(d_i)\right).$$

Since for each $k \geq 1$,

$$\phi(1) + \phi(2) + \cdots + \phi(2^{k-1}) - \phi(2^k) = 1 + 1 + 2 + 2^2 + \cdots$$
$$+ 2^{k-2} - 2^{k-1} = 0,$$

the result follows.

If n is odd, it is easy to see that the given sum is equal to $-n$.

(571) Using the Moebius Inversion Formula and the second formula of Theorem 30, we obtain

$$f(n) = \sum_{d|n} \mu(d)\frac{n}{d} = n\sum_{d|n} \frac{\mu(d)}{d} = n\prod_{p|n}\left(1 - \frac{1}{p}\right) = \phi(n).$$

(572) In each case, using the Moebius Inversion Formula, we easily find that
(a)

$$g(n) = \phi(n)\prod_{i=1}^{r} q_i^{a_i-1}(q_i + 1),$$

(b) g is the multiplicative function verifying

$$g(p^r) = \begin{cases} 1 & \text{if } r = 0 \text{ or } r = 2, \\ -2 & \text{if } r = 1, \\ 0 & \text{if } r \geq 3. \end{cases}$$

(573) The first part is trivial. For the second part, using the Moebius Inversion Formula and the fact that we only need to estimate $g(p^\alpha)$, we easily obtain that $g(n) = \mu^2(n)\sigma(n)/\phi(n)$.

(574) These relations are easily obtained by observing that each of these functions is multiplicative and therefore that the given equations only need to be verified on prime powers.

(575) Since both sides of the relation represent multiplicative functions, it is enough to prove this relation when n is a prime power, that is when $n = p^\alpha$. But

$$(1) \qquad \sum_{d|p^\alpha} \mu(d)\lambda\left(\frac{p^\alpha}{d}\right) = \mu(1)\lambda(p^\alpha) + \mu(p)\lambda(p^{\alpha-1}) = (-1)^\alpha - (-1)^{\alpha-1}$$

$$= \begin{cases} 2 & \text{if } \alpha \text{ is even}, \\ -2 & \text{if } \alpha \text{ is odd}, \end{cases}$$

so that

$$(2) \qquad (-1)^{\Omega(p^\alpha)}2^{\omega(p^\alpha)} = 2 \cdot (-1)^\alpha = \begin{cases} 2 & \text{if } \alpha \text{ is even}, \\ -2 & \text{if } \alpha \text{ is odd}. \end{cases}$$

Comparing (1) and (2), the result follows.

(576) Using the function log and then the Moebius Inversion Formula, the result follows.

(577) Since $\prod_{d|n} d^k = \prod_{d|n} \left(\frac{n}{d}\right)^k$, we have

$$\left(\prod_{d|n} d^k\right)^2 = \prod_{d|n} d^k \prod_{d|n}\left(\frac{n}{d}\right)^k = n^{k\tau(n)};$$

that is

$$\prod_{d|n} d^k = n^{k\tau(n)/2}.$$

Using Problem 576 with $f(n) = n^{k\tau(n)/2}$ and $g(d) = d^k$, we obtain the result.

(578) The answer is NO. A counter-example is provided by the functions $F(n) = \sigma(n)$ and $f(n) = n$. Indeed, it is clear that f is totally multiplicative, while F is not totally multiplicative.

(579) We first write n as $n = m^2 r$, with $(m, r) = 1$ and r squarefree, in which case

$$\Sigma_0 := \sum_{\substack{d|n \\ \mu^2(d)=1}} \mu(n/d) = \sum_{\substack{d|m^2 r \\ \mu^2(d)=1}} \mu\left(\frac{m^2 r}{d}\right)$$

$$= \sum_{\substack{d_1|m^2,\ d_2|r \\ \mu^2(d_1)=\mu^2(d_2)=1}} \mu\left(\frac{m^2}{d_1}\right) \mu\left(\frac{r}{d_2}\right).$$

But

$$\mu\left(\frac{m^2}{d_1}\right) \neq 0, \quad \text{with } \mu^2(d_1) = 1 \Longleftrightarrow d_1 = m,$$

so that $\mu\left(\dfrac{m^2}{d_1}\right) = \mu(m)$. Thus,

$$\Sigma_0 = \mu(m) \sum_{\substack{d_2|r \\ \mu^2(d_2)=1}} \mu\left(\frac{r}{d_2}\right) = \mu(m) \sum_{d_2|r} \mu\left(\frac{r}{d_2}\right)$$

$$= \mu(m) \sum_{e|r} \mu(e) = \begin{cases} \mu(m) & \text{if } r = 1, \\ 0 & \text{if } r > 1. \end{cases}$$

Since in the case $r = 1$, we have $n = m^2$, the result follows.

(580) (*AMM, Vol. 64, 1957, p. 45*). We have

$$\prod_{d|n} d^{f(d)+f(n/d)} = \prod_{d|n} d^{f(d)} \prod_{d|n} \left(\frac{n}{d}\right)^{f(d)} = n^{\sum_{d|n} f(d)}.$$

The relation (∗) then follows from the fact that $\sum_{d|n} \phi(d) = n$ (see Theorem 30).

(581) (a) We have

$$(f * g)(n) = \sum_{d|n} f(d)g(n/d) = \sum_{d|n} f(n/d)g(d) = (g * f)(n),$$

for each positive integer n, as required.

(b) It is enough to show that, for each positive integer n, $((f * g) * h)(n) = (f * (g * h))(n)$, a relation which can easily be obtained using the definition of the Dirichlet product.

(c) For each positive integer n, we have

$$(f * E)(n) = \sum_{d|n} f(d)E(n/d) = f(n),$$

since $E(n/d) = 0$ if $d < n$. Similarly, $(E * f)(n) = f(n)$ for each positive integer n.

(d) Let f be a given arithmetical function. We will show that the equation $(f * f^{-1})(n) = E(n)$ has a unique solution f^{-1}. For $n = 1$, we must solve $(f * f^{-1})(1) = E(1)$, that is $f(1)f^{-1}(1) = 1$. Since $f(1) \neq 0$, we have $f^{-1}(1) = 1/f(1)$. Assume now that the values of $f^{-1}(k)$ are uniquely determined for all $k < n$. We must then solve the equation $(f * f^{-1})(n) = E(n)$, that is

$$\sum_{d|n} f(n/d)f^{-1}(d) = 0,$$

which can be written as

$$f(n)f^{-1}(1) + \sum_{\substack{d|n \\ d<n}} f(n/d)f^{-1}(d) = 0.$$

If the values $f^{-1}(d)$ are known for all the divisors $d < n$, then the value $f^{-1}(n)$ is uniquely determined by

$$f^{-1}(n) = \frac{-1}{f(1)} \sum_{\substack{d|n \\ d<n}} f(n/d)f^{-1}(d),$$

since $f(1) \neq 0$. This therefore establishes the existence and the uniqueness of f^{-1} by induction on n.

(582) We proceed by contradiction by assuming that f is not multiplicative and then by showing that this implies that $h := f * g$ is not multiplicative. Hence assume that f is not multiplicative; that is there exists a pair of positive integers m, n such that $(m, n) = 1$ and such that $f(mn) \neq f(m)f(n)$. We choose a pair m, n in such a way that the product mn is minimal (amongst all those pairs satisfying this property).

First of all, if $mn = 1$, it means that $f(1) \neq f(1)f(1)$ so that $f(1) \neq 1$. But since $h(1) = f(1)g(1) = f(1) \neq 1$, it follows that h is not multiplicative, creating a contradiction.

If $mn > 1$, then we have, by our minimal choice, that $f(ab) = f(a)f(b)$ for all positive integers a and b such that $(a, b) = 1$ and $ab < mn$. We may therefore write

$$\begin{aligned}
h(mn) &= \sum_{\substack{a|m,\ b|n \\ ab<mn}} f(ab)g\left(\frac{mn}{ab}\right) + f(mn)g(1) \\
&= \sum_{\substack{a|m,\ b|n \\ ab<mn}} f(a)f(b)g\left(\frac{m}{a}\right)g\left(\frac{n}{b}\right) + f(mn) \\
&= \sum_{a|m} f(a)g\left(\frac{m}{a}\right)\sum_{b|n} f(b)g\left(\frac{n}{b}\right) - f(m)f(n) + f(mn) \\
&= h(m)h(n) - f(m)f(n) + f(mn).
\end{aligned}$$

Since $f(mn) \neq f(m)f(n)$, this proves that $h(mn) \neq h(m)h(n)$, thereby contradicting the fact that h is multiplicative. Thus the result.

(583) Applying the result of Problem 582 by taking $g = f^{-1}$ and observing that $f * g = f * f^{-1} = E$ is multiplicative, the result follows.

(584) (a) This follows from the fact that

$$(\mu * \iota_0)(n) = \sum_{d|n} \mu(d) = E(n).$$

(b) We have

$$(\iota_r * \iota_0)(n) = \sum_{d|n} d^r = \sigma_r(n).$$

(c) We have

$$(\iota_1 * \mu)(n) = \sum_{d|n} \frac{n}{d}\mu(d) = \phi(n).$$

(d) By definition, $\sigma(n) = \sum_{d|n} \iota_1(d)$. Therefore, using the Moebius Inversion Formula, we obtain

$$n = \iota_1(n) = \sum_{d|n} \mu(d)\sigma(n/d) = (\mu * \sigma)(n),$$

hence, the required equality.

(e) Using (a), (b) and (c), we have

$$\phi * \sigma_r = \iota_1 * \mu * \iota_r * \iota_0 = \iota_1 * \iota_r * \mu * \iota_0 = \iota_1 * \iota_r * E = \iota_1 * \iota_r.$$

(f) This follows from

$$(\iota_1 * \iota_1)(n) = \sum_{d|n} \frac{n}{d}d = n\sum_{d|n} 1 = (\iota_1\tau)(n).$$

(g) Since $\tau = \iota_0 * \iota_0$, we have

$$f(n) = \sum_{d|n} \tau(d) = (\iota_0 * \tau)(n) = (\iota_0 * \iota_0 * \iota_0)(n).$$

(585) The first part follows from the fact that $\mu * \iota_0 = E$ (see Problem 584). For the general case, we proceed as follows. Since ι_r^{-1} is a multiplicative function, we shall estimate $\iota_r^{-1}(p^k)$, $k \geq 1$. Using Problem 581, we have

$$\begin{aligned}
\iota_r^{-1}(p) &= -\iota_r(p)\iota_r^{-1}(1) = -p^r, \\
\iota_r^{-1}(p^2) &= -\iota_r(p^2)\iota_r^{-1}(1) - \iota_r(p)\iota_r^{-1}(p) = 0.
\end{aligned}$$

Using induction, one can show that $\iota_r^{-1}(p^k) = 0$ if $k > 1$, that is

$$\iota_r^{-1}(p^\alpha) = \begin{cases} -p^r & \text{if } \alpha = 1, \\ 0 & \text{if } \alpha > 1. \end{cases}$$

Hence,

$$\iota_r^{-1}(n) = \begin{cases} \prod_{p|n}(-p^r) & \text{if } n \text{ is squarefree,} \\ 0 & \text{otherwise,} \end{cases}$$

which can be written as

$$\iota_r^{-1}(n) = \mu(n)\iota_r(n) \quad \text{or} \quad \iota_r^{-1} = \mu\,\iota_r.$$

(586) (i) Since $(\mu * \iota_0)(n) = \sum_{d|n} \mu(d) = E(n)$, it follows that $\iota_0^{-1} = \mu$. For (ii), we have $(E * E)(n) = \sum_{d|n} E(n/d)E(d) = E(n)$, so that $E^{-1} = E$. Finally, for (iii), we obtain $|\mu|^{-1} = \lambda$.

(587) We have successively

$$
\begin{aligned}
(f * g)^{-1} * (f * g) &= E, \\
(f * g)^{-1} * g * f &= E, \\
(f * g)^{-1} * g &= f^{-1}, \\
(f * g)^{-1} &= f^{-1} * g^{-1}.
\end{aligned}
$$

(588) Since $\sum_{d|n} \phi(d) = n$, we have $\iota_0 * \phi = \iota_1$. Hence, $\phi = \iota_1 * \mu$ and $\phi^{-1} = \iota_1^{-1} * \iota_0$. Therefore, using Problem 585, we have

$$
\phi^{-1}(n) = \sum_{d|n} \iota_1^{-1}(d) = \sum_{d|n} \mu(d)d = \prod_{p|n}(1 - p),
$$

as required.

(589) We first prove the necessity. Assume that f is totally multiplicative and set $g(n) = \mu(n)f(n)$. We then have

$$
(g * f)(n) = \sum_{d|n} \mu(d)f(d)f(n/d) = f(n) \sum_{d|n} \mu(d) = f(n)E(n) = E(n),
$$

and the result follows.

To prove the sufficiency, we proceed as follows. Since $f^{-1}(n) = \mu(n)f(n)$, we have

$$(*)\qquad \sum_{d|n} \mu(d)f(d)f(n/d) = 0 \quad (n \geq 2).$$

Hence, if $n = p^2$,

$$\mu(1)f(1)f(p^2) + \mu(p)f(p)f(p) = 0,$$

and this shows that $f(p^2) = (f(p))^2$. Using this and again $(*)$, we obtain that $f(p^3) = (f(p))^3$, and so on. It follows that f is totally multiplicative.

(590) Since the Liouville function λ is totally multiplicative, using Problem 589, we obtain the result.

(591) Since the function $\iota_a(n) = n^a$ is totally multiplicative and since $\sigma_a = \iota_a * \iota_0$, we have, using Problems 587 and 589,

$$\sigma_a^{-1} = (\iota_a * \iota_0)^{-1} = \iota_a^{-1} * \iota_0^{-1} = \iota_a^{-1} * \mu = \iota_a\mu * \mu,$$

thus the result.

(592) If $m = p^a$ and $n = p^{a+b}$, then $(m,n) = p^a$ and

$$\sigma(m)\sigma(n) = \frac{p^{a+1} - 1}{p - 1}\frac{p^{a+b+1} - 1}{p - 1}.$$

On the other hand,

$$
\sum_{d|(m,n)} d\sigma\left(\frac{mn}{d^2}\right) = \sum_{k=0}^{a} p^k \sigma(p^{2a+b-2k})
$$

$$
= \sum_{k=0}^{a} p^k \frac{p^{2a+b-2k+1} - 1}{p - 1} = \frac{1}{p-1}\left\{ p^{a+b+1} \sum_{k=0}^{a} p^{a-k} - \sum_{k=0}^{a} p^k \right\}
$$

$$
= \frac{p^{a+1} - 1}{p - 1}\frac{p^{a+b+1} - 1}{p - 1} = \sigma(m)\sigma(n).
$$

If $m = q_1^{a_1} \cdots q_k^{a_k}$ and $n = q_1^{b_1} \cdots q_k^{a_k}$, then

$$\sigma(m)\sigma(n) = \prod_{i=1}^{k} \sigma(q_i^{a_i})\sigma(q_i^{b_i}) = \prod_{i=i}^{k} \sum_{d_i \mid (q_i^{a_i}, q_i^{b_i})} d_i \sigma \left(\frac{q_i^{a_i} q_i^{b_i}}{d_i^2} \right)$$

$$= \sum_{d \mid (m,n)} d\sigma(mn/d^2).$$

(593) (McCarthy [**23**]) It is enough to show that (a) $\Rightarrow$ (c) $\Rightarrow$ (a) $\Rightarrow$ (b) $\Rightarrow$ (a). We shall prove each of these implications one by one.

$(a) \Rightarrow (c)$ Let p be prime and $a \in \mathbb{N}$. Set $m = p^a$ and $n = p$ in (1). The result gives

$$f(p^{a+1}) = f(p)f(p^a) + f(p^{a-1})F(p).$$

For $a = 1$, we obtain $F(p) = f(p^2) - f^2(p)$, which completes the proof of this part.

$(c) \Rightarrow (a)$ Assume that (3) is true. If $(mn, m'n') = 1$, then $\big((m,n),(m',n')\big) = 1$ and $(mm', nn') = (m,n)(m',n')$. To show that (1) holds for all m and n, it is enough to show that there exists a multiplicative function F such that for each prime number p and each pair of positive integers a and b,

$$f(p^{a+b}) = \sum_{i=0}^{\min(a,b)} f(p^{a-i})f(p^{b-i})F(p^i).$$

We will show that this is the case when $F = \mu g$, where g is the totally multiplicative function defined by $g(p) = f(p^2) - f^2(p)$ for each prime number p.

Without any loss in generality, we may assume that $b \leq a$. We proceed by induction on b. The equation (3) is the one we want to establish when $b = 1$. Assume that $b > 1$ and that the equation is true when b is replaced by $b-1$, for all $a \geq b-1$. Since $F = \mu g$, we have $F(p^2) = F(p^3) = \ldots = 0$ and therefore

$$\begin{aligned} f(p^{a+b}) = f(p^{a+1+b-1}) &= f(p^{a+1})f(p^{b-1}) + f(p^a)f(p^{b-2})F(p) \\ &= \big(f(p)f(p^a) - f(p^{a-1})g(p)\big)f(p^{b-1}) - f(p^a)f(p^{b-2})g(p) \\ &= f(p^a)\big(f(p)f(p^{b-1}) - f(p^{b-2})g(p)\big) - f(p^{a-1})f(p^{b-1})g(p) \\ &= f(p^a)f(p^b) + f(p^{a-1})f(p^{b-1})F(p), \end{aligned}$$

as was to be shown.

$(a) \Rightarrow (b)$ If (1) is true for all m and n, then

$$\sum_{d|(m,n)} f(mn/d^2)g(d)$$

$$= \sum_{d|(m,n)} \sum_{D|(m/d,n/d)} f\left(\frac{m/d}{D}\right) f\left(\frac{n/d}{D}\right) \cdot \mu(D)g(D)g(d)$$

$$= \sum_{d|(m,n)} \sum_{\substack{e|(m,n)\\d|e}} f(m/e)f(n/e)\mu(e/d)g(e)$$

$$= \sum_{e|(m,n)} f(m/e)f(n/e)g(e) \sum_{d|e} \mu(e/d) = f(m)f(n).$$

$(b) \Rightarrow (a)$ We have

$$\sum_{d|(m,n)} f(m/d)f(n/d)F(d) = \sum_{d|(m,n)} F(d) \sum_{D|(m/d,n/d)} f(mn/d^2D^2)g(d),$$

and since $F = \mu\, g$, then

$$\sum_{d|(m,n)} f(m/d)f(n/d)F(d) = \sum_{d|(m,n)} \mu(d)g(d) \sum_{\substack{e|(m,n)\\d|e}} f(mn/e^2)g(e/d)$$

$$= \sum_{d|(m,n)} \mu(d) \sum_{\substack{e|(m,n)\\d|e}} f(mn/e^2)g(e) = \sum_{e|(m,n)} f(mn/d^2)g(e) \sum_{d|e} \mu(d)$$

$$= f(mn).$$

(594) It is clear that the relation we want to prove is equivalent to $\sigma * \iota_0 = \iota_1 * \tau$. Using Problem 584(b), we then have

$$\sigma * \iota_0 = \iota_1 * \iota_0 * \iota_0 = \iota_1 * \tau.$$

(595) The formula we want to prove can be written as $\phi * \tau = \sigma$. Using Problem 584(b) and (c), we then have

$$\phi * \tau = \iota_1 * \mu * \iota_0 * \iota_0 = (\iota_0 * \mu) * (\iota_1 * \iota_0) = (\iota_0 * \mu) * \sigma = E * \sigma = \sigma.$$

(596) The formula we want to prove can be written as $\phi*\sigma = \iota_1\tau$. Using Problem 584, we have

$$\phi * \sigma = \iota_1 * \mu * \iota_1 * \iota_0 = (\iota_1 * \iota_1) * (\mu * \iota_0) = \iota_1 * \iota_1 * E = \iota_1 * \iota_1 = \iota_1\tau.$$

(597) We must show that $\iota_1 * \sigma = \iota_1\tau * \iota_0$. Using Problem 584, we have

$$\iota_1 * \sigma = \iota_1 * \iota_1 * \iota_0 = \iota_1\tau * \iota_0.$$

(598) The relation we want to prove is equivalent to $\iota_1\sigma * \iota_0 = \sigma * \iota_2$. Since

$$(\iota_1\sigma)(n) = n\sigma(n) = n\sum_{d|n} d = \sum_{d|n} nd = \sum_{d|n} \frac{n}{d}d^2 = (\iota_2 * \iota_1)(n),$$

we have

$$\iota_1\sigma * \iota_0 = \iota_2 * \iota_1 * \iota_0 = \iota_2 * \sigma,$$

and since $\iota_1 * \iota_0 = \sigma$, the result follows.

(599) The relation we want to prove is equivalent to $\iota_r \sigma_{k-r} * \iota_0 = \iota_k * \sigma_r$. Since

$$(\iota_k * \iota_r)(n) = \sum_{d|n} \left(\frac{n}{d}\right)^r d^k = n^r \sum_{d|n} d^{k-r} = n^r \sigma_{k-r}(n) = (\iota_r \sigma_{k-r})(n),$$

we have

$$\iota_r \sigma_{k-r} * \iota_0 = \iota_k * \iota_r * \iota_0 = \iota_k * \sigma_r,$$

and since $(\iota_r * \iota_0)(n) = \sum_{d|n} d^r = \sigma_r(n)$, the result follows.

(600) Since $\sigma = \iota_1 * \iota_0$, $\iota_1 * \iota_1 = \iota_1 \tau$ and since $\iota_0 * \iota_0 = \tau$, it follows that

$$\sigma * \sigma = (\iota_1 * \iota_0) * (\iota_1 * \iota_0) = (\iota_1 * \iota_1) * (\iota_0 * \iota_0) = \iota_1 \tau * \tau,$$

which is the desired identity.

(601) The relation we need to prove is equivalent to

$$\sigma_r * \sigma_r = \iota_r \tau * \tau.$$

Since $\sigma_r = \iota_r * \iota_0$ and since $\iota_r * \iota_r = \iota_r \tau$, we have

$$\sigma_r * \sigma_r = (\iota_r * \iota_0) * (\iota_r * \iota_0) = \iota_r * \iota_r * \tau = \iota_r \tau * \tau,$$

which proves the identity.

(602) We must prove that $\mu * \tau = \iota_0$. But we have

$$\mu * \tau = \mu * (\iota_0 * \iota_0) = (\mu * \iota_0) * \iota_0 = E * \iota_0 = \iota_0.$$

Another way to obtain this equality is as follows. From the Moebius Inversion Formula, we must show that $\tau(n) = \sum_{d|n} 1$, which happens to be the definition of the function τ.

(603) In Problem 567, we have seen that $\sum_{d|n} \Lambda(d) = \log n$. Therefore, using the Moebius Inversion Formula, we obtain

$$\Lambda(n) = \sum_{d|n} \mu(d) \log(n/d) = (\mu * \log)(n),$$

thus the result.

(604) We have

$$\sum_{d|n} \mu^2(d) \Lambda(d) = \sum_{\substack{d|n \\ d=p^\alpha}} \mu^2(d) \Lambda(d) = \sum_{\substack{d|n \\ d=p}} \mu^2(d) \Lambda(d)$$

$$= \sum_{p|n} \mu^2(p) \log p = \sum_{p|n} \log p = \log \prod_{p|n} p = \log \delta(n).$$

(605) (*Contribution of Imre Kátai, Budapest*). Assume the contrary, that is that $f(n) = C$ (with $C = 1$ or $C = -1$) on I. Then let

$$K := \left\{ k : \frac{N+M}{k+1} - \frac{N}{k} \geq 1 \right\} \quad \text{and} \quad I_k := \left[\frac{N}{k}, \frac{N+M}{k} \right].$$

If $k \in K$, then it is clear that $I_k \cap I_{k+1}$ contains an interval of length ≥ 1, which implies that there exists an integer $\ell \in I_k \cap I_{k+1}$. But, if $m \in I_k$, then $km \in I$ and therefore $f(km) = f(k)f(m) = C$; similarly, if $r \in I_{k+1}$, then $(k+1)r \in I$ and therefore $f((k+1)r) = f(k+1)f(r) = C$. Hence, by choosing $m = r = \ell$, we obtain $f(k) = f(k+1)$. On the other hand, we observe that K contains an interval of the form $[L, 2L]$ such that $2L \geq \sqrt{N}$. Indeed, it is true that

$$(*) \qquad \frac{N+M}{k+1} - \frac{N}{k} \geq 1$$

if $(N+M)k - N(k+1) > k(k+1)$, that is if $Mk - N > k^2 + k$; that is $M > \frac{N}{k} + k + 1$. Therefore if $\frac{\sqrt{N}}{2} - 1 < k \leq \sqrt{N} + 3$, then $(*)$ holds. So let $2L$ be the largest even integer $\leq \sqrt{N} + 3$. Then, $2L > \sqrt{N} > n_0$. Therefore, $f(n) = C$ on $[L, 2L]$. We then have two possible cases:

(a) The case $C = -1$: Since there exists an integer m such that $m^2 \in [L, 2L]$ (see Problem 209), it follows that $f(m^2) = 1$, which, by our assumption, cannot occur.

(b) The case $C = 1$: If $f(2) = -1$, then $f(2L) = -f(L)$ and we are done. Otherwise, $f(2) = 1$, and then necessarily one of the elements $n_0, 2n_0, 2^2 n_0, 2^3 n_0, \dots$ belongs to the interval $[L, 2L]$, an element we shall denote by $2^\ell n_0$. But, $f(2^\ell n_0) = f(n_0) = -1$; hence, there exists an integer $n \in L, 2L]$ such that $f(n) = -1$, which contradicts the assumption $f(n) = 1$.

(606) (*Contribution of Imre Kátai, Budapest*). We easily verify the result for $N = 2, 3, 4$. So let $N \geq 5$. Then, applying the result of Problem 605 with $M = 3\sqrt{N}$ and observing that the conditions of Problem 605 are indeed fulfilled, we find the result.

(607) Using Theorem 29, we have

$$(*) \qquad \sum_{n \leq x} \sum_{d \mid n} \mu(d) = \sum_{n \leq x} E(n) = 1.$$

But the left-hand side of $(*)$ is

$$\sum_{n \leq x} \sum_{d \mid n} \mu(d) = \sum_{d \leq x} \mu(d) \sum_{\substack{n \leq x \\ d \mid n}} 1 = \sum_{d \leq x} \mu(d) \left[\frac{x}{d} \right].$$

Combining this last relation with $(*)$, we obtain the result.

(608) First of all, we observe that

$$(*) \qquad \left| \sum_{d \leq x} \mu(d) \left(\frac{x}{d} - \left[\frac{x}{d} \right] \right) \right| = \left| \sum_{d \leq x-1} \mu(d) \left(\frac{x}{d} - \left[\frac{x}{d} \right] \right) \right|$$

$$\leq \sum_{d \leq x-1} |\mu(d)| \leq x - 1.$$

On the other hand, since $\sum_{d \leq x} \mu(d)[x/d] = 1$ (see Problem 607), we derive from $(*)$ that

$$\left| x \sum_{d \leq x} \frac{\mu(d)}{d} - 1 \right| \leq x - 1.$$

Therefore,

$$\left| x \sum_{d \leq x} \frac{\mu(d)}{d} \right| = \left| x \sum_{d \leq x} \frac{\mu(d)}{d} - 1 + 1 \right| \leq x - 1 + 1 = x,$$

and the result follows.

(609) (*Putnam 1971*). It is clear that $\delta(2m+1) = 2m+1$ and that $\delta(2m) = \delta(m)$. Therefore, setting

$$S(m) = \sum_{n=1}^{m} \frac{\delta(n)}{n},$$

we obtain that $S(2k+1) = S(2k) + 1$. It follows that

$$S(2m) = \sum_{r=1}^{m} \frac{\delta(2r)}{2r} + \sum_{r=1}^{m} \frac{\delta(2r-1)}{2r-1} = \frac{1}{2}S(m) + m.$$

Letting

$$F(m) = S(m) - \frac{2m}{3},$$

we have then that

$$F(2m) = S(2m) - \frac{4m}{3} = \frac{1}{2}S(m) - \frac{m}{3} = \frac{1}{2}\left(S(m) - \frac{2m}{3}\right) = \frac{1}{2}F(m),$$

$$F(2m+1) = S(2m+1) - \frac{2(2m+1)}{3} = S(2m) + 1 - \frac{4m}{3} - \frac{2}{3}$$

$$= F(2m) + \frac{1}{3}.$$

Using induction, one easily shows that

$$0 < F(m) < \frac{2}{3}, \quad \text{for each integer } m \geq 1.$$

We have thus proved that

$$0 < S(m) - \frac{2m}{3} < \frac{2}{3} < 1,$$

and the result follows.

(610) It is enough to observe that since n can be written as $n = mr^2$, with $\mu^2(m) = 1$, then $d|r \iff d^2|n$.

(611) Using the inclusion-exclusion principle, we obtain

$$\sum_{n \leq N} f(n) \leq [N] - \left[\frac{N}{2}\right] - \left[\frac{N}{3}\right] + \left[\frac{N}{6}\right].$$

Since $[y] \leq y$ and since $-[y] \leq -y + 1$, we then have

$$\sum_{n \leq N} f(n) \leq N - \frac{N}{2} - \frac{N}{3} + \frac{N}{6} + 2 = N\left(1 - \frac{1}{2}\right)\left(1 - \frac{1}{3}\right) + 2 = \frac{N}{3} + 2,$$

and the result follows.

(612) The answer is YES. Indeed, first define f as follows:

$$f(p^k) = \begin{cases} 1 & \text{if } p > 2 \text{ and } k \geq 1, \\ 1 & \text{if } p = 2 \text{ and } k \geq 2, \\ 0 & \text{if } p = 2 \text{ and } k = 1. \end{cases}$$

Then, since each positive integer n is either odd, a multiple of 4 or else congruent to 2 modulo 4, we have

$$\sum_{n \leq N} f(n) = \sum_{\substack{n \leq N \\ n \equiv 1 \pmod 2}} f(n) + \sum_{\substack{n \leq N \\ n \equiv 0 \pmod 4}} f(n) + \sum_{\substack{n \leq N \\ n \equiv 2 \pmod 4}} f(n)$$

$$= \left[\frac{N}{2}\right] + \left[\frac{N}{4}\right] + 0 \sim \frac{3N}{4},$$

as $N \to \infty$.

(613) We have seen in Problem 610 that

$$\mu^2(n) = \sum_{d^2 | n} \mu(d).$$

We can therefore write successively

$$A(N) = \sum_{n \leq N} \mu^2(n) = \sum_{n \leq N} \sum_{d^2 | n} \mu(d) = \sum_{d^2 m \leq N} \mu(d)$$

$$= \sum_{d^2 \leq N} \mu(d) \sum_{m \leq \frac{N}{d^2}} 1 = \sum_{d^2 \leq N} \mu(d) \left[\frac{N}{d^2}\right],$$

and the result follows.

(614) We have

$$Q(n) := \frac{\phi(n)}{n\tau(n)} = \frac{1}{\tau(n)} \prod_{p|n} \left(1 - \frac{1}{p}\right).$$

Considering the infinite sequence of powers of 2, that is the numbers n of the form $n = 2^m$ with $m \in \mathbb{N}$, we have

$$0 < Q(n) = Q(2^m) = \frac{1 - \frac{1}{2}}{m+1} \to 0 \quad (m \to \infty),$$

which proves the first statement.

On the other hand, if we consider the sequence of integers $n = p$, where p is prime, we have

$$Q(n) = Q(p) = \frac{1 - \frac{1}{p}}{2} \to \frac{1}{2} \quad (p \to \infty).$$

Since, for each integer $n \geq 1$, we have $Q(n) \leq \frac{1}{2}$, the result follows.

(615) It is clear that

$$\sum_{n=1}^{\infty} \frac{1}{q_n} < \sum_{n=1}^{\infty} \frac{1}{2^n},$$

which series converges (to 1).

REMARK: Using a computer, we obtain that $\sum_{n=1}^{\infty} 1/q_n = 0.7404\ldots$.

(616) It is enough to prove that to each odd divisor d of an integer n, there corresponds a divisor of $2^n + 1$ which is larger than 1. We have

$$2^n + 1 = \left(2^{n/d}\right)^d + 1 = a^d + 1,$$

say. Therefore, since d is odd, $a + 1 | a^d + 1 = 2^n + 1$. The number $a + 1$ is therefore the required divisor.

(617) Let $d|n$ be such that $1 < d < n$; then $1 < n/d < n$. If $d \leq \sqrt{n}$, then $n/d \geq \sqrt{n}$. We derive from this that $\sigma(n) \geq n + \sqrt{n} + 1$, and the result follows. Finally, it follows from this equality that

$$\sigma(p_n + 1) - \sigma(p_n) > p_n + 1 + \sqrt{p_n + 1} = (p_n + 1) = \sqrt{p_n + 1} \to +\infty,$$

which proves the second result.

(618) If n is an odd integer, then n can be written as a product of distinct prime powers p^α, that is $n = \prod_{p^\alpha \| n} p^\alpha$, where each prime number p is odd. Since σ is a multiplicative function, we have

$$(*) \qquad \frac{\sigma(n)}{n} = \prod_{p^\alpha \| n} \frac{\sigma(p^\alpha)}{p^\alpha} = \prod_{p^\alpha \| n} \left(1 + \frac{1}{p} + \frac{1}{p^2} + \cdots + \frac{1}{p^\alpha} \right).$$

But, since each of these p's satisfies $p \geq 3$, we always have

$$1 + \frac{1}{p} + \frac{1}{p^2} + \cdots + \frac{1}{p^\alpha} \leq 1 + \frac{1}{3} + \frac{1}{3^2} + \cdots + \frac{1}{3^\alpha} < 1 + \frac{1}{3} + \frac{1}{3^2} + \cdots$$
$$= \frac{1}{1 - \frac{1}{3}} = \frac{3}{2}.$$

Using this last inequality in $(*)$, we obtain the required inequality corresponding to the case n odd. For the other case, we only need to observe that besides having $(*)$ for p odd, we have

$$1 + \frac{1}{2} + \frac{1}{2^2} + \cdots + \frac{1}{2^\alpha} < 1 + \frac{1}{2} + \frac{1}{2^2} + \cdots = 2,$$

which completes the proof of the case n even, that is

$$\frac{\sigma(n)}{n} < 2 \left(\frac{3}{2} \right)^{\omega(n)-1}.$$

(619) It is clear that

$$\sigma(n) = n + \sum_{\substack{d|n \\ d<n}} d < n + (\tau(n) - 1)n = n\tau(n).$$

(620) It is enough to take $n = p$ prime (larger than 2). We then have $\sigma(p) = 1+p$ and therefore $\sigma(p-1) \geq 1 + 2 + (p-1) = p + 2$, and the result follows.

(621) It is immediate that $\sigma(n) \geq n$. Since $\sigma(n)/n^2$ is a multiplicative function, it is enough to study this function when $n = p^a$. In this case,

$$\frac{\sigma(p^a)}{p^{2a}} = \frac{1 + p + p^2 + \cdots + p^a}{p^{2a}} = \frac{1}{p^a} \left(1 + \frac{1}{p} + \cdots + \frac{1}{p^a} \right) < \frac{1}{p^a} \frac{p}{p-1} < 1$$

which proves the result.

(622) Using the formula

$$\prod_{\substack{d|n \\ d \text{ even}}} d = (2n)^{\alpha\tau(m)/2}$$

proved in Problem 491 and observing that the number of even divisors of $n = 2^\alpha m$ is equal to $\alpha\tau(m)$, we have, comparing the geometric mean with the arithmetic mean (see Theorem 5),

$$\sqrt{2n} = \left(\prod_{\substack{d|n \\ d \text{ even}}} d \right)^{1/\alpha\tau(m)} \leq \frac{1}{\alpha\tau(m)} \sum_{\substack{d|n \\ d \text{ even}}} d = \frac{\sigma_p(n)}{\alpha\tau(m)},$$

and the required inequality is proved.

(623) First of all, it is clear that $\phi(n) \leq n-(\tau(n)-1) = n+1-\tau(n)$, since for each $d|n$, $d > 1$, we have $(d,n) > 1$. It follows that $\sigma(n) > n \geq \phi(n)+\tau(n)-1$ and therefore that $\sigma(n) \geq \phi(n)+\tau(n)$.

(624) This inequality follows from

$$\sum_{d|n} f(d)g(n/d) = f(1)g(n) + g(1)f(n) + \sum_{\substack{d|n \\ 1<d<n}} f(d)g(n/d) \geq f(n)+g(n),$$

while equality is obtained if and only if the last sum is empty, that is when n is prime.

(625) This follows from Problem 624 and the relation

$$\sum_{d|n} \sigma(d)\phi(n/d) = n\tau(n)$$

proved in Problem 596.

(626) (Schwab and Tóth [**37**]) For each integer $n \geq 1$, we have

$$(f * h)(n) + (g * h)(n) = \sum_{d|n} \Big(f(d)+g(d)\Big)h(n/d)$$

$$= 2h(n) + \sum_{\substack{d|n \\ d>1}} \Big(f(d)+g(d)\Big)h(n/d) \geq 2h(n).$$

(627) We only need to take $f = 1$, $g = \mu$ and $h(n) = n$ in Problem 626.

(628) If $\omega(n) = 1$, it is clear that $\sigma(n) > n$, and the inequality is proved. Assume that $n = q_1^{\alpha_1} \cdots q_r^{\alpha_r}$ with $r \geq 2$. We then have $q_1 < q_2 < \ldots < q_{r-1} < \sqrt{n}$, and this is why

$$\frac{n}{q_1} > \frac{n}{q_2} > \ldots > \frac{n}{q_{r-1}} > \frac{n}{\sqrt{n}} = \sqrt{n}.$$

It follows that

$$\sigma(n) > n + (r-1)\sqrt{n},$$

and the result follows.

(629) (*AMM, Vol. 79, 1972, p. 910*). We know that $\phi(n) \leq n-1$ with equality if and only if n is prime. Then,

$$\begin{aligned}\phi(n^2) + \phi((n+1)^2) &= n\phi(n) + (n+1)\phi(n+1) \\ &\leq n(n-1) + (n+1)n = 2n^2,\end{aligned}$$

with equality if and only if n and $n+1$ are both primes, that is if and only if $n = 2$.

For the general case, we proceed by induction. The induction argument follows from

$$
\begin{aligned}
\phi(n^{k+1}) + \phi((n+1)^{k+1}) &= n\phi(n^k) + (n+1)\phi((n+1)^k) \\
&< (n+1)\left(\phi(n^k) + \phi((n+1)^k)\right) \\
&< (n+1)2n^2(n+1)^{k-2} = 2n^2(n+1)^{k-1}.
\end{aligned}
$$

(630) (*AMM, Vol. 79, 1972, p. 915*). Let p_i be the i–th prime number, r a positive integer and $n = p_1 \cdots p_r$. From Dirichlet's Theorem, there exists a positive integer k such that $q = kn + 1$ is a prime number. We have $\phi(q) = kn$ and

$$
\phi(q-1) = \phi(kn) = kn \prod_{p|kn}(1-1/p) \le kn \prod_{i=1}^{r}(1-1/p_i)
$$

$$
= \phi(q)\prod_{i=1}^{r}\left(1-\frac{1}{p_i}\right).
$$

Therefore,

$$
\frac{\phi(q)}{\phi(q-1)} \ge \frac{1}{\prod_{i=1}^{r}\left(1-\frac{1}{p_i}\right)}.
$$

But, since $\prod_p(1-1/p)$ diverges to 0, it follows that, letting r tend to infinity, we have $\limsup_{n\to\infty}\phi(n)/\phi(n-1) = \infty$.

For the second part, let k be an integer such that $q = kn - 1$ is prime. Then, $\phi(q) = kn - 2$ while

$$
\phi(q+1) = \phi(kn) = kn \prod_{p|kn}\left(1-\frac{1}{p}\right) \le kn \prod_{i=1}^{r}\left(1-\frac{1}{p_i}\right).
$$

We therefore obtain that

$$
\frac{\phi(q+1)}{\phi(q)} \le \frac{kn}{kn-2}\prod_{i=1}^{r}\left(1-\frac{1}{p_i}\right).
$$

Letting r tend to infinity, we find that $\liminf_{n\to\infty}\phi(n+1)/\phi(n) = 0$.

(631) (*AMM, Vol. 84, 1977, p. 740*). The answer is NO. Let $N \ge 5$. From Bertrand's Postulate (see Theorem 15), there exists a prime number p such that

$$
\left[\frac{N+3}{2}\right] < p < 2\left[\frac{N+3}{2}\right] - 2 \le N+1.
$$

Therefore,

$$
\phi(p) = p - 1 > \frac{N+1}{2} \ge \phi(N+1) \quad \text{if } N \text{ is odd,}
$$

and

$$
\phi(p) = p - 1 > \frac{N}{2} \ge \phi(N) \quad \text{if } N \text{ is even,}
$$

meaning that N does not have the required property.

However, since $\phi(1) = \phi(2)$, $\phi(3) = \phi(4) = 2$ and $\phi(n) \ge 2$ for $n \ge 3$, the integers $1, 2, 3$ and 4 have the required property.

(632) (*AMM, Vol. 80, 1973, p. 436*). Let

$$f(n) = \frac{\phi(n)\tau^2(n)}{n^2}.$$

Then:

(*i*) We easily obtain that $f(1) = 1$, $f(2) = 1$, $f(4) = 9/8$ and $f(3) = 8/9$.

(*ii*) On the other hand, $f(p) = 4(p-1)/p^2 < 1$ if p is an odd prime number. Moreover, $f(p^{a+1})/f(p^a) = p(a+2)^2/(a+1)^2 p^2 \leq 9/4p < 1$ for $a > 0$. Hence, by induction this shows that $f(p^a) < 1$ for each positive integer a.

(*iii*) It is clear that $f(2^a) = (a+1)^2/2^{a+1} < 1$ for $a > 3$ and that $f(8) = 1$.

(*iv*) From (*i*) and (*iii*), it is clear that $f(4k) = 1$ implies that $k = 2$ where $(k, 2) = 1$. In this last case, $f(k) = 1/f(4) = 8/9$, so that $3|k$. Since $f(3^a) < 8/9$ for $a > 1$, then by (*ii*), we have $f(p^a) = 1$, so that $p = 2$, and since f is multiplicative, the only solution of $f(k) = 8/9$ is $k = 3$.

This is why the equality holds only when $n = 1, 2, 8, 12$.

(633) If $p(n)$ stands for the smallest prime factor of n, then

$$\phi(n) = n \prod_{p|n} \left(1 - \frac{1}{p}\right) \leq n \left(1 - \frac{1}{p(n)}\right) = n - \frac{n}{p(n)} \leq n - \frac{n}{n^{1/\omega(n)}},$$

and since both inequalities are strict if and only if n is not prime, the result follows.

(634) If p is the smallest prime factor of n, then, since n is composite, $p \leq \sqrt{n}$. Therefore,

$$\phi(n) \leq n \left(1 - \frac{1}{p}\right) \leq n - \sqrt{n}.$$

(635) Using the formula $\phi(n) = n \prod_{p|n} \left(1 - \frac{1}{p}\right)$, we obtain

$$\frac{\phi(n)}{n} = \prod_{p|n} \left(1 - \frac{1}{p}\right) \geq \prod_{p|n} \left(1 - \frac{1}{2}\right) = \prod_{p|n} \frac{1}{2} = \frac{1}{2^r}.$$

(636) First of all, it is clear that

$$\sigma_2(n) = \sum_{d|n} d^2 \leq \left(\sum_{d|n} d\right)^2 = (\sigma(n))^2,$$

which proves the second inequality. On the other hand, using the Cauchy-Schwarz inequality, we obtain

$$\sigma^2(n) = \left(\sum_{d|n} d \cdot 1\right)^2 \leq \sum_{d|n} d^2 \cdot \sum_{d|n} 1^2 = \sigma_2(n)\tau(n),$$

hence, the first inequality.

(637) We must prove that

$$\frac{\sigma(n)}{\tau(n)} \geq \left(\prod_{d|n} d\right)^{1/\tau(n)},$$

which is obtained by comparing the geometric mean and the arithmetic mean (see Theorem 5), if we take for the a_i's the divisors d of n and $r = \tau(n)$.

(638) From Problems 488 and 637, we have

$$\frac{\sigma(n)}{\tau(n)} \geq \left(\prod_{d|n} d\right)^{1/\tau(n)} = \left(n^{\tau(n)/2}\right)^{1/\tau(n)} = n^{1/2},$$

as was to be shown.

A second solution is the following.

Since $f(n) = \sigma(n)/\tau(n)$ is a multiplicative function, it is enough to prove that the inequality holds for $n = p^\alpha$, which is easily verified since

$$f(p^\alpha) = \frac{1 + p + p^2 + \cdots + p^\alpha}{\alpha + 1} \geq (1 \cdot p \cdots p^\alpha)^{1/(\alpha+1)} = \sqrt{p^\alpha}.$$

(639) In light of Problem 488, it follows that $n^{\tau(n)/2} = n^3$ and therefore that $\tau(n) = 6$, thus the result.

(640) To each divisor d of n such that $d \leq \sqrt{n}$ corresponds another divisor of n, that is n/d, which satisfies $n/d \geq \sqrt{n}$. Therefore, the set A of divisors of n can be written as $A = B \cup C$, where B is made up of the divisors $d \leq \sqrt{n}$ and C of the corresponding divisors n/d, observing that in the case $n = m^2$ for a certain integer m, the divisor m belongs to both sets. In any event, it is clear that $\#A \leq \#B + \#C \leq \sqrt{n} + \sqrt{n} = 2\sqrt{n}$, as required.

(641) Using the fact that $\tau(n) \leq 2\sqrt{n}$ (see Problem 640), we can write successively

$$\begin{aligned}
\sigma(n) &= \sum_{\substack{d|n \\ d \leq \sqrt{n}}} d + \sum_{\substack{d|n \\ \sqrt{n} < d \leq \frac{n}{2}}} d + n < \frac{\tau(n)}{2} \cdot \sqrt{n} + \frac{\tau(n)}{2} \cdot \frac{n}{2} + n \\
&\leq \frac{2\sqrt{n}}{2}\sqrt{n} + \frac{2\sqrt{n}}{2}\frac{n}{2} + n = n + \frac{n\sqrt{n}}{2} + n = 2n + \frac{n\sqrt{n}}{2}.
\end{aligned}$$

But

$$2n + \frac{n\sqrt{n}}{2} < n\sqrt{n} \iff \sqrt{n} > 4 \iff n > 16.$$

Finally, for $3 \leq n \leq 15$, we verify that we also have $\sigma(n) < n\sqrt{n}$.

(642) We proved in Problem 488 that $\prod_{d|n} d = n^{\tau(n)/2}$. Therefore, using the inequality comparing the geometric mean and the arithmetic mean (see Theorem 5), we obtain

$$n^{s/2} = \left(\prod_{d|n} d^s\right)^{1/\tau(n)} \leq \frac{1}{\tau(n)} \sum_{d|n} d^s.$$

This allows us to write successively

$$\int_{-\infty}^{1} \sum_{d|n} d^s \, ds \geq \int_{-\infty}^{1} \tau(n) n^{s/2} \, ds,$$

$$\sum_{d|n} \frac{d^s}{\log d} \Big|_{-\infty}^{1} \geq \frac{2\tau(n) n^{s/2}}{\log n} \Big|_{-\infty}^{1},$$

$$1 + \sum_{\substack{d|n \\ d \geq 2}} \frac{d}{\log d} \geq \frac{2\tau(n)\sqrt{n}}{\log n},$$

and the result follows.

(643) The proof is similar to that of Problem 642, except that this time we integrate from $-\infty$ to $+2$.

(644) It is clear that the inequality we want to prove is equivalent to

$$(1) \qquad \phi(n) = n \prod_{p|n} \left(1 - \frac{1}{p}\right) \leq n - (2^{\omega(n)} - 1).$$

Let $q_1 < q_2 < \ldots < q_r$ be the r distinct prime factors of n. Then,

$$\phi(n) \leq n - \sum_{\substack{q_i|n \\ 1 \leq i \leq r}} 1 - \sum_{\substack{q_i q_j|n \\ 1 \leq i < j \leq r}} 1 - \sum_{\substack{q_i q_j q_k|n \\ 1 \leq i < j < k \leq r}} 1 - \cdots - \sum_{q_1 \cdots q_r|n} 1$$

$$= n - \binom{r}{1} = \binom{r}{2} - \cdots - \binom{r}{r} = n - (2^r - 1),$$

which proves (1).

(645) By comparing the arithmetic mean with the geometric mean (see Theorem 5), we obtain

$$\frac{1}{\omega(n)} \sum_{p|n} \frac{1}{p} \geq \left(\prod_{p|n} \frac{1}{p}\right)^{1/\omega(n)} \geq \left(\frac{1}{n}\right)^{1/\omega(n)},$$

which proves the result.

(646) Writing

$$\omega(n) - 1 = \sum_{i=2}^{\omega(n)} \frac{\sqrt{q_i - q_{i-1}}}{\sqrt{q_i - q_{i-1}}}$$

and using the Cauchy-Schwarz inequality, we obtain

$$(\omega(n) - 1)^2 \leq \sum_{i=2}^{\omega(n)} (q_i - q_{i-1}) \cdot \sum_{i=2}^{\omega(n)} \frac{1}{q_i - q_{i-1}} = (P(n) - p(n)) h(n),$$

yielding the result.

(647) Writing

$$\tau(n) - 1 = \sum_{i=2}^{\tau(n)} \frac{\sqrt{d_i - d_{i-1}}}{\sqrt{d_i - d_{i-1}}},$$

and using the Cauchy-Schwarz inequality, we obtain

$$(\tau(n) - 1)^2 \leq \sum_{i=2}^{\tau(n)} (d_i - d_{i-1}) \cdot \sum_{i=2}^{\tau(n)} \frac{1}{d_i - d_{i-1}} = (n-1)H(n),$$

thus the result.

(648) If $d|n$, then $2^d - 1$ is a divisor of $2^n - 1$ because

$$2^n - 1 = 2^{dm} - 1 = (2^d - 1)(2^{d(m-1)} + 2^{d(m-2)} + \cdots + 2^d + 1).$$

(649) Using $\phi(n) = n \prod_{p|n}(1 - p^{-1})$, we get the result.

(650) Since for $n \geq 1$, $\tau(n) \leq 2\sqrt{n}$ and $\phi(n)\sigma(n) \leq n^2$, and since $\phi(mn) \leq m\phi(n)$ (see Problem 649), we have

$$\phi\left(n\left[\frac{\sigma(n)\,\tau(n)}{n^{3/2}}\right]\right) \leq \phi(n)\left[\frac{\sigma(n)\tau(n)}{n^{3/2}}\right] \leq \phi(n)\frac{\sigma(n)\tau(n)}{n^{3/2}} \leq \sqrt{n}\,\tau(n) \leq 2n.$$

(651) In light of Problem 522, we have for $d|n$, $\phi(n) \geq \phi(d)$. We thus obtain

$$\sum_{d|n} \phi(n) \geq \sum_{d|n} \phi(d) = n;$$

that is $\phi(n)\tau(n) \geq n$.

A second solution is obtained in the following way.

The inequality is verified for $n = 1$. So let $n \geq 2$. Since the expressions $\phi(n)\tau(n)$ and n both represent multiplicative functions, it is enough to show that $\phi(p^\alpha)\tau(p^\alpha) \geq p^\alpha$ for each prime number p and each positive integer α. We must therefore verify that $(p^\alpha - p^{\alpha-1})(\alpha + 1) \geq p^\alpha$; that is $(1 - \frac{1}{p})(\alpha + 1) \geq 1$. But, for each prime p and each integer $\alpha \geq 1$, we have

$$\left(1 - \frac{1}{p}\right)(\alpha + 1) \geq \left(1 - \frac{1}{2}\right)2 = 1,$$

from which the result follows.

(652) The solutions are $n = 1$ and $n = 2$. One easily checks that $n = 1$ is a solution. So let $n > 1$. From the argument used in the solution of Problem 651, we have $\phi(n)\tau(n) = n$ if and only if $(1 - \frac{1}{p})(\alpha + 1) = 1$. But if $p \geq 3$, we have $(1 - \frac{1}{p})(\alpha + 1) \geq (1 - \frac{1}{3})2 = \frac{4}{3}$. We must therefore have $p = 2$, in which case we obtain $\alpha = 1$. The only solution $n > 1$ is therefore $n = 2$.

(653) We have $\phi(n) \leq n - 1$, with equality if and only if n is prime. Assume that $m \geq n$. Then, by Problem 649,

$$
\begin{aligned}
\phi(mn) + \phi((m+1)(n+1)) &\leq m\phi(n) + (m+1)\phi(n+1) \\
&\leq m(n-1) + (m+1)n \\
&= 2mn - (m-n) \leq 2mn.
\end{aligned}
$$

We have equality if and only if n and $n + 1$ are prime numbers, that is when $n = 2$.

(654) (a) This follows from the fact that, if $p^a \| n$, then

$$1 + \frac{1}{p} \leq 1 + \frac{1}{p} + \frac{1}{p^2} + \cdots + \frac{1}{p^a}.$$

(b) If n is squarefree, it is clear that $\Psi(n) = \sigma(n)$. Assume now that $\Psi(n) = \sigma(n)$. Then,

$$\prod_{p^a \| n} \left(1 + \frac{1}{p}\right) = \prod_{p^a \| n} \left(1 + \frac{1}{p} + \frac{1}{p^2} + \cdots + \frac{1}{p^a}\right).$$

This means that if n is not squarefree, then at least one of the factors on the right-hand side is larger than the corresponding factor on the left-hand side, which leads to a contradiction.

(c) If $n = 2^a \cdot 3^b$, the result is immediate. Assume now that n is Ψ-perfect, that is that

$$\prod_{i=1}^{r} \left(1 + \frac{1}{q_i}\right) = 2,$$

where $q_1 < q_2 < \ldots < q_r$ are the r prime factors of n. First of all, it is clear that $r \geq 2$. Then,

(1) $$(q_1 + 1)(q_2 + 1) \cdots (q_r + 1) = 2q_1 q_2 \cdots q_r.$$

We must have that $q_1 = 2$, since otherwise the left-hand side of (1) is divisible by $2^r \geq 4$, while the right-hand side of (1) is divisible only by 2. Relation (1) therefore becomes

(2) $$3(q_2 + 1) \cdots (q_r + 1) = 4q_2 \cdots q_r.$$

From (2), it follows that $3 | q_2 \cdots q_r \Rightarrow q_2 = 3$. Hence, (2) becomes

$$(q_3 + 1) \cdots (q_r + 1) = q_3 \cdots q_r,$$

which is impossible, unless $r = 2$. Hence the result.

(655) (a) We have

$$\phi^*(n) = \sum_{\substack{k=1 \\ (f(k),n)=1}}^{n} 1.$$

Since $\phi^*(1) = 1$, the equation $\phi^*(mn) = \phi^*(m)\phi^*(n)$ is verified when $m = 1$ or $n = 1$. Assume that $m > 1$ and $n > 1$ are such that $(m, n) = 1$. Consider the mn consecutive integers

1	2	...	$(m-1)$	m
$m+1$	$m+2$	...	$m+(m-1)$	$m+m$
$\vdots$	$\vdots$		$\vdots$	$\vdots$
$(n-1)m+1$	$(n-1)m+2$	...	$(n-1)m+(m-1)$	nm

which constitute a complete residue system modulo mn. Clearly there are amongst these mn integers exactly $\phi^*(mn)$ integers $k \leq mn$ such that $(f(k), mn) = 1$. Each of the rows is a complete residue system modulo m, and therefore each of these rows has $\phi^*(m)$ integers k such that $(f(k), m) = 1$. Since $\{0, 1, \ldots, n-1\}$ is a complete residue system modulo n, it follows that, for $(m, n) = 1$, $\{b, m+b, \ldots, m(n-1)+b\}$ is also a complete residue system modulo n. Similarly we can argue that each of the columns contains exactly $\phi^*(n)$ integers k such that $(f(k), n) = 1$. Consequently, amongst these mn integers, there are exactly $\phi^*(n)\phi^*(m)$

integers $k \leq mn$ such that $(f(k), m) = (f(k), n) = 1$; that is $(f(k), mn) = 1$. We may therefore conclude that

$$\phi^*(mn) = \phi^*(m)\phi^*(n).$$

(b) It is enough to find the value of $\phi^*(p^a)$. Since

$$\phi^*(p^a) = \sum_{\substack{k=1 \\ (f(k),p)=1}}^{p^a} 1 = p^a - \sum_{\substack{k=1 \\ f(k)\equiv 0 \pmod p}}^{p^a} 1$$

and

$$\sum_{\substack{k=1 \\ f(k)\equiv 0 \pmod p}}^{p^a} 1 = p^{a-1}b_p,$$

where b_p is the number of values of $f(1), f(2), \ldots, f(p)$ divisible by p, the result follows.

(656) Using the result of Problem 655 with $f(k) = k(k+1)$, we obtain that the number of terms in the sequence is given by

$$n \prod_{p|n} \left(1 - \frac{2}{p}\right).$$

(657) Using the result of Problem 655, we find that the number of integers with the stated property is

$$n \prod_{p|n} \left(1 - \frac{2}{p}\right).$$

(658) If n is even, this number is clearly 0. Hence assume that n is odd. We use the result of Problem 655 with $f(k) = k(k+1)(k+2)$. In this particular case, $b_p = 3$ and therefore the number of integers with the given property is

$$n \prod_{p|n} \left(1 - \frac{3}{p}\right).$$

(659) (*Problem #316, in Barbeau, Klamkin and Moser* [**3**]) We easily observe that $f(n) = n + k$, where k is such that $k^2 < f(n) < (k+1)^2$. We will show that $k = \|\sqrt{n}\|$. But we have successively

$$k^2 < n + k < (k+1)^2,$$
$$k^2 + 1 \leq n + k \leq (k+1)^2 - 1,$$
$$k^2 - k + 1 \leq n \leq (k+1)^2 - k - 1,$$
$$k^2 - k + 1 \leq n \leq k^2 + k,$$
$$\left(k - \frac{1}{2}\right)^2 + \frac{3}{4} \leq n \leq \left(k + \frac{1}{2}\right)^2 - \frac{1}{4},$$
$$\left(k - \frac{1}{2}\right)^2 < n < \left(k + \frac{1}{2}\right)^2,$$
$$k - \frac{1}{2} < \sqrt{n} < k + \frac{1}{2},$$
$$k < \sqrt{n} + \frac{1}{2} < k + 1.$$

These inequalities imply that $k = [\sqrt{n} + \frac{1}{2}] = \|\sqrt{n}\|$.

(660) Let $f(n) := \sum_{d^k|n} \mu(d)$. First of all, we observe that f is a multiplicative function. This follows from the fact that if $(m,n) = 1$, then

$$d^k|mn \iff d = d_1 d_2, \text{ with } (d_1, d_2) = 1, \quad d_1^k|m, d_2^k|n$$

and therefore

$$f(mn) = \sum_{d^k|mn} \mu(d) = \sum_{d_1^k|m, d_2^k|n} \mu(d_1 d_2) = \sum_{d_1^k|m} \mu(d_1) \sum_{d_2^k} \mu(d_2) = f(m)f(n).$$

Finally, since

$$f(p^\alpha) = 1 + \sum_{p^k|p^\alpha} \mu(p) = \begin{cases} 1 & \text{if } \alpha < k \\ 0 & \text{if } \alpha \geq k, \end{cases}$$

the result follows.

(661) (*Contribution of Imre Kátai, Budapest*). The equation $x^2 + 1 = 2(y^2 + 1)$ being a Fermat-Pell equation (since it can be written in the form $x^2 - 2y^2 = 1$) means it must have infinitely many solutions $\{x, y\}$. Thus, given a particular solution $\{x, y\}$ of this equation, we have

$$\lambda(x^2 + 1) = \lambda(2(y^2 + 1)) = \lambda(2)\lambda(y^2 + 1) = -\lambda(y^2 + 1),$$

and the result follows.

(662) (*MMAG, Vol. 48, 1975, p. 120*). Since $\phi(n) \leq n - 1$ and $(n - 1, n) = 1$, we have

$$\phi(n) \leq n - 1 = \frac{n - 1}{n - (n - 1)} \leq \sum_{i=1}^{k} \frac{a_i}{n - a_i},$$

where $1 \leq a_i \leq n$ and $(a_i, n) = 1$.

(663) This follows from the fact that the geometric mean does not exceed the arithmetic mean (see Theorem 5). Indeed, since

$$\prod_{d|n} f(d))^{1/\tau(n)} \leq \frac{1}{\tau(n)} \sum_{d|n} f(d),$$

the result follows. For the second part, we take $f(n) = \phi(n)$ so that $F(n) = n$.

(664) We have

$$\sigma(n) = n \prod_{p^\alpha \| n} \frac{1 - p^{-\alpha - 1}}{1 - p^{-1}} \quad \text{and} \quad \phi(n) = n \prod_{p|n} (1 - p^{-1}).$$

Therefore,

$$\frac{\sigma(n)\phi(n)}{n^2} = \prod_{p^\alpha \| n} (1 - p^{-\alpha - 1}),$$

and this product is clearly located between 1 and $C := \prod_p (1 - p^{-2})$.

(665) Part (a) is immediate. To prove (b), we use the identity $\log n = \log d + \log(n/d)$ to write

$$(f * g)'(n) = \sum_{d|n} f(d)g(n/d)\log n = \sum_{d|n} f(d)\log d\, g(n/d)$$

$$+ \sum_{d|n} f(d)g(n/d)\log(n/d) = (f' * g)(n) + (f * g')(n).$$

To prove (c), we apply part (b) to the formula $E' = 0$. We then have

$$0 = E' = (f * f^{-1})' = f' * f^{-1} + f * (f^{-1})'$$

and therefore

$$f * (f^{-1})' = -f' * f^{-1}.$$

Multiplying this last equation by f^{-1}, we obtain

$$(f^{-1})' = -(f' * f^{-1}) * f^{-1} = -f' * (f^{-1} * f^{-1}).$$

Since $f^{-1} * f^{-1} = (f * f)^{-1}$, the result follows.

(666) Since $\overline{f} = (1 * f)/\tau$ and since τ is multiplicative, the result follows.
REMARK: The following ten problems are inspired by a paper of J.M. De Koninck and J. Grah [7].

(667) Let f be additive and let $(m, n) = 1$. Since

$$
\begin{aligned}
\overline{f}(mn) &= \frac{1}{\tau(mn)} \sum_{d|mn} f(d) = \frac{1}{\tau(m)\tau(n)} \sum_{\substack{d_1|n \\ d_2|m}} f(d_1 d_2) \\
&= \frac{1}{\tau(m)\tau(n)} \sum_{\substack{d_1|n \\ d_2|m}} (f(d_1) + f(d_2)) \\
&= \frac{1}{\tau(m)\tau(n)} \sum_{\substack{d_1|n \\ d_2|m}} f(d_1) + \frac{1}{\tau(m)\tau(n)} \sum_{\substack{d_1|n \\ d_2|m}} f(d_2) \\
&= \frac{1}{\tau(m)\tau(n)} \sum_{d_1|n} f(d_1) \sum_{d_2|m} 1 + \frac{1}{\tau(m)\tau(n)} \sum_{d_2|m} f(d_2) \sum_{d_1|n} 1 \\
&= \overline{f}(n) + \overline{f}(m),
\end{aligned}
$$

the result follows.

(668) We easily obtain that $\overline{1} = 1$ and $\overline{\mu} = E$.

(669) We have

$$\overline{\omega}(p^\alpha) = \frac{1}{\alpha+1}(\omega(p) + \omega(p^2) + \cdots + \omega(p^\alpha)) = \frac{\alpha}{1+\alpha}.$$

It follows that

$$\overline{\omega}(n) = \sum_{p^\alpha \| n} \frac{\alpha}{1+\alpha}.$$

(670) Since $\overline{f}$ is multiplicative, we only need to evaluate $\overline{f}(p^\alpha)$ for p prime and $\alpha \in \mathbb{N}$. Observing that

$$\overline{f}(p^\alpha) = \frac{1}{\alpha+1} \sum_{d | p^\alpha} 2^{\omega(d)} = \frac{1}{\alpha+1} \left(1 + 2^{\omega(p)} + 2^{\omega(p^2)} + \cdots + 2^{\omega(p^\alpha)} \right)$$
$$= \frac{1 + 2 + 2 + \cdots + 2}{\alpha+1} = \frac{2\alpha+1}{\alpha+1},$$

we find that

$$\overline{f}(n) = \prod_{p^\alpha \| n} \overline{f}(p^\alpha) = \prod_{p^\alpha \| n} \frac{2\alpha+1}{\alpha+1} = \frac{\tau(n^2)}{\tau(n)},$$

as required.

(671) Since $\overline{f}$ is multiplicative, we only need to evaluate $\overline{f}(p^\alpha)$ for p prime and $\alpha \in \mathbb{N}$. Since

$$\overline{f}(p^\alpha) = \frac{1}{\alpha+1} \sum_{d | p^\alpha} 2^{\Omega(d)} = \frac{1}{\alpha+1} \left(1 + 2^{\Omega(p)} + 2^{\Omega(p^2)} + \cdots + 2^{\Omega(p^\alpha)} \right)$$
$$= \frac{1 + 2 + 2^2 + \cdots + 2^\alpha}{\alpha+1} = \frac{2^{\alpha+1} - 1}{\alpha+1},$$

it follows that

$$\overline{f}(n) = \prod_{p^\alpha \| n} \overline{f}(p^\alpha) = \prod_{p^\alpha \| n} \frac{2^{\alpha+1} - 1}{\alpha+1}.$$

(672) Since $\overline{\lambda}$ is multiplicative, it follows that the function $\overline{\lambda}\tau$ is also multiplicative. Hence, in order to prove that $\overline{\lambda}\tau = \chi$, we only need to establish that

$$(\overline{\lambda}\tau)(p^\alpha) = \begin{cases} 1 & \text{if } \alpha \text{ is even,} \\ 0 & \text{otherwise.} \end{cases}$$

Since

$$(\overline{\lambda}\tau)(p^\alpha) = \sum_{d | p^\alpha} (-1)^{\Omega(d)} = (-1)^{\Omega(1)} + (-1)^{\Omega(p^2)} + \cdots + (-1)^{\Omega(p^\alpha)}$$
$$= (-1)^0 + (-1)^1 + (-1)^2 + \cdots + (-1)^\alpha,$$

a quantity which is reduced to 1 if α is even and to 0 if α is odd, the result follows.

(673) It is enough to solve the equation

$$g(n) = \frac{1}{\tau(n)} \sum_{d | n} f(d).$$

Since this equation can be written successively as

$$\tau g = 1 * f,$$
$$\mu * (\tau g) = f,$$
$$f = \mu * (\tau g)$$

and since g is multiplicative, it follows that f is also multiplicative.

(674) We know, from Problem 673, that the function f is multiplicative and is given by the equation $f = \mu * (\tau g)$. It is therefore enough to find the values of $f(p^\alpha)$ for p prime and $\alpha \in \mathbb{N}$. Clearly,

$$f(p) = \mu(1)\tau(p)g(p) + \mu(p)\tau(1)g(1) = 2g(p) - 1 = 4 - 1 = 3.$$

On the other hand, for $\alpha \geq 2$,

$$f(p^\alpha) = \mu(1)\tau(p^\alpha)g(p^\alpha) + \mu(p)\tau(p^{\alpha-1})g(p^{\alpha-1}) = (\alpha+1)\cdot 2 - \alpha\cdot 2 = 2.$$

This means that

$$f(p^\alpha) = \begin{cases} 3 & \text{if } \alpha = 1, \\ 2 & \text{if } \alpha \geq 2. \end{cases}$$

In other words,

$$f(n) = \prod_{p\|n} 3 \cdot \prod_{\substack{p^\alpha\|n \\ \alpha \geq 2}} 2.$$

(675) Since $\widehat{f} = (1*(\mu^2 f))/2^\omega$ and since μ, f and 2^ω are multiplicative functions, the result follows.

(676) We easily obtain that $\widehat{1} = 1$ and $\widehat{\lambda} = E$.

(677) Let us begin with the case $k = 3$. We have

$$
\begin{aligned}
\tau_3(n) &= \sum_{d_1 d_2 d_3 = n} 1 = \sum_{(d_1 d_2)\cdot d = n} 1 \\
&= \sum_{d|n}\sum_{d_1 d_2 = n/d} 1 = \sum_{d|n}\tau(n/d) = (1*\tau)(n) = (1*1*1)(n).
\end{aligned}
$$

If $k = 4$, we proceed in a similar manner and obtain

$$
\begin{aligned}
\tau_4(n) &= \sum_{d_1 d_2 d_3 d_4 = n} 1 = \sum_{(d_1 d_2 d_3)\cdot d = n} 1 = \sum_{d|n}\sum_{d_1 d_2 d_3 = n/d} 1 \\
&= \sum_{d|n}\tau_3(n/d) = (1*\tau_3)(n) = (1*1*1*1)(n).
\end{aligned}
$$

The result then follows by induction on k.

(678) We obtain successively

$$\sum_{k=1}^{n} F(k) = \sum_{k=1}^{n}\sum_{d|k} f(d) = \sum_{d=1}^{n} f(d) \sum_{m=1}^{[n/d]} 1 = \sum_{d=1}^{n} \left[\frac{n}{d}\right] f(d).$$

(679) (*AMM, Vol. 75, 1968, p. 77*). Setting $F = \tau$ and $f = 1$ in the equation of Problem 678, we obtain

$$\sum_{k=1}^{2n} \tau(k) = \sum_{k=1}^{2n} \left[\frac{2n}{k}\right] = \sum_{k=1}^{n} \left[\frac{2n}{k}\right] + \sum_{k=1}^{n} \left[\frac{2n}{n+k}\right] = \sum_{k=1}^{n} \left[\frac{2n}{k}\right] + n.$$

(680) This follows from Problem 678 by choosing $f(n) = \phi(n)$ and $F(n) = n$.

(681) This follows from Problem 678 by choosing $f(n) = \Lambda(n)$ and $F(n) = \log n$.

(682) In light of Problem 552, $\sum_{k|m} \lambda(k) = 1$ if m is a perfect square and 0 otherwise. This is why

$$\sum_{k=1}^{n} \lambda(k) \left[\frac{n}{k}\right] = \sum_{k=1}^{n} \lambda(k) \sum_{ki\leq n} 1 = \sum_{k=1}^{n} \lambda(k) \sum_{\substack{m\leq n \\ k|m}} 1 = \sum_{m=1}^{n} \sum_{k|m} \lambda(k)$$

$$= \sum_{\substack{m=1 \\ m \text{ perfect square}}}^{n} 1 = [\sqrt{n}].$$

(683) (*Kürschák Competition, 1983*). We shall display two such integer sequences. We first show that the number $n = 2^{k+1} + 2$ satisfies $S(n) > n$ for each positive integer k. Indeed, since n can be written as $n = 2(2^k + 1)$ with $2^k + 1 \geq 3$ and since $2^k + 1$ has at least one prime divisor $q \geq 3$, then

$$S(n) = S(2^{k+1} + 2) \geq 2^{k+1} + q^1 \geq 2^{k+1} + 3 > 2^{k+1} + 2 = n,$$

where we used the fact that 2^{k+1} is the largest power of 2 not exceeding $2^{k+1} + 2$. This therefore establishes that $n = 2^{k+1} + 2$ satisfies $S(n) > n$ for each positive integer k.

It is just as easy to see that if $n = 2p$, where p is an odd prime number, then $S(n) > n$. Indeed, let k be such that $2^k \leq n < 2^{k+1}$. Then,

$$S(n) \geq 2^k + p > \frac{n}{2} + p = p + p = 2p = n.$$

(684) We have

$$\sum_{k=1}^{n} \tau(k) = \sum_{k=1}^{n} \sum_{d|k} 1 = \sum_{d\leq n} \sum_{\substack{k\leq n \\ d|k}} 1 = \sum_{d\leq n} \left[\frac{n}{d}\right].$$

(685) We have

$$\sum_{k=1}^{n} \sigma(k) = \sum_{k=1}^{n} \sum_{d|k} d = \sum_{d=1}^{n} d \sum_{m\leq n/d} 1 = \sum_{d=1}^{n} d \left[\frac{n}{d}\right].$$

(686) We may first write

$$\sum_{k=1}^{n} \phi(k) = \sum_{k=1}^{n} k \sum_{d|k} \frac{\mu(d)}{d} = \sum_{d=1}^{n} \mu(d) \sum_{j\leq n/d} j.$$

Using the fact that

$$\sum_{j=1}^{m} j = \frac{m(m+1)}{2},$$

we obtain the result.

(687) This is an immediate consequence of Problems 1 and 552.

(688) (*AMM, Vol. 94, 1987, p. 795*). Let $\{x\} = x - [x]$ stand for the fractional part of x. Then $k \in S(n)$ if and only if $\{n/k\} \geq 1/2$. If $k > 2n$, then

$n/k < 1/2$ and therefore $k \notin S(n)$. Hence, $k \in S(n)$ implies $k \leq 2n$. Since $[2x] - 2[x] = 0$ if $\{x\} < 1/2$ and 1 otherwise, we have

$$
\begin{aligned}
\sum_{k \in S(n)} f(k) &= \sum_{k=1}^{2n} f(k) \left(\left[\frac{2n}{k} \right] - 2 \left[\frac{n}{k} \right] \right) \\
&= \sum_{k=1}^{2n} f(k) \left[\frac{2n}{k} \right] - 2 \sum_{k=1}^{n} f(k) \left[\frac{n}{k} \right] \\
&= g(2n) - 2g(n),
\end{aligned}
$$

where we used the fact that $[n/k] = 0$ when $k > n$.

For the particular cases, it is enough to use the identities

$$
\sum_{k=1}^{n} \phi(k) \left[\frac{n}{k} \right] = \frac{n(n+1)}{2}, \qquad \sum_{k=1}^{n} \mu(k) \left[\frac{n}{k} \right] = 1,
$$

$$
\sum_{k=1}^{n} \Lambda(k) \left[\frac{n}{k} \right] = \log n!, \qquad \sum_{k=1}^{n} \lambda(k) \left[\frac{n}{k} \right] = [\sqrt{n}],
$$

which are respectively the subjects of Problems 680, 607, 681 and 682.

(689) Using the relation $\sum_{d|k} \phi(d) = k$, we have

$$
\begin{aligned}
\sum_{n=1}^{\infty} \frac{\phi(n) x^n}{1 - x^n} &= \sum_{n=1}^{\infty} \sum_{j=1}^{\infty} \phi(n) x^{jn} = \sum_{k=1}^{\infty} \left(\sum_{d|k} \phi(d) \right) x^k \\
&= \sum_{k=1}^{\infty} k x^k = \frac{x}{(1-x)^2}.
\end{aligned}
$$

(690) We obtain successively

$$
\begin{aligned}
\sum_{n=1}^{\infty} f(n) x^n &= \sum_{n=1}^{\infty} \left(\sum_{d|n} g(d) \right) x^n = \sum_{d_1=1}^{\infty} g(d_1) \sum_{d_2=1}^{\infty} x^{d_1 d_2} \\
&= \sum_{d_1=1}^{\infty} g(d_1) \left(x^{d_1} + x^{2d_1} + x^{3d_1} + \cdots \right) \\
&= \sum_{n=1}^{\infty} g(n) \left(x^n + x^{2n} + x^{3n} + \cdots \right) \\
&= \sum_{n=1}^{\infty} g(n) x^n \left(1 + x^n + x^{2n} + \cdots \right) \\
&= \sum_{n=1}^{\infty} g(n) \frac{x^n}{1 - x^n}.
\end{aligned}
$$

(691) (Niven, Zuckerman and Montgomery [**25**], page 313). In fact, we will show a much more general result, namely the following: *"Let f be a multiplicative function and consider the matrix $M_{n \times n} = (b_{ij})_{n \times n}$, where*

$b_{i,j} = f((i,j))$, that is the value of the function f evaluated at the GCD of i and j. Then,

$$\det M = g(1)g(2)\cdots g(n),$$

where g is defined by $g(n) = \sum_{d|n} \mu(d)f(n/d)$." It is clear that by taking $f(n) = n$ and using the fact that $\sum_{d|n} \mu(d)\frac{n}{d} = \phi(n)$, we find as a particular case the original problem.

Let $A = (a_{ij})_{n\times n}$ be the matrix for which each element a_{ij} is defined by $a_{ij} = 1$ if $j|i$ and $a_{ij} = 0$ otherwise. Observe that the matrix A is triangular with 1's on the diagonal and 0's above the diagonal, so that $\det A = 1$ and moreover that

(1) $$\det A^T = 1,$$

where A^T stands for the transpose of matrix A. Finally, let $H = (h_{ij})_{n\times n}$ be the matrix defined by $h_{ij} = g(j)a_{ij}$. We easily see that H is also a triangular matrix, with the elements $h_{jj} = g(j)a_{jj} = g(j)$ on its diagonal and 0's above its diagonal. It follows that

(2) $$\det H = \prod_{j=1}^{n} g(j).$$

Then consider the matrix HA^T. We obtain that $HA^T = (\ell_{ij})_{n\times n}$, where

(3) $$\ell_{ij} = \sum_{k=1}^{n} g(k)a_{ik}a_{jk} = \sum_{k|(i,j)} g(k) = \sum_{k|(i,j)} \sum_{d|k} \mu(k/d)f(d)$$

$$= \sum_{d|(i,j)} \sum_{r|\frac{(i,j)}{d}} \mu(r)f(d) = \sum_{d|(i,j)} f(d) \sum_{r|\frac{(i,j)}{d}} \mu(r) = f((i,j)),$$

where we used the fact that $\sum_{d|z} \mu(d) = 1$ if $z = 1$ and 0 if $z > 1$. It then follows from (1) and (2) that

(4) $$\det HA^T = \det H \cdot \det A^T = \prod_{j=1}^{n} g(j).$$

In light of (3) and (4), the result follows.

REMARK: The determinant M is often called *Smith's determinant* (see Shapiro [**38**], page 75).

(692) Setting $g(m) = 1$ in the solution of Problem 691, in which case $f(m) = \tau(m)$, we have

$$\det M = \prod_{j=1}^{n} g(j) = 1.$$

(693) Setting $g(m) = m$ in the solution of Problem 691, in which case $f(m) = \sigma(m)$, we have

$$\det M = \prod_{j=1}^{n} g(j) = \prod_{j=1}^{n} j = n!$$

(694) Setting $f(m) = \mu(m)$ in the solution of Problem 691, in which case the corresponding function is multiplicative with, for each prime number p,

$$g(p) = -2, \qquad g(p^2) = 1, \qquad g(p^\alpha) = 0 \quad \text{if } \alpha \geq 3,$$

we get that

$$
\begin{aligned}
\det M_{1\times 1} &= 1, \\
\det M_{2\times 2} &= 1\cdot(-2) = -2, \\
\det M_{3\times 3} &= 1\cdot(-2)\cdot(-2) = 4, \\
\det M_{4\times 4} &= 1\cdot(-2)\cdot(-2)\cdot 1 = 4, \\
\det M_{5\times 5} &= 1\cdot(-2)\cdot(-2)\cdot 1\cdot(-2) = -8, \\
\det M_{6\times 6} &= 1\cdot(-2)\cdot(-2)\cdot 1\cdot(-2)\cdot 4 = -32, \\
\det M_{7\times 7} &= 1\cdot(-2)\cdot(-2)\cdot 1\cdot(-2)\cdot 4\cdot(-2) = 64,
\end{aligned}
$$

while for each integer $n \geq 8$,

$$
\det M_{n\times n} = 1\cdot(-2)\cdot(-2)\cdot 1\cdot(-2)\cdot 4\cdot(-2)\cdot 0\cdot\ldots = 0.
$$

(695) Using the solution of Problem 691 in order to first evaluate the determinant of the matrix $M_0 = (a_{ij})_{n\times n}$, where $a_{ij} = 1/(i,j)$, that is by setting $f(m) = 1/m$ so that

$$
g(m) = \sum_{d\mid m} \mu(d) f(m/d) = \frac{1}{m}\sum_{d\mid m}\mu(d)d = \frac{1}{m}\prod_{p\mid m}(1-p) = \frac{\phi(m)}{m^2}\prod_{p\mid m}(-p),
$$

we have that

$$
(1) \qquad \det M_0 = \prod_{j=1}^{n} g(j) = \prod_{j=1}^{n}\frac{\phi(j)}{j^2}\prod_{p\mid j}(-p) = \prod_{j=1}^{n}\frac{\phi(j)}{j^2}(-1)^{\omega(j)}\gamma(j)
$$

$$
= \frac{1}{(n!)^2}\prod_{j=1}^{n}\phi(j)(-1)^{\omega(j)}\gamma(j).
$$

Using the fact that a common factor of the elements of a row or of a column can be factored, we obtain

$$
(2) \qquad \det M = \det\left([i,j]\right)_{n\times n} = \det\left(\frac{ij}{(i,j)}\right)_{n\times n} = n!\det\left(\frac{j}{(i,j)}\right)_{n\times n}
$$

$$
= n!n!\det\left(\frac{1}{(i,j)}\right)_{n\times n} = (n!)^2\det M_0.
$$

Combining (1) and (2) gives us the result.

(696) We have

$$
\begin{aligned}
\sum_{\substack{n\leq x \\ (n,d)=1}} \mu(n)g(x/n) &= \sum_{\substack{n\leq x \\ (n,d)=1}} \mu(n)\sum_{\substack{m\leq x/n \\ (m,d)=1}} f(x/mn) \\
&= \sum_{\substack{mn\leq x \\ (mn,d)=1}} \mu(n)f(x/mn) \\
&= \sum_{\substack{r\leq x \\ (r,d)=1}} f(x/r)\sum_{n\mid r}\mu(r) = f(x).
\end{aligned}
$$

(697) The result follows from the fact that

$$\sum_{d|k} \sum_{m \le N/d} \mu(d)f(md) = \sum_{\substack{md \le N \\ d|k}} \mu(d)f(md) = \sum_{n \le N} \left(\sum_{\substack{d|k \\ d|n}} \mu(d) \right) f(n)$$

$$= \sum_{\substack{n \le N \\ (n,k)=1}} f(n).$$

(698) We obtain the result by writing

$$\sum_{n \le x} M(x/n) = \sum_{n \le x} \sum_{m \le x/n} \mu(m) = \sum_{mn \le x} \mu(m) = \sum_{r \le x} \sum_{d|r} \mu(d)$$

$$= 1 + \sum_{2 \le r \le x} \sum_{d|r} \mu(d) = 1 + 0 = 1.$$

(699) This result is trivial, since $p(n(n+1)) = 2$ for each positive integer n.

(700) Since $10^{c(n)-1} \le n < 10^{c(n)}$, we have $c(n) = [\log_{10} n] + 1$. Set

$$\Sigma_1 = \sum_{1 \le i \le c(n)} \ell_i^2 \quad \text{and} \quad \Sigma_2 = \sum_{\substack{d|n \\ 1<d<n}} d.$$

Let us show that for $n > 10^6$, we cannot have $\Sigma_1 = \Sigma_2$, because Σ_1 is much to small with respect to Σ_2. Indeed,

$$(*) \qquad \Sigma_1 \le 81c(n) \le 81(\log_{10} n + 1).$$

On the other hand, since each composite number has at least one proper divisor $> \sqrt{n}$,

$$(**) \qquad \Sigma_2 > \sqrt{n}.$$

Thus, combining $(*)$ and $(**)$, we would have

$$\sqrt{n} < \Sigma_2 = \Sigma_1 \le 81(\log_{10} n + 1),$$

which is not true if $n > 10^6$.

(701) (a) Using MAPLE, we can write
```
> for n to 10000 do
> p(n) := 4*tau(n + 2)−phi(n);
> if p(n) = 0 then print(n) else fi;
> od:
```

This procedure generates the numbers 15, 32, 60, 64, 68, 90, 102, 110 and 130.

REMARK: In fact, one can prove that these are the only solutions of the equation $4\tau(n+2) = \phi(n)$. Indeed, first observe that it follows from Problem 207 that

$$\sum_{p \le n} \log \left(1 - \frac{1}{p} \right) = -\sum_{p \le n} \frac{1}{p} - \sum_{\alpha \ge 2} \sum_{p \le n} \frac{1}{\alpha p^\alpha} > -\sum_{p \le n} \frac{1}{p} - \sum_{\alpha \ge 2} \sum_{p} \frac{1}{p^\alpha}$$

$$> -\sum_{p \le n} \frac{1}{p} - 1.$$

Using this inequality, we get that

$$(i) \quad \phi(n) = n \prod_{p|n} \left(1 - \frac{1}{p}\right) \geq n \prod_{p \leq n} \left(1 - \frac{1}{p}\right) = n \, e^{\sum_{p \leq n} \log(1 - 1/p)}$$

$$> \frac{n}{e \cdot e^{\sum_{p \leq n} \frac{1}{p}}}.$$

On the other hand, it follows from relation (1) of Problem 220 that

$$(ii) \qquad\qquad \sum_{p \leq n} \frac{1}{p} < 2 \log \log n \qquad (n \geq 100).$$

Combining (i) and (ii), we obtain that

$$(iii) \qquad\qquad \phi(n) > n \, e^{-2 \log \log n} = \frac{n}{\log^2 n} \qquad (n \geq 100).$$

But, we have seen in Problem 640 that

$$(iv) \qquad\qquad \tau(n) \leq 2\sqrt{n} \qquad (n \geq 1).$$

Thus, combining (iii) and (iv) and observing that

$$8\sqrt{n+2} < \frac{n}{\log^2 n}$$

for at least all $n > 3\,000\,000$, we obtain that

$$4\tau(n+2) \leq 8\sqrt{n+2} < \frac{n}{\log^2 n} < \phi(n),$$

as soon as $n > 3 \cdot 10^6$. Therefore, we only need to check, using a computer, that $4\tau(n+2) = \phi(n)$ has no solution $n \leq 3 \cdot 10^6$, and the problem is solved!

(b) Similarly, with MAPLE, we can write

```
> for n from 1 to 2000 do p(n):=sigma(n)-((2*n)-1);
> if p(n)=0 then print( n) else fi;
> od:
```

This procedure generates all powers of 2 smaller than 2000.

(702) It seems natural to first examine the possible solutions of the form $n = 2^k$. For such a number to be a solution, we must have that $\phi(\sigma(2^k)) = 2^k$ and therefore that

$$(*) \qquad\qquad \phi(2^{k+1} - 1) = 2^k.$$

We observe that such is the case if $k = 0, 1, 3, 7$. This suggests that we should examine the cases where $k = 2^\alpha - 1$, that is when

$$\begin{aligned}
2^{k+1} - 1 = 2^{2^\alpha} - 1 &= (2^{2^{\alpha-1}} - 1)(2^{2^{\alpha-1}} + 1) \\
&= (2^{2^{\alpha-2}} - 1)(2^{2^{\alpha-2}} + 1)(2^{2^{\alpha-1}} + 1) \\
&= (2^{2^{\alpha-3}} - 1)(2^{2^{\alpha-3}} + 1)(2^{2^{\alpha-2}} + 1)(2^{2^{\alpha-1}} + 1) \\
&= (2^2 - 1)(2^2 + 1)(2^3 + 1)(2^4 + 1) \cdots (2^{2^{\alpha-1}} + 1) \\
&= 3 \cdot 5 \cdot 17 \cdot 257 \cdots (2^{2^{\alpha-1}} + 1).
\end{aligned}$$

Hence, if $2^{2^i}+1$ is a prime number for each integer i such that $1 \le i \le \alpha-1$, then substituting $k = 2^\alpha - 1$ in $(*)$, we do indeed obtain

$$\phi(2^{2^\alpha} - 1) = \phi(3)\phi(5)\phi(17)\cdots\phi(2^{2^{\alpha-1}} + 1) = 2^{2^\alpha - 1},$$

since

$$2 \cdot 2^2 \cdot 2^4 \cdots 2^{2^{\alpha-1}} = 2^{2^\alpha - 1}.$$

It is known that $2^{2^\beta}+1$ is (a Fermat) prime if $\beta = 0, 1, 2, 3, 4$, thus yielding the numbers $k = 1, 3, 7, 15, 31$. Taking into account the trivial solution $n = 2^0 = 1$, we have thus found six solutions of $\phi(\sigma(n)) = n$, namely

$$n = 1, 2, 2^3, 2^7, 2^{15}, 2^{31}.$$

REMARK: Only 24 solutions are known for this equation: besides the six mentioned above, J.L. Selfridge, F. Hoffman and R. Schroeppel (see [16], p. 99) have found the 18 solutions shown in the following table:

$$
\begin{aligned}
12 &= 2^2 \cdot 3 \\
240 &= 2^4 \cdot 3 \cdot 5 \\
720 &= 2^4 \cdot 3^2 \cdot 5 \\
6912 &= 2^8 \cdot 3^3 \\
142\,560 &= 2^5 \cdot 3^4 \cdot 5 \cdot 11 \\
712\,800 &= 2^5 \cdot 3^4 \cdot 5^2 \cdot 11 \\
1\,140\,480 &= 2^8 \cdot 3^4 \cdot 5 \cdot 11 \\
1\,190\,400 &= 2^9 \cdot 3 \cdot 5^2 \cdot 11 \\
3\,345\,408 &= 2^{10} \cdot 3^3 \cdot 11^2 \\
3\,571\,200 &= 2^9 \cdot 3^2 \cdot 5^2 \cdot 31 \\
5\,702\,400 &= 2^8 \cdot 3^4 \cdot 5^2 \cdot 11 \\
14\,859\,936 &= 2^5 \cdot 3^6 \cdot 7^2 \cdot 13 \\
29\,719\,872 &= 2^6 \cdot 3^6 \cdot 7^2 \cdot 13 \\
50\,319\,360 &= 2^{12} \cdot 3^3 \cdot 5 \cdot 7 \cdot 13 \\
4\,389\,396\,480 &= 2^{13} \cdot 3^7 \cdot 5 \cdot 7^2 \\
21\,946\,982\,400 &= 2^{13} \cdot 3^7 \cdot 5^2 \cdot 7^2 \\
11\,681\,629\,470\,720 &= 2^{21} \cdot 3^3 \cdot 5 \cdot 11^3 \cdot 31 \\
58\,408\,147\,353\,600 &= 2^{21} \cdot 3^3 \cdot 5^2 \cdot 11^3 \cdot 31
\end{aligned}
$$

(703) Let $n = n_0$ be a solution of $\phi(\sigma(n)) = n$. Set $m_0 = \sigma(n_0)$. We indeed have that $m = m_0$ is a solution of $\sigma(\phi(m)) = m$, since

$$\sigma(\phi(m_0)) = \sigma(\phi(\sigma(n_0))) = \sigma(n_0) = m_0.$$

In Problem 702, we showed that $n = 1$, 2, 8, 2^7, 2^{15} and 2^{31} are solutions of $\phi(\sigma(n)) = n$; we derive from this that $m = \sigma(1) = 1$, $m = \sigma(2) = 3$, $m = \sigma(8) = 15$, $m = \sigma(2^7) = 2^8 - 1 = 255$, $m = \sigma(2^{15}) = 2^{16} - 1 = 65535$ and $m = \sigma(2^{31}) = 2^{32} - 1 = 4\,294\,967\,295$ are solutions of $\sigma(\phi(m)) = m$.

REMARK: It is clear that one can generalize this result and state that if f and g are two arithmetic functions and that if $n = n_0$ is a solution of the equation $f(g(n)) = n$, then $m = m_0 = g(n_0)$ is a solution of $g(f(m)) = m$.

(704) (Golomb [15]). Let $n = 3^{p-1}$, where p and $(3^p - 1)/2$ are two prime numbers. We then have

$$\sigma(\phi(n)) = \sigma(2 \cdot 3^{p-2}) = 3 \cdot \frac{3^{p-1} - 1}{2} = \frac{3^p - 3}{2},$$

while

$$\phi(\sigma(n)) = \phi\left(\frac{3^p - 1}{2}\right) = \frac{3^p - 1}{2} - 1 = \frac{3^p - 3}{2},$$

which proves the first statement.

Using the MATHEMATICA program

```
Do[p=Prime[i];If[PrimeQ[q=(3^p-1)/2], Print[p," ",q," ",
n=3^(p-1)]],{i,1,100}]
```

we obtain the following values for p: 3, 7, 13, 71, 103, 541. The first three corresponding values of n are then 9, 729 and 531 441.

One can also find these values using the following MAPLE program:

```
> for i to 20 do if isprime( (3^(ithprime(i))-1)/2) then
> print('n'=3^(ithprime(i)-1) ) fi; od;
```

There exist other solutions apart from those listed above, namely 225, 242, 516, 3 872, and many others.

(705) It seems natural to consider the numbers n of the form $n = 2^k$, k a positive integer. For such a number to be a solution of $\phi(\tau(n)) = \tau(\phi(n))$, we must have $\phi(k+1) = \tau(2^{k-1})$; that is $\phi(k+1) = k$, an equation which is solvable only when $k+1$ is a prime number, that is when $k = p - 1$, with p prime. This is why the equation $\phi(\tau(n)) = \tau(\phi(n))$ has infinitely many solutions.

(706) We will show that n is a solution of $\tau(\gamma(n)) = \gamma(\tau(n))$ if and only if $n = p^\alpha$ with p prime and α a positive integer such that $\alpha + 1$ is a power of 2.

Let us first show that the condition is sufficient. So let $n = p^\alpha$, where p is prime and $\alpha = 2^k - 1$ for a certain positive integer k. We then have

$$\tau(\gamma(p^\alpha)) = \tau(p) = 2,$$

while

$$\gamma(\tau(p^\alpha)) = \gamma(\alpha + 1) = \gamma(2^k) = 2.$$

To prove that the condition is necessary, we proceed by contradiction. Two situations may occur:

(i) $n = p^\alpha$ with $\alpha \neq 2^k - 1$ for each integer $k \geq 1$.

(ii) $\omega(n) \geq 2$.

In case (i), we have $\tau(\gamma(p^\alpha)) = \tau(p) = 2$, while $\gamma(\tau(p^\alpha)) = \gamma(\alpha + 1) > 2$. In case (ii), we have $n = q_1^{\alpha_1} q_2^{\alpha_2} \cdots q_r^{\alpha_r}$ for certain prime numbers $q_1 < q_2 < \ldots < q_r$ and certain positive integers α_i's, $i = 1, 2, \ldots, r$, with $r \geq 2$. It follows that $\tau(\gamma(n)) = 2^r$, while $\gamma(\tau(n)) = \gamma((\alpha_1+1)(\alpha_2+1)\cdots(\alpha_r+1))$ is either equal to 2 or else divisible by a prime number $p > 2$, and therefore in both cases, $\gamma(\tau(n))$ cannot be equal to 2^r with $r \geq 2$.

(707) The following is the list of all the solutions $n < 10^6$, n nonsquarefree, of (∗): 49, 1681, 18 490, 23 762, 39 325, 57 121 and 182 182. The fact that (∗) has infinitely many nonsquarefree solutions n follows from the fact that the Fermat-Pell equation (∗∗) $2x^2 - y^2 = 1$ has infinitely many solutions: indeed, to every solution (x, y) of (∗∗), one can associate the solution n of (∗) given by $n = y^2$ and $n + 1 = 2x^2$, in which case $\delta(n + 1) - \delta(n) = 2 - 1 = 1$.

(708) We will show that the only two solutions are $n = 1$ and $n = 6$. Taking into account the remark following the solution of Problem 703, we get a

bijection between the solutions of $\gamma(\sigma(n)) = n$ and those of

(1) $$\sigma(\gamma(n)) = n.$$

We will therefore look for the solutions of this last equation. First of all, it is clear that $n = 1$ is a solution of (1). So let $n > 1$ be a solution of (1). If n is a prime power, that is $n = p^\alpha$, then (1) implies that $\sigma(p) = p^\alpha$; that is $p + 1 = p^\alpha$, which is impossible. We then have that n has at least two distinct prime factors. We write n as a product of distinct prime powers, that is

$$n = q_1^{\alpha_1} q_2^{\alpha_2} \cdots q_r^{\alpha_r} \qquad (r \geq 2).$$

From (1), we have

(2) $$(q_1 + 1)(q_2 + 1) \cdots (q_r + 1) = q_1^{\alpha_1} q_2^{\alpha_2} \cdots q_r^{\alpha_r}.$$

If $q_1 \geq 3$, then the left-hand side of (2) is even, while its right-hand side is odd, which is impossible. Hence, $q_1 = 2$. Therefore, (2) can be written as

(3) $$3(q_2 + 1) \cdots (q_r + 1) = 2^{\alpha_1} q_2^{\alpha_2} \cdots q_r^{\alpha_r}.$$

Since q_r divides the left-hand side of (2), it must divide 3 or one of the factors $q_i + 1$ for a certain $2 \leq i \leq r$. Since this last alternative is impossible, we have that $q_r | 3$, so that $q_r = q_2 = 3$. It follows that equation (3) becomes

$$\begin{aligned} (2+1)(3+1) &= 2^{\alpha_1} 3^{\alpha_2}, \\ 2^2 \cdot 3 &= 2^{\alpha_1} 3^{\alpha_2}, \end{aligned}$$

and therefore by the uniqueness of factorization, we may conclude that $\alpha_1 = 2$ and $\alpha_2 = 1$, which gives rise to the solution $n = 12$ of (1). Hence, we easily obtain that the only two solutions of $\gamma(\sigma(n)) = n$ are $n = 1$ and $n = 6$.

(709) Since 8 does not divide k, it is clear that k is either odd or of the form $4\ell + 2$, $\ell = 0, 1, 2, \ldots$, or else of the form $8\ell + 4$, $\ell = 0, 1, 2, \ldots$.

In the first case, we have that $\sigma_k(6) = 1^k + 2^k + 3^k + 6^k$ is divisible by 6, since

$$1^k + 2^k + 3^k + 6^k \equiv 1 + 3^k \equiv 1 + 1 \equiv 0 \pmod 2$$

and

$$1^k + 2^k + 3^k + 6^k \equiv 1 + 2^k \equiv 1 - 1^k \equiv 0 \pmod 3,$$

while $\sigma_k(2) = 1 + 2^k \not\equiv 0 \pmod 2$, $\sigma_k(3) = 1 + 3^k \not\equiv 0 \pmod 3$, $\sigma_k(4) = 1 + 2^k + 4^k \not\equiv 0 \pmod 4$ and $\sigma_k(5) = 1 + 5^k \not\equiv 0 \pmod 5$.

In the second case, we have that $\sigma_k(10) = 1^k + 2^k + 5^k + 10^k$ is divisible by 10, since

$$1^k + 2^k + 5^k + 10^k \equiv 1 + 5^k \equiv 1 + 1 \equiv 0 \pmod 2$$

and

$$1^k + 2^k + 5^k + 10^k \equiv 1 + 2^k \equiv 1 + 2^{4\ell} \cdot 2^2 \equiv 1 + 1 \cdot 4 \equiv 0 \pmod 5,$$

while $\sigma_k(n) \not\equiv 0 \pmod{10}$ for each integer $2 \leq n \leq 9$.

In the third case, $\sigma_k(34) = 1^k + 2^k + 17^k + 34^k$ is divisible by 34, since

$$1^k + 2^k + 17^k + 34^k \equiv 1 + 17^k \equiv 1 + 1^k \equiv 0 \pmod 2$$

and

$$1^k + 2^k + 17^k + 34^k \equiv 1 + 2^k = 1 + 2^{8\ell} \cdot 2^4$$
$$\equiv 1 + (-1)^{2\ell} \cdot (-1) \equiv 0 \pmod{17},$$

while $\sigma_k(n) \not\equiv 0 \pmod{34}$ for each integer $2 \le n \le 33$.

(710) We will establish that

$$(*) \qquad \frac{\phi(n)}{n} = \frac{2}{3} \Longleftrightarrow n = 3^k, \quad k = 1, 2, \dots .$$

The implication ($\Leftarrow$) is easy to verify. So assume that $\frac{\phi(n)}{n} = \frac{2}{3}$. Since $3\phi(n) = 2n$, we have that $3|n$, and this is why there exist two positive integers k and r with $(r,3) = 1$ such that $n = 3^k r$. We then obtain successively

$$3\phi(3^k)\phi(r) = 2 \cdot 3^k \cdot r$$
$$3 \cdot 3^{k-1}(3-1)\phi(r) = 2 \cdot 3^k \cdot r$$
$$\phi(r) = r,$$

which implies that $r = 1$, thereby completing the proof of $(*)$.

(711) By hypothesis, we have $7\phi(n) = 4n$. It follows that $7|n$ and therefore that there exist two positive integers β and r, with $(r,7) = 1$, such that $n = 7^\beta r$. We then obtain successively

$$7\phi(7^\beta)\phi(r) = 4 \cdot 7^\beta \cdot r,$$
$$7 \cdot 7^{\beta-1}(7-1)\phi(r) = 4 \cdot 7^\beta \cdot r,$$
$$6\phi(r) = 4r,$$
$$\frac{\phi(r)}{r} = \frac{2}{3}.$$

Using the result of Problem 710, we obtain that $r = 3^\alpha$ for a certain positive integer α, and this is why $n = 7^k r = 3^\alpha 7^\beta$, as was to be shown.

(712) (*AMM, Vol. 88, 1981, p. 764*). First assume that $n = p^r$; we then have

$$(*) \qquad S(p^r) = \frac{(r+1)(r+2)}{2}$$

so that $n = 3$ if $n = S(n)$. Since τ is a multiplicative function, the function S is also multiplicative. However, if $n = S(n)$ and $n \ne 1$ and 3, then there exist prime numbers q_1 and q_2 and also positive integers a and b such that $q_1^a|n$, $q_2^b|n$ and $(q_1^a, n/q_1^a) = 1$, $(q_2^b, n/q_2^b) = 1$, $S(q_1^a)/q_1^a > 1$ and $S(q_2^b)/q_2^b < 1$. From relation $(*)$, it follows that q_1^a must be 2, 4 or 8. Therefore, $1 < S(q_1^a)/q_1^a \le 3/2$; this is why q_2^b is not a power of 2 and must satisfy $2/3 < S(q_2^b)/q_2^b < 1$. Again using $(*)$, we find that the only possible value of q_2^b is 9. Therefore,

$$\frac{S(q_2^b)}{q_2^b} = \frac{2}{3}, \qquad \frac{S(q_1^a)}{q_1^a} = \frac{3}{2},$$

so that no prime numbers other than 2 and 3 can appear in the representation of n as a product of distinct prime powers. We conclude that the only solutions are $n = 1, 3, 18$ and 36.

(713) (a) The only such numbers are $n = 14$, 15 and 23. Since the divisors of 24 are 1, 2, 3, 4, 6, 8, 12 and 24 and since $\sigma(n) = \prod_{i=1}^{r}(1 + q_i + \cdots + q_i^{a_i})$ when $n = q_1^{a_1} \cdots q_r^{a_r}$, it follows that the numbers 1 and 2 cannot be written as $1 + q_i + \cdots + q_i^{a_i}$, while $3 = 1+2$, $4 = 1+3$, $6 = 1+5$, $8 = 1+7$, $12 = 1+11$ and $24 = 1 + 23$. Let us now examine all possible ways of writing 24 as a product: first $3 \cdot 8$ gives $n = 2 \cdot 7 = 14$; then, $4 \cdot 6$ gives $n = 3 \cdot 5 = 15$; finally, $1 \cdot 24$ gives $n = 23$.

(b) Since $57 = 3 \cdot 19$ and since it is not possible to write 3, 19 and 57 as $1 + p + \cdots + p^a$, we may conclude that no integer n such that $\sigma(n) = 57$ exists.

(714) It is clearly $n = 1$, with $x = 1$.

(715) Building a table of values of $\sigma(x)$ for $x = 1, 2, 3, \ldots$, we easily obtain that $\sigma(x) = 12$ has exactly two solutions, namely $x = 6$ and $x = 11$, and that there are no other numbers smaller than 12 with this property.

(716) It is $n = 24$, with $x = 14, 15, 23$.

(717) Since $\sigma(x) \geq x$, it is clear that given any positive integer n, $\sigma(x) = n$ has only a finite number of solutions x. But the integers of the form p^{n-1}, where p is a prime number, are solutions of $\tau(x) = n$. Therefore, $\tau(x) = n$ has infinitely many integer solutions x.

(718) The answer is YES. If n is prime, then n has only two divisors, 1 and n, in which case $\sigma(n) = n + 1$. Conversely, if n is not prime, then it has a proper divisor d, in which case $\sigma(n) \geq n + d + 1 > n + 1$.

(719) Assume the contrary, that is that n has at most three distinct prime factors. First consider the case when there are exactly three, say $3 \leq q_1 < q_2 < q_3$. Then, since

$$\frac{\sigma(p^\alpha)}{p^\alpha} = 1 + \frac{1}{p} + \cdots + \frac{1}{p^\alpha} < 1 + \frac{1}{p} + \frac{1}{p^2} + \cdots = \frac{1}{1 - \frac{1}{p}} = \frac{p}{p-1},$$

we will have

$$\frac{35}{16} \leq a = \frac{\sigma(n)}{n} = \prod_{p^\alpha \| n} \frac{\sigma(p^\alpha)}{p^\alpha} < \frac{q_1}{q_1 - 1} \frac{q_2}{q_2 - 1} \frac{q_3}{q_3 - 1}$$

$$\leq \frac{3}{3 - 1} \frac{5}{5 - 1} \frac{7}{7 - 1} = \frac{35}{16},$$

a contradiction. From this argument, it is clear that we again obtain a contradiction if we assume that n has exactly one prime factor or exactly three prime factors.

(720) Assume the contrary, that is that n has at most three distinct prime factors. First consider the case when there are exactly three prime factors, say $2 \leq q_1 < q_2 < q_3$. Then, since

$$\frac{\sigma(p^\alpha)}{p^\alpha} = 1 + \frac{1}{p} + \cdots + \frac{1}{p^\alpha} < 1 + \frac{1}{p} + \frac{1}{p^2} + \cdots = \frac{1}{1 - \frac{1}{p}} = \frac{p}{p-1},$$

we will have

$$\frac{15}{4} \le a = \frac{\sigma(n)}{n} = \prod_{p^\alpha \| n} \frac{\sigma(p^\alpha)}{p^\alpha} < \frac{q_1}{q_1 - 1} \frac{q_2}{q_2 - 1} \frac{q_3}{q_3 - 1}$$

$$\le \frac{2}{2 - 1} \frac{3}{3 - 1} \frac{5}{5 - 1} = \frac{15}{4},$$

a contradiction. From this argument, it is clear that we obtain a contradiction if we assume that n has exactly one prime factor or exactly three prime factors.

(721) Since $4\sigma(n) = 9n$ and $(4, 9) = 1$, it is clear that $4 | n$. Assume that $n = 2^a p^b$ where $(2, p) = 1$, $a \ge 2$. We will establish that the only solutions of this type are $n = 40$ and $n = 224$. First of all, we have

(1) $\qquad 4\sigma(n) = 4\sigma(2^a)\sigma(p^b) = 4(2^{a+1} - 1)(1 + p + \cdots + p^b),$

while

(2) $\qquad\qquad\qquad 9n = 9 \cdot 2^a \cdot p^b.$

Combining (1) and (2), we obtain

(3) $\qquad (2^{a+1} - 1)(1 + p + \cdots + p^b) = 9 \cdot 2^{a-2} \cdot p^b.$

Since $(p^b, 1 + p + \cdots + p^b) = 1$, it follows that $p^b | 2^{a+1} - 1$. Similarly, since $(2^{a-2}, 2^{a+1} - 1) = 1$, then $2^{a-2} | (1 + p + \cdots + p^b)$. Therefore, relation (3) can be written as

(4) $\qquad \dfrac{2^{a+1} - 1}{p^b} \cdot \dfrac{1 + p + \cdots + p^b}{2^{a-2}} = 9.$

Three cases are then possible:
    Case #1:

$$\frac{2^{a+1} - 1}{p^b} = 1 \quad \text{and} \quad \frac{1 + p + \cdots + p^b}{2^{a-2}} = 9.$$

If $a = 2$, then $7 = p^b$; hence, $b = 1$ and $p = 7$. Substituting these values in the second equation, we obtain $1 + p = 8 = 9 \cdot 2^{a-2}$, which is not possible. It follows that we must have $a \ge 3$, which implies, in view of the second relation, that b must be odd. If $b = 1$, then $p = 2^{a+1} - 1$ and $p = 9 \cdot 2^{a-2} - 1$, which implies that $2^{a+1} = 9 \cdot 2^{a-2}$, which is impossible. If $b \ge 3$, then, since b is odd,

$$2^{a+1} = p^b + 1 = (p + 1)(p^{b-1} - p^{b-2} + \cdots - p + 1) = (p + 1)Q,$$

where $Q > 1$ is odd, because it is the sum of b odd numbers, which is also impossible.
    Case #2:

$$\frac{2^{a+1} - 1}{p^b} = 3 \quad \text{and} \quad \frac{1 + p + \cdots + p^b}{2^{a-2}} = 3.$$

As in the preceding case, we must have $a \ge 3$ and b odd. If $b = 1$, then the above relations bring us to the solution $p = 5$ and $a = 3$, that is $n = 40$ and no other solutions. One easily checks that the cases $a = 4$ and $a = 5$ generate no solutions. If $b \ge 3$, then setting $\beta = a + 1$, we have $\beta \ge 6$ and we obtain $2^\beta - 1 = 3p^b$, an equation which has no solutions. Indeed, if β is odd, then

$$2^\beta - 1 \equiv (-1)^\beta - 1 = -2 \equiv 1 \not\equiv 0 \pmod 3,$$

a contradiction. Similarly, if β is even, then

$$2^\beta - 1 = (2^{\beta/2} - 1)(2^{\beta/2} + 1),$$

which means that $3|(2^{\beta/2} - 1) > 3$ or that $3|(2^{\beta/2} + 1) > 3$, and this is why we necessarily have that $p|(2^{\beta/2} - 1) > 3$ or that $p|(2^{\beta/2} + 1) > 3$, which would mean that $p|2$, which is nonsense.

Case #3:

$$\frac{2^{a+1} - 1}{p^b} = 9 \quad \text{and} \quad \frac{1 + p + \cdots + p^b}{2^{a-2}} = 1.$$

As in the preceding case, we must have $a \geq 3$ and b odd. If $b = 1$, then the above equations lead to the solution $p = 7$ and $a = 5$, that is $n = 224$, and no other solutions.

REMARK: The equation $\sigma(n)/n = 9/4$ has other solutions, in particular $n = 174\,592$ and $n = 492\,101\,632$. It is quite possible that this equation has infinitely many solutions, but no one knows how to prove it.

(722) (Sierpinski [**39**], p. 176) Consider the numbers $m = 2(13^k - 1)$, where $k = 1, 2, \ldots$. Since

$$\sigma(14 \cdot 13^{k-1}) = \sigma(15 \cdot 13^{k-1}) = \sigma(23 \cdot 13^{k-1}) = 24 \cdot \frac{13^k - 1}{13 - 1}$$
$$= 2 \cdot (13^k - 1) = m,$$

the result follows.

(723) Setting $n = 2^k p$ in the equation $\tau(n) + \sigma(n) = 2n$, we easily find that

(*) $$p = 2^{k+1} + 2k + 1.$$

Therefore, to each prime number p of the form (*), we can associate the dihedral perfect number $n = 2^k p$. The five smallest numbers n of this kind are 14, 52, 184, 656 and 34688.

(724) Let $x = \prod_{i=1}^r q_i^{a_i}$, so that

$$\phi(x) = \prod_{i=1}^r q_i^{a_i - 1}(q_i - 1) = 24;$$

this is why the largest prime factor of x must be ≤ 23. Hence, the only prime factors of x are 2, 3, 5, 7, 11, 13, 17, 19 and 23. Since $10 \nmid 24$, $16 \nmid 24$, $18 \nmid 24$ and $22 \nmid 24$, it follows that the prime numbers 11, 17, 19 and 23 cannot divide x. Hence, the only possibilities for the prime numbers are: $p = 2, 3, 5, 7$ and 13. Let $x = 2^a \cdot 3^b \cdot 5^c \cdot 7^d \cdot 13^f$. Since $\phi(x) = 24$, we derive that $0 \leq a \leq 5$, $0 \leq b \leq 2$ and $0 \leq c, d, e, f \leq 1$. We may therefore conclude that the solutions of $\phi(x) = 24$ are 35, 39, 45, 52, 56, 70, 72, 78, 84 and 90.

(725) Let $n = q_1^{a_1} \cdots q_k^{a_k}$. Then,

$$\phi(n) = q_1^{a_1 - 1} \cdots q_k^{a_k - 1}(q_1 - 1) \cdots (q_k - 1).$$

If q_i is an odd prime number, then the term $q_i - 1$ contributes to a factor 2, in which case x cannot have more than r distinct odd prime factors.

(726) First of all, assume that $n = 2^r$ with $r \geq 1$, so that $\phi(n) = 2^{r-1}$. Since $4 \nmid \phi(n)$, it is clear that r must be equal to 1 or 2; that is $n = 2$ or 4. Now, assume that $n = 2^r N$ where $N > 1$ is an odd number. If $\phi(n)$ is not divisible by 4, then from Problem 725, n cannot have more than one odd prime factor. Let $n = 2^r p^k$, where $k \geq 1$ and p is an odd prime number. Since $4 | p - 1$ if $p \equiv 1 \pmod 4$, we have that p must be of the form $4M + 3$. In this case, $\phi(p^k)$ is divisible by 2 and not by 4. Hence, we must choose r so that $\phi(2^r)$ is odd; that is we must choose $r = 0$ or $r = 1$. Finally, the only integers n for which $\phi(n)$ is not divisible by 4 are 1, 2, 4 and the numbers of the form p^k or $2p^k$, where $p \equiv 3 \pmod 4$.

(727) First assume that $n = 3^{6k+2}$ or $2 \cdot 3^{6k+2}$. It is clear that

$$\frac{\phi(3^{6k+2})}{3^{6k+2}} = \frac{2}{3},$$

so that

$(*)$ $\qquad\qquad \phi(3^{6k+2}) = 2 \cdot 3^{6k+1} = m.$

It follows that we also have

$$\phi(2 \cdot 3^{6k+2}) = \phi(3^{6k+2}) = 2 \cdot 3^{6k+1} = m.$$

Reciprocally, if $\phi(n) = m = 2 \cdot 3^{6k+1}$, then setting $n = r \cdot 3^{6k+2}$ with $(r, 3^{6k+2}) = 1$, we will have

$$\phi(n) = \phi(r)\phi(3^{6k+2}) = 2 \cdot 3^{6k+1} = m,$$

which implies by $(*)$ that $\phi(r) = 1$ and therefore that $r = 1$ or 2, as required. Finally, since to each integer $k \geq 1$ we can associate two solutions n (that is $n = 3^{6k+2}$ and $n = 2 \cdot 3^{6k+2}$) of $\phi(n) = 2 \cdot 3^{6k+1}$, the second part follows immediately.

(728) Since $2 \cdot 7^m$ is divisible by 2 and not by 4, it follows from Problem 726 that $\phi(n) = 2 \cdot 7^m$, which in turn implies that n must be of the form p^k or $2p^k$, where $p \equiv 3 \pmod 4$. In each of these cases, $\phi(n) = p^{k-1}(p-1)$ and $\phi(n) = 2 \cdot 7^m$ implies that $k = 1$ and $p - 1 = 2 \cdot 7^m$, that is $p = 2 \cdot 7^m + 1$. Since $7^m \equiv 1 \pmod 3$, it follows that $2 \cdot 7^m + 1 \equiv 0 \pmod 3$, and this is why $2 \cdot 7^m + 1$ cannot be a prime number if $m \geq 1$.

(729) If n is prime, the result is immediate. Conversely, if n is not prime, then n must have a proper divisor d which cannot be relatively prime with n. Hence, $\phi(n) \leq n - 2$.

(730) Let $x = q_1^{a_1} \cdots q_r^{a_r}$ be the representation of x as a product of distinct prime powers; then

$$\phi(x) = \prod_{i=1}^r q_i^{a_i - 1}(q_i - 1) = 2p.$$

Consequently, $q_i - 1 | 2p$. But, since the divisors of $2p$ are 1, 2, p and $2p$ and since $2p + 1$ is a composite number, we conclude that the only possibilities are $q_i - 1 = 1$ and $q_i - 1 = 2$. This implies that x must be of the form $x = 2^a 3^b$. But $\phi(x) = 2p$ implies that $2^{a-1} 3^{b-1} = p$. Hence, $p = 2$ or $p = 3$, and since $2p + 1$ is a composite number, we conclude that no such x exists.

(731) If $n = 2^k$ with $k \geq 1$, the result is immediate. Assume now that $n = 2^k N$, $(2, N) = 1$. In this case, $\phi(n) = 2^{k-1}\phi(N)$ and the condition $\phi(n) = n/2$ implies $\phi(N) = N$; that is $N = 1$ so that $n = 2^k$.

(732) Let $n = 2^r 5^s N$, where $(N, 10) = 1$. Using the condition $\phi(n) = 2n/5$, one can easily show that it implies that $\phi(N) = N$ and therefore $N = 1$. Conversely, $\phi(2^r 5^s) = 2^{r+1} 5^{s-1} = 2n/5$.

(733) Let $n = 3^a N$, where $(N, 3) = 1$. Then, $\phi(n) = 3^{a-1} 2\phi(N) = 3^{a-1} N$ and therefore $\phi(N) = N/2$. Assume now that $N = 2^b M$, where $(2, M) = 1$. Substituting, we find $\phi(M) = M$; that is $M = 1$. Therefore, there are infinitely many values of n satisfying $\phi(n) = n/3$, namely $n = 2^b 3^a$, $b \geq 0$, $a > 0$.

(734) Since
$$\frac{\phi(n)}{n} = \prod_{p|n} \left(1 - \frac{1}{p}\right) = \frac{1}{4},$$
we must have
$$4 \prod_{p|n}(p-1) = \prod_{p|n} p.$$
Since $4 \nmid \prod_{p|n} p$, we conclude that there exist no integers n verifying the relation of the statement.

(735) (*TYCM, Vol. 26, 1995, p. 298*). If $p = 12k + 11$ and a is an even number, then
$$\phi(p^a) = p^{a-1}(p-1) = (12k+11)^{a-1}(12k+10)$$
$$= 2\big((12k+11)^{a-1}(6k+5)\big) = 2(6n+1),$$

for a certain positive integer n, since $11^{a-1} \cdot 5 \equiv (-1)^{a-1}(-1) \equiv 1 \pmod 6$. Reciprocally, if $\phi(p^a) = 2(6n+1)$,
$$\phi(p^a) = p^{a-1}(p-1) = 2(6n+1) = 2 \prod_i p_i^{\alpha_i} \prod_j q_j^{\beta_j},$$

where the p_i's are prime numbers of the form $6k+1$ and the q_j's are prime numbers of the form $6k+5$. The prime p must be one of the p_i's or one of the q_j's. Assume that $p = p_k$, so that $a - 1 = \alpha_k$. The above equation can therefore be reduced to
$$p_k - 1 = 2 \prod_{i \neq k} p_i^{\alpha_i} \prod_j q_j^{\beta_j}.$$

The left-hand side is congruent to 0 modulo 6, while the right-hand side is congruent to $2(-1)^{\sum \beta_j} \equiv \pm 2$ modulo 6. This contradiction implies that p must be a q_j, say $p = q_k$. Therefore, $a - 1 = \beta_k$ and the above equation becomes
$$q_k - 1 = 2 \prod_i p_i^{\alpha_i} \prod_{j \neq k} q_j^{\beta_j}.$$

The left-hand side is congruent to 4 modulo 6, while the right-hand side is congruent to $2(-1)^r$ modulo 6, where $r = \sum_{j \neq k} \beta_j$. Clearly, r must be odd. However, we also have
$$6n + 1 = \prod_i p_i^{\alpha_i} \prod_j q_j^{\beta_j} \equiv (-1)^{r+\beta_k} \pmod 6.$$

Since the left-hand side is congruent to 1 modulo 6, it is clear $r + \beta_k$ must be even and therefore that β_k must be odd. It follows that $a = \beta_k + 1$ is even. We have thus shown that $p \equiv 5 \pmod 6$ and therefore that p is congruent to 5 or 11 modulo 12. If $p = 12n + 5$, then

$$\phi(p^\alpha) = 2\Big((12n + 5)^{a-1}(6n + 2)\Big) = 2m,$$

where m is congruent to 4 modulo 6. This is a contradiction, and this is why we must have $p = 12n + 11$, as required.

(736) (a) It is immediate that $\phi(n) \le n$. Hence, we only need to prove the left inequality. The result is immediate for $n = 1$. So, assume that $n > 1$, and write $n = 2^{a_0} q_1^{a_1} \cdots q_r^{a_r}$, where $2 < q_1 < \ldots < q_r$ are prime numbers and the a_i's positive integers. We then have

$$\phi(n) = 2^{a_0 - 1} \prod_{i=1}^{r} q_i^{a_i - 1}(q_i - 1).$$

Since $p(p - 3) + 1 = (p - 1)^2 - p > 0$ for $p \ge 3$, we have $p - 1 > \sqrt{p}$. On the other hand, for each positive integer a_i, we have $a_i - \frac{1}{2} \ge \frac{1}{2} a_i$. This is why

$$\begin{aligned}
\phi(n) &= 2^{a_0 - 1} q_1^{a_1 - 1} \cdots q_r^{a_r - 1}(q_1 - 1) \cdots (q_r - 1) \\
&\ge 2^{a_0 - 1} q_1^{a_1 - \frac{1}{2}} \cdots q_r^{a_r - \frac{1}{2}} \\
&\ge 2^{a_0 - 1} q_1^{\frac{1}{2} a_1} \cdots q_r^{\frac{1}{2} a_r} \ge \frac{1}{2} \sqrt{n},
\end{aligned}$$

and the result follows.

(b) Since $\phi(x) = n \ge \frac{1}{2} \sqrt{x}$, the result is immediate.

(737) To these three questions, we easily find the answers $n = 3$, $n = 2$ and $n = 1$, respectively.

(738) First let $n = p$, a prime number. Then the equation $\sum_{d|n} f(d) = \tau(n) f(n)$ implies that $f(1) + f(p) = 2f(p)$, so that $f(p) = f(1)$. Similarly, we show that $f(p^a) = f(1)$ for each positive integer a. Then we show that $f(pq) = f(1)$, and so on. The result then follows with $c = f(1)$.

(739) The result follows from Problem 738 using the fact that, since f is multiplicative, $f(1) = 1$.

(740) We shall prove that n is a solution of $(*)$ if and only if $n = p^p$, where p is a prime number, thereby establishing in particular that $(*)$ has infinitely many solutions. Now, first of all, it is clear that if p is prime, then by setting $n = p^p$, we have $\Omega(n) = p$, so that $\Omega(n)^{\Omega(n)} = p^p = n$. Let us assume that there exists a composite number $r \ge 4$ such that $n = r^r$ is a solution of $(*)$. In that case, it will follow that $\Omega(n) = \Omega(r^r) \ge 2r$, meaning that

$$\Omega(n)^{\Omega(n)} \ge (2r)^{2r} > r^r = n,$$

thereby contradicting $(*)$. We have thus established that the set of solutions $(*)$ is the set $\{p^p : p \text{ prime}\}$, thus proving our claim.

(741) We shall examine successively the possible solutions n of $(*)$ according to the positive integer β such that $2^\beta \| n$.

We first observe that equation $(*)$ has no odd solution. Indeed, assume that such a solution $n = q_1^{\alpha_1} q_2^{\alpha_2} \cdots q_r^{\alpha_r}$, with $3 \le q_1 < q_2 < \ldots < q_r$, exists. In this case, because the function $\sum_{d|n} \gamma(d)$ is multiplicative, (∗) can be written as

(2) $\qquad q_1^{\alpha_1} q_2^{\alpha_2} \cdots q_r^{\alpha_r} = (1 + \alpha_1 q_1)(1 + \alpha_2 q_2) \cdots (1 + \alpha_r q_r).$

If each exponent α_i is larger than 1, then the left-hand side of (2) is larger than its right-hand side, a contradiction. It follows that there exists at least one exponent α_i equal to 1, say $\alpha_{i_0} = 1$ ($1 \le i_0 \le r$). We then have $1 + \alpha_{i_0} q_{i_0} = 1 + q_{i_0}$, so that the left-hand side of (2) is odd, while its right-hand side is even, a contradiction. This shows that any solution of (∗) must be even.

Let $\beta \ge 1$ be the unique integer such that $2^\beta \| n$. We write n as

$$n = 2^\beta q_1 \cdots q_r p_1^{\alpha_1} \cdots p_s^{\alpha_s},$$

where the α_i's are greater than 1 and the q_i's and p_i's are the odd prime factors of n. We allow the possibility that there are no p_i's or no q_i's. Observe that, since $q_i + 1 | n$ and $q_i + 1$ is even, we can assume that $r \le \beta$. We will first show that $\beta \le 4$. Indeed, equation (∗) can be written as

$$\frac{1 + 2\beta}{2^\beta} \frac{q_1 + 1}{q_1} \cdots \frac{q_r + 1}{q_r} = \frac{p_1^{\alpha_1}}{1 + \alpha_1 p_1} \cdots \frac{p_s^{\alpha_s}}{1 + \alpha_s p_s}.$$

It follows that

$$\frac{1 + 2\beta}{2^\beta} \frac{q_1 + 1}{q_1} \cdots \frac{q_r + 1}{q_r} \ge 1.$$

Hence, assuming that $\beta \ge 5$, we would get

$$\frac{1 + 2\beta}{2^\beta} \frac{q_1 + 1}{q_1} \cdots \frac{q_r + 1}{q_r} \le \frac{11}{32} \frac{4}{3} \frac{6}{5} \frac{8}{7} \frac{12}{11} \frac{14}{13} < 1,$$

a contradiction, which proves that $\beta \le 4$. We will now consider separately the four cases $\beta = 1$, $\beta = 2$, $\beta = 3$ and $\beta = 4$.

First assume that $\beta = 4$. Then $1 + 4 \cdot 2 = 9$ divides n, which implies that

$$\frac{p_1^{\alpha_1}}{1 + \alpha_1 p_1} \cdots \frac{p_s^{\alpha_s}}{1 + \alpha_s p_s} \ge \frac{3^2}{1 + 2 \cdot 3} = \frac{9}{7}.$$

On the other hand,

$$\frac{1 + 2\beta}{2^\beta} \frac{q_1 + 1}{q_1} \cdots \frac{q_r + 1}{q_r} \le \frac{9}{16} \frac{4}{3} \frac{6}{5} \frac{8}{7} \frac{12}{11} < \frac{9}{7},$$

a contradiction. It follows that $\beta = 1$, 2 or 3.

Now assume that $\beta = 3$. We then have that $1 + 3 \cdot 2 = 7$ divides n. Clearly, either $7^2 | n$ or $7 \| n$. If $7^2 | n$, we have

$$\frac{p_1^{\alpha_1}}{1 + \alpha_1 p_1} \cdots \frac{p_s^{\alpha_s}}{1 + \alpha_s p_s} \ge \frac{7^2}{1 + 2 \cdot 7} = \frac{49}{15}.$$

But since $\beta = 3$, we must have

$$\frac{1 + 2\beta}{2^\beta} \frac{q_1 + 1}{q_1} \cdots \frac{q_r + 1}{q_r} \le \frac{7}{8} \frac{4}{3} \frac{6}{5} \frac{8}{7} = \frac{8}{5} < \frac{49}{15},$$

a contradiction. Therefore $7 \| n$, in which case we obtain that $2^3 \cdot 7 = 56$ is a solution. Now, n cannot have any other odd prime factor q such that $q \| n$ since otherwise it would imply that $(1 + 7)(q + 1)|n$, forcing β to be larger than 3. Finally, n cannot have any odd prime p such that $p^2|n$ because it would imply

$$\frac{p^2}{p+1} > 1 = \frac{1+3\cdot2}{2^3}\frac{1+7}{7},$$

again a contradiction. We have thus established that $n = 56$ is the only solution of $(*)$ provided by the case $\beta = 3$.

Now assume that $\beta = 2$. In this case, $1 + 2 \cdot 2 = 5$ divides n. Either $5^2|n$ or $5\|n$. If $5^2|n$, we get

$$\frac{p_1^{\alpha_1}}{1+\alpha_1 p_1} \cdots \frac{p_s^{\alpha_s}}{1+\alpha_s p_s} \geq \frac{25}{11}.$$

On the other hand, since $\beta = 2$, we have

$$\frac{1+2\beta}{2^\beta}\frac{q_1+1}{q_1} \cdots \frac{q_r+1}{q_r} \leq \frac{5}{4}\frac{4}{3}\frac{6}{5} = 2 < \frac{25}{11},$$

a contradiction. Hence $5\|n$, which implies that $5 + 1 = 6$ divides n, so that $3|n$. Assuming that $3\|n$, we would have that $(3+1)(5+1)|n$, so that $8|n$, thereby contradicting the fact that $\beta = 2$. Hence, either $3^3|n$ or $3^2\|n$. If $3^3|n$, then

$$\frac{p_1^{\alpha_1}}{1+\alpha_1 p_1} \cdots \frac{p_s^{\alpha_s}}{1+\alpha_s p_s} \geq \frac{3^3}{1+3\cdot3} = \frac{27}{10} > 2 \geq \frac{1+2\beta}{2^\beta}\frac{q_1+1}{q_1} \cdots \frac{q_r+1}{q_r},$$

a contradiction. We can therefore assume that $3^2\|n$, in which case $1 + 2 \cdot 3 = 7$ divides n. Therefore, either $7\|n$ or $7^2|n$. If $7\|n$, then $(7+1)(5+1)|n$, so that $16|n$, which contradicts the fact that $\beta = 2$. Hence, $7^2|n$, so that

$$\frac{p_1^{\alpha_1}}{1+\alpha_1 p_1} \cdots \frac{p_s^{\alpha_s}}{1+\alpha_s p_s} \geq \frac{49}{15} > 2,$$

also a contradiction.

It remains to consider the case $\beta = 1$. This case implies that $1+2 = 3$ divides n. Then either $3\|n$ or $3^2\|n$ or $3^3|n$. If $3\|n$, it implies that $3+1 = 4$ divides n, contradicting the fact that $\beta = 1$. On the other hand, if $3^3|n$, we get

$$\frac{p_1^{\alpha_1}}{1+\alpha_1 p_1} \cdots \frac{p_s^{\alpha_s}}{1+\alpha_s p_s} \geq \frac{3^3}{1+3\cdot3} = \frac{27}{10}.$$

But on the other hand,

$$\frac{1+2\beta}{2^\beta}\frac{q_1+1}{q_1} \cdots \frac{q_r+1}{q_r} \leq \frac{3}{2}\frac{4}{3} = 2 < \frac{27}{10},$$

a contradiction. We must therefore have that $3^2\|n$, which implies that $1 + 2 \cdot 3 = 7$ divides n. Under the hypothesis that $7\|n$ we get $7 + 1 = 8$ divides n, contradicting $\beta = 1$. But, on the other hand, if $7^2|n$, we obtain

$$\frac{p_1^{\alpha_1}}{1+\alpha_1 p_1} \cdots \frac{p_s^{\alpha_s}}{1+\alpha_s p_s} \geq \frac{49}{15} > 2 \geq \frac{1+2\beta}{2^\beta}\frac{q_1+1}{q_1} \cdots \frac{q_r+1}{q_r},$$

also a contradiction.

We can therefore conclude that the only solution of $(*)$ is $n = 56$.

(742) (*TYCM, Vol. 29, 1998, p. 242*). Since $\sigma(n) \geq n$ and since $\phi(n) \leq n$, it follows that $\sigma(n) - \phi(n) = (-1)^n \tau(n)$ has no solutions when n is odd. We only need to find all the positive even integers n such that $\sigma(n) - \phi(n) = \tau(n)$. Using Problem 623, we find that this is possible if and only if n is prime. Since n is even, the only possibility is $n = 2$.

(743) (*AMM, Vol. 79, 1972, p. 911*). The only solutions are $(m,n) = (2,2), (3,4)$ and $(4,3)$. Indeed, from relation $\phi(mn) = d\phi(m)\phi(n)/\phi(d)$, where $d = (m,n)$ (see Problem 522), we can write the given equation as $1/a + 1/b = d$, where $a = \phi(m)/\phi(d)$ and $b = \phi(n)/\phi(d)$. Since a and b are positive integers, it follows that $d = 2$ and $a = b = 1$ or else that $d = 1$ and $a = b = 2$. The first case leads to $\phi(m) = \phi(n) = 1$, hence, $m = n = 2$, and the second case leads to $\phi(m) = \phi(n) = 2$, which gives that one of the integers m and n is equal to 3 while the other is equal to 4.

(744) We have $\phi(1) = \gamma(1) = 1$ and $1 = \phi(2) \neq 2 = \gamma(2)$. So let $n \geq 3$, in which case $\phi(n)$ is even. We may therefore conclude that any other solution n of $\phi(n) = \gamma(n)$ must be even. If $n = 2^\alpha$ with $\alpha \geq 1$, we have $2^{\alpha-1} = 2$ so that $\alpha = 2$. We have thus found the solution $n = 4$. On the other hand, since for each even number n we have $2\|\gamma(n)$, it follows that if n is not a power of 2, then $n = 2 \cdot p^\alpha$ for a certain odd prime number p and a positive integer α. If $\alpha \geq 3$, then $p^2|\phi(n)$ and $p\|\gamma(n)$, a contradiction. Hence, $\alpha = 1$ or 2. If $\alpha = 1$, we obtain $p - 1 = 2p$, which is nonsense. If $\alpha = 2$, we have $p(p-1) = 2p$, which gives $p = 3$. We have thus found the solution $n = 18$ while at the same time showing that there is no other one.

(745) First of all, $\phi(1) = \gamma(1) = 1$ and $1 = \phi(2) \neq 4 = \gamma(2)^2$. So let $n \geq 3$, in which case $\phi(n)$ is even. We may therefore conclude that any other solution n of $\phi(n) = \gamma(n)^2$ must be even. If $n = 2^\alpha$ with $\alpha \geq 1$, we have $2^{\alpha-1} = 2^2$ so that $\alpha = 3$. We have thus found the solution $n = 8$. On the other hand, since for each even number n we have $4\|\gamma(n)^2$, it follows that if n is not a power of 2, there are three types of possible solutions: (i) $n = 4 \cdot p^\alpha$ for a certain odd prime number $p \equiv 3 \pmod 4$ and a certain positive integer α, (ii) $n = 2 \cdot p^\alpha$ for a certain prime number $p \equiv 1 \pmod 4$ and a certain positive integer α, (iii) $n = 2 \cdot p^\alpha \cdot q^\beta$ for certain odd prime numbers $p < q$ and certain positive integers α, β.

In case (i), we have $2p^{\alpha-1}(p-1) = 4p^2$; that is $p^{\alpha-1}(p-1) = 2p^2$. Since we must have $\alpha = 3$, it follows that $p = 3$, which gives rise to the solution $n = 2^2 \cdot 3^3 = 108$.

In case (ii), we have $p^{\alpha-1}(p-1) = 4p^2$. Since we must then have $\alpha = 3$, it follows that $p = 5$, which gives rise to the solution $n = 2 \cdot 5^3 = 250$.

In case (iii), we have

$$(*) \qquad p^{\alpha-1} \cdot \frac{p-1}{2} \cdot q^{\beta-1} \cdot \frac{q-1}{2} = p^2 q^2,$$

in which case $1 \leq \alpha \leq 3$ and $1 \leq \beta \leq 3$. We first consider the case $\beta = 3$. In this case, it follows from $(*)$ that

$$p^{\alpha-1} \cdot \frac{p-1}{2} \cdot \frac{q-1}{2} = p^2.$$

If $\alpha = 3$, it is easy to see that $p = q = 3$, which contradicts the fact that $p < q$. If $\alpha = 2$, then we must have $\frac{q-1}{2} = p$ and $p = 3$, so that $q = 7$, which gives rise to the solution $n = 2 \cdot 3^2 \cdot 7^3 = 6174$. If $\alpha = 1$, then we must have $\frac{q-1}{2} = p^2$ and $p = 3$, so that $q = 2 \cdot 3^2 + 1 = 19$, which gives rise to the solution $n = 2 \cdot 3 \cdot 19^3 = 41\,154$.

It remains to consider the cases $\beta = 2$ and $\beta = 1$. In the first case, $(*)$ becomes

$$p^{\alpha-1} \cdot \frac{p-1}{2} \cdot \frac{q-1}{2} = p^2 q.$$

We then observe that each of the cases $\alpha = 3$, $\alpha = 2$ and $\alpha = 1$ leads to a contradiction. Similarly, the case $\beta = 1$ generates no solutions.

We have thus proved that the only solutions of $\phi(n) = \gamma(n)^2$ are the six solutions mentioned in the statement.

(746) Property (a) is easy to obtain. Indeed, if $n > 1$ is odd, then $\gamma(n)^2 = \sigma(n)$ is also odd, so that $n = m^2$ for a certain integer m, in which case $n < \sigma(n) = \gamma(n)^2 = \gamma(m^2)^2 = \gamma(m)^2 \le m^2$, a contradiction. To verify property (b), we first observe that if n is squarefree, say $n = q_1 q_2 \cdots q_r$, for certain prime numbers $q_1 < q_2 < \ldots < q_r$, then it follows from $(*)$ that

$$(q_1 + 1)(q_2 + 1) \cdots (q_r + 1) = q_1^2 q_2^2 \cdots q_r^2,$$

which is impossible since, for each prime number q, we have $q + 1 < q^2$. It therefore follows that any solution of $(*)$ must be squarefree.

Finally, using a computer, we find that $n = 1782 = 2 \cdot 3^4 \cdot 11$ is a solution of $(*)$ and in fact that it is the only solution $n > 1$ which is smaller than 10^9.

REMARK: To this day, we do not know if equation $(*)$ has a solution $n > 1782$. The first author has raised this question in the Problem Section of the AMM (Problem #10966, **109** (2002), 759).

(747) Let k be a fixed positive integer. Let $n = 2^\alpha \cdot 3^\beta$, where α and β are positive integers yet to be determined. We will show that an appropriate choice of α and β will provide infinitely many integers n satisfying the desired property. Indeed, with n as above, we have

$$\frac{\sigma(n)}{\gamma(n)^k} = \frac{(2^{\alpha+1} - 1) \cdot \frac{3^{\beta+1}-1}{2}}{2^k \cdot 3^k} = \frac{(2^{\alpha+1} - 1)(3^{\beta+1} - 1)}{2^{k+1} \cdot 3^k}.$$

This last quantity will be an integer if

$$2^{\alpha+1} \equiv 1 \pmod{3^k} \qquad \text{and} \qquad 3^{\beta+1} \equiv 1 \pmod{2^{k+1}}.$$

Now, in light of Euler's Theorem, these congruences will be satisfied if $\alpha + 1$ is a multiple of $\phi(3^k)$ and if $\beta + 1$ is a multiple of $\phi(2^{k+1})$. Clearly, these requirements will be fulfilled if $\alpha = 2r \cdot 3^{k-1} - 1$ and if $\beta = s \cdot 2^k - 1$, where r and s are any positive integers. Hence, by choosing

$$n = 2^{2r \cdot 3^{k+1} - 1} \cdot 3^{s \cdot 2^k - 1}, \qquad r, s \quad \text{positive integers,}$$

we get the result.

(748) We will prove that the only solution of

$$(*) \qquad\qquad \phi(n) + \gamma(n) = \sigma(n)$$

is $n = 2$.

It is clear that $n = 1$ is not a solution of $(*)$.

First assume that $n \geq 3$ is an odd integer solution of $(*)$. Since $\phi(n)$ is even for each integer $n \geq 2$, it follows from $(*)$ that

$$\text{even} + \text{odd} = \text{odd},$$

in which case there exists an odd integer $a \geq 3$ such that $n = a^2$. We derive from $(*)$ that

$$a^2 + a > \phi(a^2) + \gamma(a^2) = \sigma(a^2) \geq a^2 + a + 1,$$

which is nonsense.

We have thus established that if n is a solution of $(*)$, then n must be even. So let $n = 2^\alpha m$, where α is a positive integer and m is an odd positive integer. If $m = 1$, then $n = 2^\alpha$, in which case equation $(*)$ gives

$$2^{\alpha-1} + 2 = 2^{\alpha+1} - 1,$$

which is only possible if $\alpha = 1$, thereby providing the solution $n = 2$.

On the other hand, if $m > 1$, it follows from $(*)$ that

$$(2^{\alpha-1} + 2)m = 2^{\alpha-1}m + 2m > 2^{\alpha-1}\phi(m) + 2\gamma(m)$$
$$= (2^{\alpha-1} - 1)\sigma(m) > (2^{\alpha-1} - 1)m,$$

which implies that

$$2^{\alpha-1} + 2 > 2^{\alpha+1} - 1,$$

which is nonsense for any positive integer α.

All this proves that the only solution of $(*)$ is $n = 2$.

(749) We will show that the numbers $n = 32 \cdot 3^{2r+1}$, $r = 1, 2, 3, \ldots$, serve our purpose. To do so, it is enough to show that if $n = 32 \cdot 3^{2r+1}$, then

$$(*) \qquad\qquad \phi(n) + \sigma(n) = 16 \cdot 3^{2r} \cdot 2 + 63 \cdot \frac{3^{2r+2} - 1}{2}$$

is indeed a multiple of 36 $(= \gamma(n)^2)$. Since 36 divides the first term on the right-hand side of $(*)$ and since $9|63$, the only difficulty rests in the proof that $\frac{3^{2r+2}-1}{2}$ is a multiple of 4, which boils down to showing that $3^{2r+2} - 1$ is divisible by 8. This last claim follows from the fact that

$$3^{2r+2} = 9^{r+1} \equiv 1 \pmod 8.$$

This proves that $\dfrac{\phi(n) + \sigma(n)}{\gamma(n)^2}$ is an integer infinitely many times.

REMARK: The ten smallest integers n such that $\dfrac{\phi(n) + \sigma(n)}{\gamma(n)^2}$ is an integer are 288, 864, 2430, 7776, 27 000, 55 296, 69 984, 82 134, 215 622 and 432 000.

(750) Since $\phi(n) \geq \sqrt{n}/2$ (see Problem 736), we have $2^{\phi(n)} \geq 2^{\sqrt{n}/2}$. Since for $x \geq 9.5$, the function $2^x - 8x^2$ is increasing, it follows that for $n \geq 361$,

$$2^{\sqrt{n}/2} \geq 8 \left(\frac{\sqrt{n}}{2} \right)^2 = 2n.$$

Hence, the inequality is true provided $1 \leq n \leq 360$. Using a computer, we quickly obtain the following values: $n = 1, 2, 3, 4, 6, 8, 10$ and 12.

With MAPLE, we obtain these values by typing in the program (after having opened the library numtheory, namely by typing the instruction with(numtheory))

```
> for to 360 do if evalf(2^phi(n)-2*n)<=0 then print(n)
> else fi; od;
```

(751) With MAPLE, the program can be written as follows:
```
> carre:=proc(N::integer)
> local n;
> for n from 1 to N do
> if type(sqrt(return(n)),integer)=true then print(n)
> else fi; od; end:
```
where the procedure return is

```
> return:=proc(n::integer)
> local m,s;
> m:=n; s:=0;
> while m<>0 do
> s:=10*s+irem(m,10);
> m:=iquo(m,10) od; s end:
```

(752) With MAPLE, the program can be written as follows:
```
> Niven:=proc(N)
> local n,k,r,s;
> for n from 12476 to N do
> k:=n; r:=0; while k<>0 do
> s:=irem(k,10); r:=s+r; k:=iquo(k,10); od;
> if irem(n,r)=0 then print(n); fi; od; end:
```
Using this procedure with $N = 12645$, that is by writing Niven(12645), we obtain that the Niven numbers $n \in [12476, 12645]$ are: 12480, 12492, 12495, 12496, 12501, 12504, 12510, 12520, 12525, 12528, 12532, 12540, 12544, 12546, 12558, 12563, 12564, 12570, 12582, 12600, 12610, 12612, 12614, 12615, 12618, 12636.

(753) With MAPLE, the program can be written as follows:
```
> for n to 1000 do
> if return(n)<>n and return(n^2)=n^2 then
> print(n) else fi; od;
```

(754) The only Cullen prime number < 1000 is 141, and to establish this result, we type with MAPLE the program
```
> for n from 2 to 1000 do
> if isprime(n*2^n+1) then print(n) else fi od;
```

(755) With MAPLE, the program can be written as follows:
```
> deuxcarres:=proc(x)
> local i,j, value , matable, tab;
> matable:=array(0..x,1..2);
> for i from 0 by 1 to x do
> for j from 1 by 1 to 2 do
> matable[i,j]:=i+j-1; tab[i]:=0;
> od; od;
> for i from 0 by 1 to trunc(sqrt(x)) do
> j:=i;
> for j from 0 by 1 to trunc(sqrt(x)) do
> value :=i^2+j^2;
```

```
> if (value <x) then tab[value ]:=1 fi; if (value <x) then
> matable[value ,2]:=i^'2' + j^'2' fi;
> od;
> od;
> print('The numbers smaller than', x, 'which can');
> print('be written as the sum of two squares are:');
> for i from 1 by 1 to x do
> if tab[i]=1 then
> print(i=matable[i,2]);
> fi;
> od;
> end:
> deuxcarres(30);
```

We thus obtain the numbers 1, 2, 4, 5, 8, 9, 10, 13, 16, 17, 18, 20, 25, 26, 29.

(756) With MAPLE, the program can be written as follows:

```
> Euler:=proc(N)
> local m;
> for m to 3*N do if m<>N and
> phi(m)=phi(N) then print(' m'=m,phi(' m')=phi(m))
> else fi; od; end:
```

(757) Using MAPLE, we can write the program as follows:

```
> silv:=proc(N) local n,i,pi;
> for n to N do
> for i from 2 to trunc(sqrt(N)) do
> pi:=ithprime(i);
> if isprime(n-pi)=true then print(n,pi,n-pi)
> else fi; od; od; end:
> silv(25);
```

Taking $N = 25$, that is by typing in silv(25), we obtain that 22 and 24 are the only Silverbach numbers ≤ 25.

(758) Using MAPLE, we can write the program as follows:

```
> for i to 1000 do
> if (ithprime(i)-1)!+1mod(ithprime(i)^2)=0
> then print(ithprime(i)) else fi;od;
```

In this way, we find the Wilson primes 5, 13 and 563. It is known that there are no other Wilson primes smaller than $5 \cdot 10^8$.

(759) We have

(a) By definition,

$$n = 1 + k \sum_{\substack{d|n \\ 1<d<n}} d = 1 + k(\sigma(n) - 1 - n),$$

an equation which gives after simplification

$$k\sigma(n) = (k+1)n + k - 1.$$

(b) Dividing both sides of the equation $k\sigma(n) = (k+1)n + k - 1$ by k, we get the result.

(c) This congruence follows immediately from the definition, since the expression $\sum_{d|n,\ 1<d<n} d$ is an integer.

(d) Assume the contrary, that is that $p|n$ with $p \le k$. We then have that n/p is a proper divisor of n, in which case, using part (b) above, we have

$$\sigma(n) \ge n + 1 + \frac{n}{p} \ge n + 1 + \frac{n}{k} > n + 1 + \frac{n-1}{k} = \sigma(n),$$

a contradiction.

(e) Assume that there exists a prime number p and integers $\alpha \ge 1$ and $k \ge 2$ such that p^α is k-hyperperfect. If $\alpha = 1$, then we have $p = 1$, which makes no sense. On the other hand, if $\alpha \ge 2$, then

$$p^\alpha = 1 + k(p + p^2 + \cdots + p^{\alpha-1}),$$

which implies that $p|1$, again a contradiction.

(f) Using the MATHEMATICA program
$v = \{\}; \text{Do}[n = 2*m + 1; \text{If}[2*\text{DivisorSigma}[1,n]==3*n+1,$
$\text{Print}[n]],$
$\{m,1,10^5\}, v=\text{Append}[v,n]], \{m,1,500000\}; \text{Print}[v]$
we find the numbers 21, 2133, 19521 and 176661.
With the MAPLE software, one can type:
$>$ with(numtheory):
to use the library numtheory which contains the arithmetical functions. Thereafter, we build the following program:
$>$ for n to 1000000 do p(n):=2*σ(n)-(3*n+1); if p(n)=0
$>$ then print(n) else fi; od;

(g) Substituting $n = 3^\alpha \cdot p$ in $2\sigma(n) = 3n + 1$, we obtain

$$2\frac{3^{\alpha+1} - 1}{2}(p + 1) = 3^{\alpha+1}p + 1,$$

an equation which is easily reduced to

$$p = 3^{\alpha+1} - 2.$$

It follows that if for a certain positive integer α, the number $p = 3^{\alpha+1} - 2$ is prime, then the number $n = 3^\alpha \cdot p$ is 2-hyperperfect. Thus, with the program
$\text{Do}[\text{If}[\text{PrimeQ}[(p = 3\text{^}(k+1)-2)], \text{Print}[k," ",p," ",$
$3\text{^}k * p]], \{k,1,30\}]$
we obtain the following table of 2-hyperperfect numbers:

α	p	$3^\alpha \cdot p$
1	7	21
3	79	2133
4	241	19521
5	727	176661
8	19681	129127041

REMARK: In fact, the next number α such that $3^{\alpha+1} - 2$ is prime is $\alpha = 21$.

Using the MAPLE program
$>$ for k to 50 do if isprime(3$\wedge$(k+1)-2) then
$>$print(3$\wedge$k*(3$\wedge$(k+1)-2),

> 'is a 2-hyperperfect number'), else fi; od;

(760) (Chris K. Caldwell: www.utm.edu/research/primes/notes/proofs/
Theorem3.html). Given a positive integer n, denote by $s(n)$ the sum
of its digits. It is easy to see that $s(n) \equiv n \pmod 9$. Therefore, it is
enough to show that if n is an even perfect number, then $n \equiv 1 \pmod 9$.
But it is known that if n is an even perfect number, then there exists a
prime number p (> 2, because $n > 6$) such that $n = 2^{p-1}(2^p - 1)$, where
$p = 3$ or $p \equiv 1 \pmod 6$ or $p \equiv 5 \pmod 6$. Since $2^6 = 64 \equiv 1 \pmod 9$,
we have respectively

$$n = 2^{p-1}(2^p - 1) \equiv 2^{3-1}(2^3 - 1),\ 2^{1-1}(2^1 - 1) \text{ and } 2^{5-1}(2^5 - 1) \pmod 9.$$

Since each of these three expressions is $\equiv 1 \pmod 9$, the proof is complete.

(761) Using the equality $\sum_{k=1}^{n} t_k = \frac{1}{2} \sum_{k=1}^{n} (k^2 + k)$ and the identities of Problem 1 (a) and (b), we obtain the result.

(762) If $n = 0$ or $n = 1$, the result is true. If $n > 1$, it is enough to show that
$n + 1$ divides $\dfrac{(2n)!}{n!n!} = \dbinom{2n}{n}$. Since

$$(2n + 1)\binom{2n}{n} = (n + 1)\binom{2n + 1}{n + 1}$$

and since $(n + 1, 2n + 1) = 1$, the result follows.

(763) We must show that $\sum\limits_{\substack{d \mid M \\ d < M}} d = N$ and that $\sum\limits_{\substack{d \mid N \\ d < N}} d = M$. Since

$$\sum_{\substack{d \mid M \\ d < M}} d \ =\ \sigma(M) - M = (2^{k+1} - 1)(p + 1)(q + 1) - 2^k pq$$

$$=\ (2^{k+1} - 1) \cdot 3 \cdot 2^{k-1} \cdot 3 \cdot 2^k - 2^k(3 \cdot 2^{k-1} - 1)(3 \cdot 2^k - 1)$$

is an expression which gives, after simplification, $3^2 \cdot 2^{3k-1} - 2^k$, we have
obtained N. In a similar way, one can show that $\sum\limits_{\substack{d \mid N \\ d < N}} d = M$.

(764) Assume the contrary, that there exist positive integers x and y such that

$$\frac{x(x + 1)}{2} = 4\frac{y(y + 1)}{2};$$

that is

$$x(x + 1) = 4y(y + 1).$$

It follows successively that

$$\begin{aligned}
x^2 + x + 1 &= 4y^2 + 4y + 1 = (2y + 1)^2, \\
4x^2 + 4x + 1 + 3 &= 4(2y + 1)^2, \\
(2x + 1)^2 + 3 &= 4(2y + 1)^2 = [2(2y + 1)]^2, \\
3 &= [2(2y + 1)]^2 - (2x + 1)^2, \\
3 &= (4y + 2 - 2x - 1)(4y + 2 + 2x + 1), \\
3 &= (4y - 2x + 1)(4y + 2x + 3).
\end{aligned}$$

This means that the positive integer $4y + 2x + 3$ divides 3, which makes
no sense, since $4y + 2x + 3 \geq 9$.

(765) The smallest odd abundant number is 945. To prove that there exist infinitely many abundant numbers, we first set $n = 945m$, where m is a positive integer relatively prime with 2, 3, 5 and 7. Since $945 = 3^3 \cdot 5 \cdot 7$, we have $(m, 945) = 1$ and therefore

$$\sigma(n) = \sigma(945)\sigma(m) \geq \sigma(945)m = 1920m > 2 \cdot 945m = 2n.$$

Since m is odd, n is also odd. Finally, since there exist infinitely many integers m which are relatively prime with 2, 3, 5 and 7, the result follows.

(766) The result is certainly true for $n = 1$. So let $n \geq 2$ and $n = q_1^{\alpha_1} q_2^{\alpha_2} \cdots q_r^{\alpha_r}$ be its representation as a product of distinct prime powers. If $d|n$, then $d = q_1^{\beta_1} q_2^{\beta_2} \cdots q_r^{\beta_r}$, for some nonnegative integers $\beta_i \leq \alpha_i$ for $i = 1, 2, \ldots, r$. Therefore, we have

$$\frac{\sigma(n)}{n} = \prod_{i=1}^{r} \left(1 + \frac{1}{p} + \cdots + \frac{1}{p^{\alpha_i}}\right) \geq \prod_{i=1}^{r} \left(1 + \frac{1}{p} + \cdots + \frac{1}{p^{\beta_i}}\right) = \frac{\sigma(d)}{d},$$

as was to be shown.

(767) Let S be the set of positive integers n such that $n|(2^n + 1)$. We will show that 3^k belongs to S for each integer $k \geq 1$. We proceed by induction. First of all, it is clear that $3 \in S$. Assume that $3^k \in S$ and let us show that $3^{k+1} \in S$. To do so, we must show that $3^{k+1}|m$, where $m = 2^{3^{k+1}} + 1$. Setting $x = 2^{3^k}$, we have

$$m = (2^{3^k})^3 + 1 = x^3 + 1 = (x + 1)(x^2 - x + 1).$$

Using the induction hypothesis, we find that $3^k|(x + 1)$. Hence, we only need to prove that $3|(x^2 - x + 1)$. Since

$$x = 2^{3^k} \equiv (-1)^{3^k} = -1 \pmod{3},$$

it follows that

$$x^2 - x + 1 \equiv (-1)^2 - (-1) + 1 \equiv 0 \pmod{3},$$

which establishes that $3^{k+1}|m$ and therefore that $3^{k+1} \in S$, as required.
REMARKS: The sequence of numbers n such that $n|(2^n + 1)$ is the subject of Problem #16 of the book by Sierpinski [**39**]. In this problem, the author also observes that if $n \in S$, then $2^n + 1 \in S$, which serves as a second way of establishing that S is an infinite set. Let us also mention that it is obvious that if $n \in S$, then n is odd. Hence, using a computer, one easily obtains that the 20 first elements of S are 1, 3, 9, 27, 81, 171, 243, 513, 729, 1539, 2187, 3249, 4617, 6561, 9747, 13203, 13851, 19683, 29241 and 39609.

(768) (*Contribution of Nicolas Doyon*). Let S be the set of positive integers n such that $n|(2^n + 1)$. If $n \in S$, the following observations are immediate:
 (a) n is odd;
 (b) $2^n \equiv -1 \pmod{n}$;
 (c) $2^{2n} \equiv 1 \pmod{n}$;
 (d) if a is the smallest positive integer such that $2^a \equiv 1 \pmod{n}$, then $a \nmid n$ and $a|2n$.
Set

$$n = q_1^{\alpha_1} q_2^{\alpha_2} \cdots q_k^{\alpha_k} \qquad (3 \leq q_1 < q_2 < \ldots < q_k; \ \alpha_i \in \mathbb{N} \text{ for } i = 1, 2, \ldots, k).$$

It follows from (d) that

$$a = 2q_1^{\beta_1} q_2^{\beta_2} \cdots q_k^{\beta_k} \qquad (0 \le \beta_i \le \alpha_i, \text{ for } i = 1, 2, \ldots, k; \ \beta_1 \ge 1).$$

Let b be the smallest positive integer such that $2^b \equiv 1 \pmod{q_1}$. Then, since $2^{q_1 - 1} \equiv 1 \pmod{q_1}$, we have

$$b | a \qquad \text{and} \qquad b | (q_1 - 1),$$

so that $b | d$ where $d = \mathrm{GCD}(a, q_1 - 1)$.

Since a and $q_1 - 1$ are even, we have $d \ge 2$. If an odd number p divides d, then since it divides a, it follows that $p \ge q_1$. On the other hand, since it divides $q_1 - 1$, it must satisfy $p < q_1$, which is contradictory. It follows that d is a power of 2; because 4 does not divide a, it follows that $d = 2$.

Hence, since $b | 2$, we have $b = 1$ or $b = 2$. If $b = 1$, then $2 \equiv 1 \pmod{q_1}$, which is impossible. Therefore, we must have $b = 2$, in which case $2^2 \equiv 1 \pmod{q_1}$, which implies that $q_1 = 3$. We may thus conclude that n is a multiple of 3.

(769) It is enough to show that the equation $4^n = p^k - 1$ has no solutions in integers $n \ge 2$, $k \ge 2$, p an odd prime. We shall examine separately the cases "k even" and "k odd".

First of all, if k is even, we have

$$4^n = (p^{k/2} - 1)(p^{k/2} + 1),$$

so that there exist integers $r > s > 0$ such that

$$p^{k/2} - 1 = 2^s, \qquad p^{k/2} + 1 = 2^r,$$

in which case, by subtracting the first equation from the second, we find

$$2^r - 2^s = 2, \qquad \text{that is } 2^{r-1} - 2^{s-1} = 1,$$

which is nonsense.

On the other hand, if k is odd (and therefore ≥ 3), we have

$$4^n = (p - 1)(p^{k-1} + p^{k-2} + \cdots + p + 1),$$

so that there exist two integers $r > s > 0$ such that

$$p - 1 = 2^s, \qquad p^{k-1} + p^{k-2} + \cdots + p + 1 = 2^r,$$

in which case, by subtracting the first equation from the second, we find

$$\begin{aligned} p^{k-1} + p^{k-2} + \cdots + p^2 + 2 &= 2^r - 2^s, \\ p^2(p^{k-3} + p^{k-4} + \cdots + p + 1) &= 2^r - 2^s - 2. \end{aligned}$$

Since this last equation makes no sense, its left-hand side being an odd number while its right-hand side is an even number, the result follows.

(770) The answer is NO. Indeed, assume that q_1 and q_2 are two distinct prime numbers such that $p | 2^{q_1} - 1$ and $p | 2^{q_2} - 1$ for a certain prime number p. Let r be the smallest positive integer such that $2^r \equiv 1 \pmod{p}$. It follows that $r | q_1$, so that $r = 1$ or $r = q_1$. Since $r > 1$, we have that $r = q_1$. Similarly, we obtain that $r = q_2$, so that $q_1 = q_2$, thereby contradicting our assumption.

(771) We use the following MATHEMATICA program:

$r = 1$; $n = 2$; While[s =DivisorSigma[1,n]/n; $(r < 2)$||$(s < 2)$,
$(n++;r = s)$]; Print[$n-1$," ",n]

We then obtain $n - 1 = 5775$ and $n = 5776$ as the required numbers.

With MAPLE, the following program will do:
> with(numtheory):
> for n from 2 to 10000 do if sigma(n-1)/(n-1)>=2
> and sigma(n)/n>=2 then print(n-1,n) else fi; od;
which yields the pair $(5775, 5776)$.

To find the three smallest nondeficient consecutive numbers, we may write a similar program with MATHEMATICA, for instance:

$r = 1$; $s = 1$; $n = 3$; While[t =DivisorSigma[1,n]/n;
$(r < 2)$||$(s < 2)$||$(t < 2)$,$(n++;r = s;s = t)$];
Print[$n-2$," ",$n-1$," ",n]

We then obtain $n - 2 = 171\,078\,830$, $n - 1 = 171\,078\,831$ and $n = 171\,078\,832$ for the required numbers.

For the third part of the problem, we first observe that

$$\frac{\sigma(6)}{6} = \frac{\sigma(2)}{2} \cdot \frac{\sigma(3)}{3} = \left(1 + \frac{1}{2}\right)\left(1 + \frac{1}{3}\right) = 2$$

and that

$$\left(1 + \frac{1}{5}\right)\left(1 + \frac{1}{7}\right) \cdots \left(1 + \frac{1}{31}\right) = 2.00097 \ldots > 2.$$

We then set $p_{r_1} = p_2 = 3$ and $p_{r_2} = p_{11} = 31$.
Then let p_{r_3} be the smallest prime number such that

$$\left(1 + \frac{1}{p_{12}}\right)\left(1 + \frac{1}{p_{13}}\right) \cdots \left(1 + \frac{1}{p_{r_3}}\right) \geq 2.$$

More generally, for each positive integer $i < k$, having determined p_{r_i}, we set $p_{r_{i+1}}$ as the smallest prime number such that

$$\left(1 + \frac{1}{p_{r_i+1}}\right)\left(1 + \frac{1}{p_{r_i+2}}\right) \cdots \left(1 + \frac{1}{p_{r_{i+1}}}\right) \geq 2.$$

This process has no end since $\prod_p (1 + 1/p)$ diverges (see Theorem 16).

Hence, since for each divisor d of a positive integer m, we have $\sigma(m)/m \geq \sigma(d)/d$ (see Problem 766), we only need to find a positive integer n satisfying the system of congruences

$$\begin{cases} n \equiv 0 \pmod{p_1 p_2}, \\ n \equiv -1 \pmod{p_3 \cdots p_{11}}, \\ n \equiv -2 \pmod{p_{12} \cdots p_{r_3}}, \\ \vdots \\ n \equiv -k+1 \pmod{p_{r_{k-1}+1} \cdots p_{r_k}}. \end{cases}$$

By the Chinese Remainder Theorem, the above system of congruences has a solution n. Summarizing, the nature of the congruences of this system and the fact that $\sigma(n)/n \geq \sigma(d)/d$ if $d|n$ guarantees that

$$\frac{\sigma(n)}{n} \geq \frac{\sigma(p_1 p_2)}{p_1 p_2} = 2,$$

$$\frac{\sigma(n+1)}{n+1} \geq \frac{\sigma(p_3 p_4 \cdots p_{11})}{p_1 p_2 \cdots p_{11}} \geq 2,$$

$$\vdots \qquad \vdots$$

$$\frac{\sigma(n+k-1)}{n+k-1} \geq \frac{\sigma(p_{r_{k-1}+1} \cdots p_{r_k})}{p_{r_{k-1}+1} \cdots p_{r_k}} \geq 2,$$

thereby producing k consecutive nondeficient numbers.

(772) We already know that $\dfrac{\sigma(120)}{120} = 3$ so that, since $\sigma(m)/m \geq \sigma(d)/d$ when $d|m$ (see Problem 766), we have

$$\frac{\sigma(120m)}{120m} \geq \frac{\sigma(120)}{120} = 3 \qquad (m = 1, 2, \ldots).$$

On the other hand, since

$$\left(1 + \frac{1}{7}\right)\left(1 + \frac{1}{11}\right) \cdots \left(1 + \frac{1}{743}\right) = 3.00179 \ldots > 3,$$

it is enough to find a positive integer n such that

$$\begin{cases} n \equiv 0 \pmod{120}, \\ n \equiv -1 \pmod{7 \cdot 11 \cdots 743}. \end{cases}$$

Using the Chinese Remainder Theorem, we may therefore conclude that such a number n exists (even if it is certainly very large), as was to be shown.

(773) First observe that, for each prime number q and positive integer α,

$$(1) \qquad \frac{\sigma(q^\alpha)}{q^\alpha} = 1 + \frac{1}{q} + \frac{1}{q^2} + \cdots + \frac{1}{q^\alpha} < 1 + \frac{1}{q} + \frac{1}{q^2} + \cdots = \frac{q}{q-1}.$$

Therefore, for each integer $n > 1$, written in its canonical form $n = q_1^{\alpha_1} q_2^{\alpha_2} \cdots q_r^{\alpha_r}$, it follows from (1) that

$$(2) \qquad \frac{\sigma(n)}{n} = \prod_{i=1}^{r}\left(1 + \frac{1}{q_i} + \frac{1}{q_i^2} + \cdots + \frac{1}{q_i^{\alpha_i}}\right) < \prod_{i=1}^{r}\left(1 + \frac{1}{q_i} + \frac{1}{q_i^2} + \cdots\right)$$

$$= \prod_{i=1}^{r} \frac{q_i}{q_i - 1}.$$

Now, if n is tri-perfect, $\frac{\sigma(n)}{n} = 3$, in which case (2) implies that

$$(3) \qquad \prod_{i=1}^{r} \frac{q_i}{q_i - 1} > 3,$$

which can only take place if the left-hand side of (3) has at least eight prime factors, this minimum being attained when

$$\{q_1, q_2, \ldots, q_8\} = \{3, 5, 7, 11, 13, 17, 19, 23\}.$$

(774) We only need to verify that the $\binom{4}{2} = 6$ pairs of elements of A satisfy the stated property. Therefore, since $ab + 1 = c^2$ for a certain integer c, we have

$$a \cdot (a + b + 2\sqrt{ab + 1}) + 1 = a^2 + ab + 2ac + 1$$
$$= a^2 + c^2 - 1 + 2ac + 1 = (a + c)^2,$$

as required. The five other pairs can easily be handled. Finally, setting $a = 1$, $b = 3$, we obtain $A = \{1, 3, 8, 120\}$; to $a = 2$, $b = 4$ corresponds the set $A = \{2, 4, 12, 420\}$.

REMARK: This result was obtained by Euler. For an extensive study of this problem, see L. Jones [21].

(775) Let n be such an integer; then there exists a positive integer m such that $n = (m - 1)m = m^2 - m$. We then have $n + m = m^2$, so that we have successively

$$\sqrt{n + m} = m, \quad \sqrt{n + \sqrt{n + m}} = m, \quad \sqrt{n + \sqrt{n + \sqrt{n + m}}} = m,$$

and so on. It follows that

$$\sqrt{n + \sqrt{n + \sqrt{n + \sqrt{n + \cdots}}}} = m,$$

as required.

(776) We use induction. First of all, the result is true for $n = 2$, since $F_1 F_3 - F_2^2 = 1 \cdot 2 - 1^2 = 1 = (-1)^2$. Assuming that the identity is true for $n = k$, that is that

$$F_{k-1} F_{k+1} - F_k^2 = (-1)^k,$$

we will prove that it is also true for $n = k + 1$. Using the induction hypothesis, we have successively

$$\begin{aligned}
F_{k-1} + F_k &= F_{k+1}, \\
F_{k-1} F_{k+1} + F_k F_{k+1} &= F_{k+1}^2, \\
F_{k-1} F_{k+1} - (-1)^k + F_k F_{k+1} &= F_{k+1}^2 + (-1)^{k+1}, \\
F_k^2 + F_k F_{k+1} - F_{k+1}^2 &= (-1)^{k+1}, \\
F_k(F_k + F_{k+1}) - F_{k+1}^2 &= (-1)^{k+1}, \\
F_k F_{k+2} - F_{k+1}^2 &= (-1)^{k+1},
\end{aligned}$$

as required.

(777) There are $\binom{4}{2} = 6$ verifications to be made. We will show only two, namely that $F_{2n} F_{2n+2}$ and that $F_{2n} F_{2n+4}$ are both squares. First of all, from the Cassiny identity (see Problem 776), we have $F_{2n} F_{2n+2} - (-1)^{2n+1} = F_{2n+1}^2$, and therefore $F_{2n} F_{2n+2} + 1 = F_{2n+1}^2$, proving our claim.

On the other hand, again calling upon the Cassiny identity, we have

$$
\begin{aligned}
F_{2n}F_{2n+4} + 1 &= F_{2n}(F_{2n+3} + F_{2n+2}) + 1 = F_{2n}F_{2n+3} + F_{2n}F_{2n+2} + 1 \\
&= F_{2n}(F_{2n+2} + F_{2n+1}) + F_{2n}F_{2n+2} + 1 \\
&= 2F_{2n}F_{2n+2} + F_{2n}F_{2n+1} + 1 \\
&= 2(F_{2n+1}^2 - 1) + F_{2n}F_{2n+1} + 1 \\
&= F_{2n+1}^2 + F_{2n+1}^2 + F_{2n}F_{2n+1} - 1 \\
&= F_{2n+1}^2 + F_{2n+1}F_{2n+2} - 1 = F_{2n+1}(F_{2n+1} + F_{2n+2}) - 1 \\
&= F_{2n+1}F_{2n+3} - 1 = F_{2n+2}^2,
\end{aligned}
$$

as required.

(778) The choice $k = n(n+3)/2$ implies that

$$
\frac{n(n+1)(n+2)(n+3)}{8} = \frac{k(k+1)}{2},
$$

which proves the result.

(779) By Dirichlet's Theorem, any arithmetic progression $an + b$, where $(a,b) = 1$, contains infinitely many primes. So let $a = 10\,000$ and $b = 7777$. Clearly, $(a,b) = 1$. Hence, there exist infinitely many prime numbers of the form

$$(*) \qquad 10\,000n + 7777,$$

which proves our claim. Using a computer, we easily find that the smallest five prime numbers of the form $(*)$ are $47\,777$, $67\,777$, $97\,777$, $107\,777$ and $137\,777$.

(780) With a computer, we find the numbers 111, $11\,112$ and $1\,122\,112$. Let S be the set of numbers with this property. First of all, since it is easy to see that $9 \mid 111\,111\,111$, it is clear that this last number belongs to S. In fact, using a computer, we observe that $111\,111\,111$ is the fourth smallest element of S, while the fifth one is $122\,121\,216$. Finally, setting $n_0 = 1\,122\,112$, we find that the number

$$
\underbrace{11\ldots1}_{16} \cdot 10^7 + n_0 = \underbrace{11\ldots1}_{16} 1\,122\,112
$$

also belongs to S.

REMARK: For an extended study of these numbers, one may consult the paper of J.M. De Koninck and N. Doyon [6].

(781) To show that the composite number n is a Carmichael number, it is enough to show that

$$b^{n-1} \equiv 1 \pmod{n}, \quad \text{for each positive integer } b \text{ such that } (b,n) = 1.$$

But, if we assume that $\phi(n)$ is a proper divisor of $n - 1$, there exists a positive integer r such that $n - 1 = r\phi(n)$. We then have

$$b^{n-1} = b^{r\phi(n)} = \left(b^{\phi(n)}\right)^r \equiv 1^r = 1 \pmod{n},$$

and the result follows.

(782) If $n = d_1 d_2 \cdots d_r$ satisfies the equation

$$(*) \qquad n = d_1^1 + d_2^2 + d_3^3 + \cdots + d_r^r,$$

then

$$n = d_1^1 + d_2^2 + \cdots + d_r^r \le 9 + 9^2 + \cdots + 9^r$$
$$= 9(1 + 9 + \cdots + 9^{r-1}) = \frac{9}{8}(9^r - 1).$$

On the other hand, since $n > 10^{r-1}$, we must have

$$10^{r-1} < \frac{9}{8}(9^r - 1),$$

an inequality which holds only if $r \ge 23$. Hence, any number satisfying $(*)$ must be smaller than 10^{22}.

Using a computer, we find that the numbers 89, 135, 175, 518, 598, 1306, 1676 and 2427 satisfy $(*)$.

(783) If $n = d_1 d_2 \cdots d_r$ satisfies the equation

$$(*) \qquad\qquad n = d_1^r + d_2^{r-1} + d_3^{r-2} + \cdots + d_r^1,$$

then

$$n = 10^{r-1} d_1 + 10^{r-2} d_2 + \cdots + 10 d_{r-1} + d_r$$
$$= d_1^r + d_2^{r-1} + d_3^{r-2} + \cdots + d_{r-1}^2 + d_r,$$

in which case

$$0 = d_1(10^{r-1} - d_1^{r-1}) + d_2(10^{r-2} - d_2^{r-2}) + \cdots + d_{r-1}(10 - d_{r-1}) > 0,$$

a contradiction. Hence, there exist no integers n satisfying $(*)$.

(784) Let $n = 2^\alpha$, where α is a positive integer. Since

$$\sigma(n) + 1 = 2^{\alpha+1} - 1 + 1 = 2 \cdot 2^\alpha = 2n,$$

the sequence of powers of 2 provides infinitely many solutions to the equation $\sigma(n) = 2n - 1$.

REMARK: Besides the numbers $n = 2^\alpha$, $\alpha = 1, 2, \ldots$, no other solution of the equation $\sigma(n) = 2n - 1$ is known. Also, no number n such that $\sigma(n) = 2n + 1$ is known.

(785) (*Contribution of Claude Levesque, Québec*). Let $q \mid n^2 + n + 1$, q an odd prime number. Then,

$$4q \mid 4(n^2 + n + 1) = (2n + 1)^2 + 3,$$

which implies that $q \mid (2n+1)^2 + 3$, so that the congruence $x^2 \equiv -3 \pmod{q}$ is solvable. We have thus proved that

$$(1) \qquad\qquad \left(\frac{-3}{q}\right) = 1.$$

Then, using the law of quadratic reciprocity, we have

$$(2) \qquad \left(\frac{-3}{q}\right) = \left(\frac{-1}{q}\right)\left(\frac{3}{q}\right) = \left(\frac{-1}{q}\right)\left(\frac{q}{3}\right)(-1)^{\frac{q-1}{2} \cdot \frac{3-1}{2}}$$
$$= (-1)^{\frac{q-1}{2}}(-1)^{\frac{q-1}{2}}\left(\frac{q}{3}\right) = \left(\frac{q}{3}\right).$$

We want to prove that $q \equiv 1 \pmod 3$. Assume that this is not the case. Since $n^2 + n + 1 \not\equiv 0 \pmod 3$, it means that $q \equiv 2 \pmod 3$, in which case $\left(\frac{q}{3}\right) = -1$, which implies, in light of (2), that $\left(\frac{-3}{q}\right) = -1$, thus contradicting (1). This proves that we must have $q \equiv 1 \pmod 3$, as was to be shown.

(786) (Sierpinski [**39**], Problem #176) There is only one, namely $x = 3$. Setting $x = t + 3$, (∗) is reduced to

$$(**)\qquad\qquad 2t(t^2 + 9t + 21) = 0.$$

Since the quadratic polynomial $t^2 + 9t + 21$ has no real roots, the only solution of (∗∗) is $t = 0$. It follows that the only solution of (∗) is $x = 3$.

(787) Assume that such a solution $\{x, y\}$ exists. Since $9|117$, we must have that $9|x^3 + 5$, which is impossible, since $x^3 + 5 \equiv 4, 5, 6 \pmod 9$.

REMARK: In his book [**32**], Joe Roberts makes the following interesting observation:

> In the chapter "Diophantine Equations: p-adic Methods" in *Studies in Number Theory*, [**22**] D.J. Lewis states on page 26 that "The equation $x^3 - 117y^3 = 5$ is known to have at most 18 integral solutions but the exact number is not known." Finkelstein and London (1971) [**12**] made use of the field $\mathbf{Q}(\sqrt[3]{117})$, where the cube root is real, to show that, in fact, the equation has no solutions in integers. Halter-Koch (1973) [**17**] and Udrescu (1973) [**33**] independently observed that by considering the equation modulo 9 we get $x^3 \equiv 5 \pmod 9$ and this congruence clearly has no solutions. Consequently we immediately see that the equation has no solutions.

(788) It is important to make sure that each of the terms $(\ldots)^3$ is positive. To do so, if we take $a \geq 3$ and $b = a + 1$, it is easy to see that each of the four expressions $(\ldots)^3$ is positive. Since the Ramanujan identity holds for each integer $a \geq 3$, the first result is proved. On the other hand, to find the "double" representation of 1729, we first set $a = 3$ and $b = 4$ in (∗), in which case we obtain

$$7^3 + 84^3 = 63^3 + 70^3.$$

Dividing each of the four terms of this last identity by 7^3, we obtain the double representation of 1729 noticed by Ramanujan.

(789) If $ax + by = b + c$ is solvable, then $d = (a, b)|(b + c)$. Since $d|b$, it follows that $d|c$, which implies that $ax + by = c$ is solvable. The other implication can be handled in a similar manner.

(790) We know that $ax + by = c$ is solvable if and only if $d = (a, b)|c$, which is equivalent to $d|(a, b, c)$. This shows that $(a, b) = (a, b, c)$.

(791) Since $(a, b) = 1$, there exist integers x^* and y^* such that $ax^* + by^* = 1$. The solutions of $ax + by = n$ are then given by $x = nx^* - bk$ and $y = ny^* + ak$, where $ax^* + by^* = 1$. Hence, the equation has positive solutions if $nx^* - bk > 0$ and $ny^* + ak > 0$, that is if

$$-\frac{y^*n}{a} < k < \frac{x^*n}{b}.$$

To show that there exists at least one such a value of k, we only need to show that

$$\frac{-y^*n}{a} + 1 < \frac{x^*n}{b},$$

an inequality which is equivalent to $n(ax^* + by^*) > ab$, that is $n > ab$.

Finally, if $n = ab$, then $-y^*b < k < x^*a$; and since $ax^* + by^* = 1$, we obtain $ax^* - 1 < k < ax^*$, which is impossible.

(792) The solution is $x = 19$, $y = 11$, $z = 70$. Indeed, if we multiply the first equation by 2 and subtract this new equation from the second one, we obtain

$$(*) \qquad\qquad 30y - 19z = -1000.$$

Reducing modulo 19, we obtain $11y \equiv 7 \pmod{19}$ and therefore (multiplying by 7) we have

$$y \equiv 11 \pmod{19} \quad \text{that is} \quad y = 11 + 19k, \ k \in \mathbb{Z}.$$

Substituting this value in $(*)$, we find $z = 70 + 30k$ and finally $x = 19 - 49k$. For these solutions to be positive, we must choose $k = 0$, which gives the solution stated above.

(793) (Marco Carmosini, Queen's University, Canadian Congress of Students in Mathematics, May 1999). If P and A stand respectively for the perimeter and the area of such a triangle, then using Heron's formula $A = \sqrt{\frac{P}{2}\left(\frac{P}{2} - a\right)\left(\frac{P}{2} - b\right)\left(\frac{P}{2} - c\right)}$ where $P = a + b + c$, we are led to the equation

$$a + b + c = \sqrt{\frac{a+b+c}{2} \cdot \frac{-a+b+c}{2} \cdot \frac{a-b+c}{2} \cdot \frac{a+b-c}{2}},$$

that is

$$(*) \qquad 16(a + b + c) = (-a + b + c)(a - b + c)(a + b - c).$$

Since the left-hand side of $(*)$ is even, it follows that $(-a + b + c)$ or $(a - b + c)$ or $(a + b - c)$ must be even. It is easy to see that if any of these three quantities is even, each of the other two will also be even. It follows that there exist three integers $m \le n \le k$ such that

$$-a + b + c = 2m, \qquad a - b + c = 2n, \qquad a + b - c = 2k,$$

so that

$$a = n + k, \qquad b = m + k, \qquad c = m + n.$$

Substituting these values in $(*)$, we obtain that

$$mnk = 4(m + n + k).$$

We will treat separately the following four cases:

$$m = 1, \qquad m = 2, \qquad m = 3, \qquad m \ge 4.$$

If $m = 1$, we obtain successively $nk = 4(1 + n + k)$, $nk - 4n - 4k = 4$, $nk - 4n - 4k + 16 = 4 + 16$ and $(n - 4)(k - 4) = 20$, in which case the only possible values of (n, k) are $(n, k) = (5, 24)$, $(6, 14)$ and $(8, 9)$. To these values correspond the three triangles whose sides a, b, c are $(a, b, c) = (20, 15, 7)$, $(17, 10, 9)$ and $(29, 25, 6)$.

In the case $m = 2$, we have $2nk = 4(2 + n + k)$; that is $(n - 2)(k - 2) = 8$, which gives the only possible values $(n, k) = (3, 10)$ and $(4, 6)$, thus yielding the two triangles of lengths a, b, c given by $(a, b, c) = (13, 12, 5)$ and $(10, 8, 6)$.

In the case $m = 3$, we obtain successively $3nk = 4(3 + n + k)$, $(3n - 4)(3k-4) = 52$, and hence the pair $(n, k) = (2, 10)$, which must be rejected since $n = 2 < 4 = m$.

It remains to consider the case $m \geq 4$. Let us assume that there exists a solution (m, n, k), with $m \geq 4$, to the equation $mnk = 4(m + n + k)$. We would then successively have

$$m = \frac{4n + 4k}{nk - 4} \geq 4, \quad n + k \geq nk - 4, \quad k \leq \frac{5}{n - 1} + 1 \qquad \text{for all } n \geq 4.$$

In particular, we would have

$$k \leq \frac{5}{4 - 1} + 1 = \frac{8}{3} < 3, \text{ and therefore } 4 \leq m \leq n \leq k < 3,$$

a contradiction.

To sum up, the only solutions (a, b, c) are given by the five triples

$$(13, 12, 5), \quad (10, 8, 6), \quad (29, 25, 6), \quad (20, 15, 7), \quad (17, 10, 9).$$

(794) There are none. Indeed, if x, y is a solution, since x is not a multiple of 3, then $x^2 \equiv 1 \pmod 3$, in which case $x^2 + 3y \equiv 1 \pmod 3$, while $5 \equiv 2 \pmod 3$.

(795) We may assume that $x^2 + y^2 = z^2$. We proceed by contradiction by assuming that $xyz \not\equiv 0 \pmod 5$, in which case $x^2, y^2, z^2 \equiv 1, 4 \pmod 5$. The only three possible values modulo 5 of $x^2 + y^2$ are therefore $1 + 1$, $1 + 4$ and $4 + 4$, that is 2, 0 and 3 modulo 5, while we should have 1 or 4.

(796) If $x > 1$, then using the first equation, we have that $y < 1$, which in turn implies that $z > 1$. But "$x > 1$, $z > 1$" contradicts the third equation. Hence, $x \leq 1$. By a similar argument, one can show that $x \geq 1$. Hence, $x = 1$. We can then do the same reasoning with each unknown, allowing us to conclude that $x = y = z = 1$.

(797) (*TYCM, Vol. 13, 1982, p. 263*). Assume that there exist nonnegative integers x and y such that $ax + by = ab - a - b$. In this case, we have $a(x + 1) = b(a - y - 1)$. Since a and b are relatively prime, it is clear that

$$b | x + 1 \qquad \text{and} \qquad a | a - y - 1,$$

which implies that $a | y + 1$. Hence, $y + 1 \geq a$, $x + 1 \geq b$ and therefore $ab = (x + 1)a + (y + 1)b \geq 2ab$, which is impossible, since a and b are positive.

(798) We know that the solutions of $n = ax + by$ are of the form $x = x_0 + bt$, $y = y_0 - at$, where $ax_0 + by_0 = n$, $t \in \mathbb{Z}$. We must choose t so that $y_0 - at \geq 0$ and $x_0 + bt \geq 0$, which is equivalent to $-(x_0/b) \leq t \leq (y_0/a)$. The number of solutions is therefore $[y_0/a] - [-x_0/b]$, and since $[a] - [b] = [a - b]$ or $[a - b] + 1$, we obtain the result.

(799) Say Peter has paid \$1.04. The only way this can happen is if Peter has bought x apples and y oranges, with x and y such that $5x + 7y = 104$. We can express this situation as $(x, y) = (x_0, y_0) = (4, 12)$. All the integer solutions of $5x + 7y = 104$ are given by $x = x_0 + 7t = 4 + 7t$ and $y = y_0 - 5t = 12 - 5t$. Since we must have $4 + 7t > 0$ (that is $t > -4/7$) and $12 - 5t > 0$ (that is $t < 12/5$), it follows that $0 \leq t \leq 2$. The only three suitable values of t are therefore 0, 1 and 2. Since Peter's purchase corresponds to the value $t = 0$, Paul's purchase must necessarily

correspond to $t = 1$ or to $t = 2$, that is $x = 11$ and $y = 7$ or $x = 18$ and $y = 2$. Since by hypothesis $y \geq 3$, we may conclude that Paul has bought 11 apples and 7 oranges.

(800) First of all, since $(3, 7) = 1 | 11$, the given Diophantine equation has integer solutions. We easily establish that $(x_0, y_0) = (6, -1)$ is a particular solution of this Diophantine equation. The set of all the solutions is therefore given by

$$x = 6 + 7t, \qquad y = -1 - 3t, \qquad \text{where } t \in \mathbb{Z}.$$

The solutions located in the second quadrant are those corresponding to the points (x, y) satisfying

$$x = 6 + 7t < 0 \quad \text{and} \quad y = -1 - 3t > 0,$$

that is when $t < -\frac{6}{7}$ and $t < -\frac{1}{3}$. This means that we must have $t \leq -1$. The set A of integer points (x, y) which are solutions of $3x + 7y = 11$ and which are located in the second quadrant is therefore given by

$$A = \{(x, y) : x = 6 + 7t \text{ and } y = -1 - 3t, \text{ where } t = -1, -2, -3, \ldots\}.$$

(801) We first establish that $(x_0, y_0) = (5, -2)$ is a particular solution of this equation. This point generates the solutions

$$x = 5 + 7t, \qquad y = -2 - 5t, \qquad \text{where } t \in \mathbb{Z}.$$

Since we are interested in the points (x, y) such that $y > x$, we need to establish the integer values of t for which

$$-2 - 5t > 5 + 7t, \qquad \text{that is} \quad t < -7/12,$$

which is only possible if $t \leq -1$. The required set E is therefore

$$E = \{(x, y) : x = 5 + 7t \text{ and } y = -2 - 5t, \text{ where } t = -1, -2, -3, \ldots\}.$$

(802) Taking $t = 0$, we obtain that $(x, y) = (5, 1)$ is a solution of (∗). Choosing $t = 1$, we obtain the solution $(x, y) = (1, -2)$. These two solutions give rise to the system

$$\begin{cases} 5a + b = 11, \\ a - 2b = 11, \end{cases}$$

a solution of which is $a = 3$ and $b = -4$, which produces the required numbers a and b.

(803) Since $(x, y, z) = 1$, we have $(x, y) = (x, z) = (y, z) = 1$, and therefore only one of the terms x, y and z can be even. If x is even, then $x^2 \equiv 0$ (mod 4). The fact that x is even implies that y and z are both odd and $z^2 - 3y^2 \equiv 2$ (mod 4). It follows that x must be odd and that y or z is even.

I) If y is even, then $(z + x, z - x) = 2$ and therefore $z + x = 2u$ and $z - x = 2v$, where $(u, v) = 1$. We then have $3y^2 = (z - x)(z + x) = 4uv$. Hence, $(y/2)^2 = uv/3$, and since $(u, v) = 1$, we may assume that $3 | u$, so that there exists a positive integer m such that $u = 3m$. It follows that there exist positive integers r and s such that $v = r^2$ and $m = s^2$, in which case

$$\frac{z + x}{2} = u = 3m = 3s^2 \quad \text{and} \quad \frac{z - x}{2} = v = r^2.$$

We easily see that in this case, we must have

$$(r, s) = 1, \quad s > r, 3 \nmid r, \quad y = 2rs, \quad x = 3s^2 - r^2, \quad z = 3s^2 + r^2.$$

II) If z is even, then $(z + x, z - x) = 1$, in which case for r and s odd and $(r, s) = 1$, $\quad s > r, 3 \nmid r$, we have

$$y = rs, \quad x = \frac{3s^2 - r^2}{2}, \quad z = \frac{3s^2 + r^2}{2}.$$

(804) (Sierpinski [**39**], Problem #170). We begin with the identity

(∗) $$(x + y + z)^3 - (x^3 + y^3 + z^3) = 3(x + y)(x + z)(y + z).$$

It follows that if x, y and z are integers such that $x + y + z = 3$ and $x^3 + y^3 + z^3 = 3$, then, by (∗), we have

(∗∗) $$8 = (x + y)(x + z)(y + z) = (3 - x)(3 - y)(3 - z),$$

so that, in light of $x + y + z = 3$, we have

(∗ ∗ ∗) $$8 = (3 - x)(3 - y)(3 - z).$$

Relation (∗ ∗ ∗) implies that either the three numbers $3 - x$, $3 - y$, $3 - z$ are even or else only one of the three is even. In the first case, in light of (∗∗), they are all in absolute value equal to 2; therefore, by (∗ ∗ ∗), they are equal to 2, in which case $x = y = z = 1$. In the second case, in light of (∗∗), one of the numbers $3 - x$, $3 - y$, $3 - z$ is in absolute value equal to 8, while the others are in absolute value equal to 1; thus, by (∗ ∗ ∗), one of the two is equal to 8, while the others are equal to -1. This finally yields $x = -5$ and $y = z = 4$, or $x = y = 4$ and $z = -5$ or $x = 4$, $y = -5$ and $z = 4$. We can therefore conclude that the system of equations has only four integer solutions, namely $(1, 1, 1)$, $(-5, 4, 4)$, $(4, -5, 4)$ and $(4, 4, -5)$.

(805) We only need to consider, for each positive integer n, the triples $\{x, y, z\}$, where

$$\begin{aligned} x &= n^{10}(n + 1)^8, \\ y &= n^7(n + 1)^5, \\ z &= n^4(n + 1)^3. \end{aligned}$$

(806) Consider the equation $5x + 7y = 136$. Reducing this equation modulo 5, we obtain $y = 3 + 5k$. Substituting in the equation, we obtain $x = 23 - 7k$. The condition "$x > 0$ and $y > 0$" allows us to conclude that solutions are possible for $k = 0, 1, 2, 3$, that is $136 = 115 + 21 = 80 + 56 = 45 + 91 = 10 + 126$.

(807) Setting $x = a - r$, $y = a$ and $z = a + r$, we find the equation $x^2 + y^2 = z^2$ becomes $a(a - 4r) = 0$. Hence, $a = 4r$ and therefore $x = 3r$, $y = 4r$ and $z = 5r$, where $r \in \mathbb{N}$.

(808) Setting $r = 16$ and $s = 5$ in Theorem 34, we obtain $z = 281$, $y = 160$ and $x = 231$.

(809) Since $3|xy$ and $4|xy$, we have that $12|xyz$. Hence, we only need to show that $5|xyz$. We first observe that if $5 \nmid m$, then $m = 5k \pm 1$ or $m = 5k \pm 2$, for a certain integer k. In the first case, $m^2 = 5(5k^2 \pm 2k) + 1$ and in the second case, $m^2 = 5(5k^2 \pm 4k) + 4$. Using this observation, we see that if none of the numbers x, y, z are divisible by 5, then $x^2 + y^2$ gives, after dividing by 5, the remainders 2, 3 or 0. Since $x^2 + y^2 = z^2$, the first

two cases are clearly impossible. The only possibility is the third one, in which case z^2 is divisible by 5, so that z is divisible by 5.

(810) Since $x = r^2 - s^2$, $y = 2rs$ and $z = r^2 + s^2$, we have $s(r - s) = 6$. Solving for s, we find

$$s = \frac{r \pm \sqrt{r^2 - 24}}{2}.$$

Hence, in order for s to exist, we must have $r^2 \geq 24$, in which case $\sqrt{r^2 - 24}$ is an integer. Therefore, there exists an integer u such that $r^2 - 24 = u^2$, in which case

$$(r - u)(r + u) = 24 = 1 \cdot 24 = 2 \cdot 12 = 3 \cdot 8 = 4 \cdot 6.$$

From this, we derive the values $r = 7$ and $r = 5$. We thus obtain the Pythagorean triples $(20, 21, 29)$, $(16, 30, 34)$, $(13, 84, 85)$ and $(48, 14, 50)$.

(811) The equations $x^2 + y^2 = z^2$ and $x + y + z = xy$ allow us to obtain the equation $(x - 2)(y - 2) = 2$. Hence, $x = 3$, $y = 4$ and $z = 5$ are the dimensions of the required triangle.

(812) The relation $(n - 1)^2 + n^2 = (n + 1)^2$ implies that $n^2 = 4n$; that is $n = 4$.

(813) Whatever the parity of n, the left-hand side is always odd, while the right-hand side is always even, a contradiction.

(814) Since $x^2, y^2 \equiv 0, 1 \pmod 4$, we have $x^2 + y^2 \not\equiv 3 \pmod 4$, while $4x + 7 \equiv 3 \pmod 4$.

(815) First observe that the primitive solutions of $x^2 + y^2 = (z^2)^2$ are given by $x = r^2 - s^2$, $y = 2rs$ and $z^2 = r^2 + s^2$, with $r > s > 0$, $(r, s) = 1$, r, s of opposite parity. Since the primitive solutions of $z^2 = r^2 + s^2$ are in turn given by $r = m^2 - n^2$, $s = 2mn$ and $z = m^2 + n^2$, with $m > n > 0$, $(m, n) = 1$, m, n of opposite parity, we may conclude that all primitive solutions of $x^2 + y^2 = z^4$ are given by $y = 4mn(m^2 - n^2)$, $x = m^4 + n^4 - 6m^2 n^2$ and $z = m^2 + n^2$, with $m > n > 0$, $(m, n) = 1$, m, n of opposite parity.

(816) First of all, it is clear that $(x_0, y_0) = (2, -1)$ is a solution of the Diophantine equation $x + y = 1$. All the integer solutions (x, y) of the equation are therefore given by

$$x = 2 + t, \quad y = -1 - t \quad (t \in \mathbb{Z}).$$

Hence, we are looking for the values of x and y such that $x^2 + y^2 \leq 9$. But

$$x^2 + y^2 = (2 + t)^2 + (-1 - t)^2 = 2t^2 + 6t + 5.$$

This means that we must have

$$2t^2 + 6t + 5 \leq 9,$$

an inequality which is verified for the integers $t = -3, -2, -1, 0$, yielding the integer solutions

$$(-1, 2), \quad (0, 1), \quad (1, 0), \quad (2, -1).$$

(817) First observe that $3136 = 56^2$. We are therefore looking for the primitive solutions of

$$(*) \qquad\qquad x^2 + 56^2 = z^2.$$

Since all the primitive solutions of $X^2 + Y^2 = Z^2$ are given by

$$X = r^2 - s^2, \ Y = 2rs, \ Z = r^2 + s^2,$$

where $r > s > 0$, $(r, s) = 1$, r, s of opposite parity, we must look for integers r and s such that

$(\ast\ast)$ $Y = 56 = 2rs$, where $r > s > 0$, $(r, s) = 1$, r, s of opposite parity.

Hence, we only need to search for the solutions of $(\ast\ast)$. There are two of them, namely $(r, s) = (28, 1)$ and $(r, s) = (7, 4)$, these in turn giving rise to the solutions $(X, Y, Z) = (x, 56, z) = (783, 56, 785)$ and $(X, Y, Z) = (x, 56, z) = (33, 56, 65)$.

(818) Comparing the geometric mean with the arithmetic mean (see Theorem 5) we have, for any positive real numbers x and y,

$$\frac{x^2 + y^2}{2} \geq (x^2 y^2)^{1/2} = xy, \text{ and therefore } x^2 + y^2 \geq 2xy.$$

Hence, we cannot have $x^2 + y^2 = xy$ unless $x = y = 0$. This is why the only integer solution of $x^2 + y^2 = xy$ is $(x, y) = (0, 0)$.

(819) It is obvious that x must be even. Setting $x = 2u$, we have

$$2u^2 + y^2 = 2z^2.$$

It is then clear that y must be even, in which case setting $y = 2v$, we obtain

$$u^2 + 2v^2 = z^2.$$

Reducing modulo 4, we see that v must be even. Setting $v = 2w$, we then have

$$u^2 + 8w^2 = z^2.$$

This equation has infinitely many solutions for each fixed value of u. Indeed, $w = u$ and $z = 3u$ is a solution for each positive integer u. Moreover, for each solution of $(\ast)$ we can write

$$8w^2 + u^2 = (3w - a)^2,$$

for some integer a. Thus, we have

$$w = 3a \pm \sqrt{8a^2 + u^2}.$$

Since $w_1 = u$ is a solution, it follows that

$$w_2 = 3u + \sqrt{8u^2 + u^2} = 6u$$

is also a solution. Other solutions are given by

$$w_3 = 35u,$$

$$w_4 = 204u,$$

and more generally by

$$w_n = 6w_{n-1} - w_{n-2}.$$

To find all the solutions for a fixed u, we only need to search for the solutions such that w is between 0 and u inclusively and then to iterate from these solutions.

(820) (*Contribution of John Brillhart, Arizona*). Let a, b, c be the lengths of the three sides of the required triangle and let α and 2α be the angles opposite to the sides a and b. Calling upon the law of sines and thereafter to the law of cosines, we obtain successively

$$\frac{b}{a} = \frac{\sin 2\alpha}{\sin \alpha} = \frac{2\sin\alpha\cos\alpha}{\sin\alpha} = 2\cos\alpha = \frac{b^2 + c^2 - a^2}{bc},$$

so that

$$
\begin{aligned}
b^2 c &= ab^2 + ac^2 - a^3, \\
b^2 c - b^2 a &= a(c^2 - a^2), \\
b^2(c - a) &= a(c^2 - a^2), \\
b^2 &= a(c + a), \quad \text{since } c \neq a.
\end{aligned}
$$

It is clear that the choice $a = 4$, $c = 5$, $b = 6$ serves our purpose.

(821) It is easy to check that the only solution is $x = 1$ and $y = 3$. As for the other equation, it has no integer solution.

(822) Since $X^2 + Y^2 = Z^2$ implies $\left(\dfrac{X}{Z}\right)^2 + \left(\dfrac{Y}{Z}\right)^2 = 1$, we have

$$x = \frac{X}{Z} = \frac{2rs}{r^2 + s^2}, \qquad y = \frac{Y}{Z} = \frac{r^2 - s^2}{r^2 + s^2}.$$

Dividing the numerator and the denominator by r^2, we obtain by setting $t = s/r$,

$$x = \frac{2t}{1 + t^2}, \qquad y = \frac{1 - t^2}{1 + t^2}, \qquad 0 \le t \le 1.$$

(823) Let $y = 2rs = 24$, so that $rs = 12 = 12 \cdot 1 = 6 \cdot 2 = 4 \cdot 3$. We thus find $(r, s) = (12, 1) = (6, 2) = (4, 3)$, and this is why the primitive Pythagorean triangles are obtained when $(r, s) = (12, 1)$ and $(r, s) = (4, 3)$.

(824) Assume that x, y and z is a primitive solution of $x^2 + y^2 = z^2$. Hence, $x = t^2 - s^2$, $y = 2ts$ and $z = t^2 + s^2$, so that letting A be the area of the triangle and letting r be the radius of the inscribed circle, we have

$$A = \frac{xy}{2} = \frac{rx}{2} + \frac{ry}{2} + \frac{rz}{2},$$

and therefore

$$r = \frac{xy}{x + y + z} = \frac{2ts(t^2 - s^2)}{2ts + (t^2 - s^2) + (t^2 + s^2)} = s(t - s),$$

an integer.

(825) One only needs to reduce the equation modulo 8, thereby obtaining a contradiction.

(826) We will show that at least two of the numbers x, y, z must be even. Assume the contrary, that is that the three numbers x, y, z are odd. Then t^2 is a number of the form $8k + 3$ and therefore must be odd, which contradicts the fact that t is even. If only one of the numbers x, y, z is even, the sum $x^2 + y^2 + z^2 = t^2$ is of the form $4k + 2$, which is impossible since the square of an even number must be of the form $4k$.

(827) If $(x, y) = (y, z) = (x, z) = 1$, then x and z are odd and y is even. Set $z - x = 2u$, $z + x = 2v$, where $(u, v) = 1$ and u and v are of opposite parity. Substituting in the equation, we find that $y^2 = 2uv$. Assuming that u is even, set $u = 2M$, that is $y^2 = 4Mv$, in which case we must have that $M = r^2$ and $v = s^2$, $(r, s) = 1$. By substitution, we obtain $y = 2rs$, $z = s^2 + 2r^2$ and $x = s^2 - 2r^2$, with $(r, s) = 1$.

(828) (*AMM, Vol. 65, 1958, p. 43*). The first equation becomes

$$(a - b - c)\left(a^2 + (b - c)^2 + ab + bc + ca\right) = 0.$$

Since $a^2 + (b - c)^2 + ab + bc + ca \neq 0$ (because $a, b, c > 0$), we have $a - b - c = 0$, which implies that $a = b + c = a^2/2$ and allows us to conclude that $a = 2$ and $b = c = 1$.

(829) (*AMM, Vol. 73, 1966, p. 895*). Multiplying the equation of the statement by 4 and adding 1, we obtain

$$4x^4 + 4x^3 + 4x^2 + 4x + 1 = (2y + 1)^2.$$

For $x = -1$, we find $y = -1$ or 0; for $x = 0$, we find $y = -1$ or 0; for $x = 2$, we find $y = -6$ or 5; finally, for $x = 1$, y is not an integer. On the other hand, for $x < -1$ or $x > 2$, the left-hand side of the above equation is larger than $(2x^2 + x)^2$ but smaller than $(2x^2 + x + 1)^2$ and therefore cannot be the square of an integer for integer values of x, while the right-hand side is the square of an integer for all integer values of y. It follows that the six solutions listed above are the only integer solutions of the given equation.

(830) (*AMM, Vol. 75, 1968, p. 193*). If $x^2 + ry^2 = p$, then $x^2 \equiv -ry^2$ (mod p) and therefore $-ry^2$; that is $-r$ is a quadratic residue of p. Hence, the required prime number must satisfy $\left(\dfrac{-r}{p}\right) = 1$ for $1 \leq r \leq 10$. This will be satisfied if $-1, 2, 3, 5$ and 7 are quadratic residues of p. It follows that the congruences $p \equiv 1$ (mod 8), $p \equiv 1$ (mod 3), $p \equiv 1$ or -1 (mod 5), and $p \equiv 1, 2$, or 4 (mod 7). The required prime number p must therefore satisfy

$$p \equiv 1, 121, 169, 289, 361 \text{ or } 529 \quad (\text{mod } 840),$$

and this is why the smallest prime number satisfying the conditions is $p = 1009$. We have therefore obtained

$$\begin{aligned}
1009 &= 15^2 + 28^2 = 19^2 + 2 \cdot 18^2 = 31^2 + 3 \cdot 4^2 = 15^2 + 4 \cdot 14^2 \\
&= 17^2 + 5 \cdot 12^2 = 25^2 + 6 \cdot 8^2 = 1^2 + 7 \cdot 12^2 = 19^2 + 8 \cdot 9^2 \\
&= 28^2 + 9 \cdot 5^2 = 3^2 + 10 \cdot 10^2.
\end{aligned}$$

(831) (*AMM, Vol. 75, 1968, p. 685*). Setting $x = a + 3d$, $y = a + 4d$, $z = a + 5d$ and $w = a + 6d$, we find the given equation becomes $a(a^2 + 9ad + 21d^2) = 0$. The only integer solution a of this equation is $a = 0$, and therefore the only solution $\{x, y, z, w\}$ of the given equation is $\{3d, 4d, 5d, 6d\}$.

(832) We have $x_{n+1} = x_1 + 6n(n+2)$, $n \geq 1$, and therefore $x_{n+1} = (n+2)^3 - n^3$. Hence, if x_{n+1} is a cube, say A^3, then we have $x_{n+1} = A^3 = (n+2)^3 - n^3$. Since no integer satisfies such an equation, the result is proved.

(833) (*AMM, Vol. 85, 1978, p. 118*). Assume that there exists a solution $\{x, y, n\}$. Since

$$x^n = y^{n+1} - 1 = (y-1)(y^n + y^{n-1} + \cdots + 1),$$

it is clear that any prime divisor p of $y-1$ divides x, and since $(x, n+1) = 1$, we have $p \nmid (n+1)$. Since $y \equiv 1 \pmod{y-1}$, it follows that

$$1 + y + y^2 + \cdots + y^n \equiv n + 1 \pmod{y-1}$$

and therefore that the numbers $y-1$ and $1 + y + \cdots + y^n$ are relatively prime. Hence, we may write

$$x^n = (y-1)(1 + y + \cdots + y^n),$$

which implies that $1 + y + \cdots + y^n$ is an n–th power of an integer. But this is impossible since

$$y^n < 1 + y + \cdots + y^n < (y+1)^n.$$

(834) (*AMM, Vol. 87, 1980, p. 138*). If the exponent of 3 is not zero, it is easy to see that none of these equations are solvable modulo 3.

(835) (*AMM, Vol. 76, 1969, p. 308*). If $x \le y \le z$, then $4^x + 4^y + 4^z$ is a perfect square under the condition that there exists a positive integer m and a positive odd integer t such that

$$1 + 4^{y-x} + 4^{z-x} = (1 + 2^m t)^2.$$

Therefore,

$$(*) \qquad\qquad 4^{y-x}(1 + 4^{z-y}) = 2^{m+1} t (1 + 2^{m-1} t),$$

so that we must have $m = 2y - 2x - 1$. Substituting this value in $(*)$, we obtain

$$\begin{aligned} t - 1 &= 4^{y-x-1}(4^{z-2y+x+1} - t^2) \\ &= 4^{y-x-1}(2^{z-2y+x+1} + t)(2^{z-2y+x+1} - t). \end{aligned}$$

Since t is odd, this last equation is possible when $t = 1$, and consequently $z = 2y - x - 1$. Therefore, the only integer solutions are $\{x, y, 2y - x - 1\}$, with arbitrary x and y. Finally, these values produce the square $(2^x + 2^{2y-x-1})^2$.

(836) (*AMM, Vol. 76, 1969, p. 84*). Setting $a = 3d$, $c = 2b - 3d$, we obtain $x + y = 3b$, and the second equation boils down to

$$(x - y)^2 = (b - 8d)^2 - 40d^2,$$

of which a solution is given by

$$x - y = m^2 - 10n^2, \quad b - 8d = m^2 + 10n^2, \quad d = mn,$$

where $m, n \in \mathbb{N}$. Hence, the solutions are: $a = 3d$, $b = 8d + m^2 + 10n^2$, $c = 2b - 3d$, $x = 12d + 2m^2 + 10n^2$, $y = 12d + m^2 + 20n^2$. To obtain infinitely many solutions when a, b, c are in arithmetic progression, it is enough to choose

$$a = 3mn, \quad b = m^2 + 8mn + 10n^2, \quad c = 2m^2 + 13mn + 20n^2$$

and

$$x = 2m^2 + 12mn + 10n^2, \quad y = m^2 + 12mn + 20n^2.$$

(837) *(AMM, Vol. 83, 1976, p. 569)*. First consider the equation $(*)$, $x^m(x^2 + y) = y^{m+1}$. If $m = 0$, the solutions are $x = 0$ and y arbitrary. If $m \geq 1$, the solutions are given by $x = b(b^m - 1)$, $y = b^2(b^m - 1)$, where $b \in \mathbb{Z}$. It is easy to verify that these are indeed solutions. Let (x, y) be a nontrivial solution; that is $xy \neq 0$. Then, we can write $x = ac$, $y = bc$ where a and b are relatively prime and $a \geq 1$. From $(*)$, we derive

$$a^m(a^2 c + b) = b^{m+1}.$$

This implies $a = 1$ and $c = b(b^m - 1)$, hence the solutions x and y. The only solutions are therefore $(m, x, y) = (0, 0, y)$ where y is arbitrary, and $(m, x, y) = (m, b(b^m - 1), b^2(b^m - 1))$, where $m \geq 1$ and $b \in \mathbb{Z}$.

Let us now examine the equation $(**)$, $x^m(x^2 + y^2) = y^{m+1}$. If $m = 0$, we have the only solution $x = 0$, $y = 1$. If $m \geq 1$, we only have the trivial solution $x = y = 0$. Indeed, assume that there exists a nontrivial solution (x, y). Then $xy \neq 0$, and we write again $x = ac$, $y = bc$, where a and b are relatively prime and $a \geq 1$. From $(**)$ we derive

$$a^m c(a^2 + b^2) = b^{m+1}.$$

This implies $a = 1$, so that $1 + b^2$ divides b^{m+1}. Since $b \neq 0$, we obtain a contradiction. The only solutions are therefore $(m, x, y) = (0, 0, 1)$ and $(m, 0, 0)$ where $m \geq 1$.

(838) *(AMM, Vol. 95, 1988, p. 141)*. These equations cannot be satisfied by integers. Indeed, for each integer h, we have

$$h^2 \equiv \begin{cases} 0 \pmod 8 & \text{if } h \equiv 0 \pmod 4, \\ 1 \pmod 8 & \text{if } h \equiv 1 \text{ or } 3 \pmod 4, \\ 4 \pmod 8 & \text{if } h \equiv 2 \pmod 4. \end{cases}$$

We therefore have

$$h^2 + k^2 \equiv \begin{cases} 0, 1 \text{ or } 4 \pmod 8 & \text{if } h \equiv 0 \pmod 4, \\ 1, 2 \text{ or } 5 \pmod 8 & \text{if } h \equiv 1 \text{ or } 3 \pmod 4, \\ 0, 4 \text{ or } 5 \pmod 8 & \text{if } h \equiv 2 \pmod 4, \end{cases}$$

for each integer h and k. Hence, since $\{x + 1, x + 2, x + 3, x + 4\}$ forms a complete residue system modulo 4, the congruences

$$(x+1)^2 + a^2 \equiv (x+2)^2 + b^2 \equiv (x+3)^2 + c^2 \equiv (x+4)^2 + d^2$$
$$\equiv n \pmod 8$$

are satisfied only if $n \in \{0, 1, 4\} \cap \{1, 2, 5\} \cap \{0, 4, 5\}$, which is impossible.

(839) Verifying the parity, we easily notice that two of the three integers are even, while the other is odd. Setting $x = 2m$, $y = 2n$ and $z = 2r + 1$, the equation becomes

$$4m^2 + 4n^2 + 4r^2 + 4r + 2 = 4mn(2r + 1),$$

which would mean that $4|2$, which is nonsense. Hence, there are no solutions.

(840) If $x = 0$, then from $(*)$, $y = \pm 1$; hence, because of $(**)$, we have $y = 1$. It follows that $(x, y) = (0, 1)$ is a solution of the system. Similarly, $(x, y) = (1, 0)$ is a solution of the system. Assume that $x \neq 0$ and $y \neq 0$. If $x > 1$, then $2x^3 - x^2 + y^2 = x^2(2x - 1) + y^2 > x^2 + y^2 > 1 + y^2 > 1$, which contradicts $(*)$; hence $x \leq 1$. Similarly, $y \leq 1$. By adding $(*)$ and $(**)$,

we derive that $x^3 + y^3 = 1$. If $x < 0$, then $y > 1$, which contradicts $y \leq 1$; hence $x > 0$. Similarly, $y > 0$. We then have $0 < x < 1$ and $0 < y < 1$. By hypothesis, $2x^3 - x^2 + y^2 = 2y^3 - y^2 + x^2$, so that $x^3 - x^2 = y^3 - y^2$. Let $u = y/x$. It follows that $x^3 - x^2 = u^3 x^3 - u^2 x^2$ and therefore that

(1) $$x - 1 = u^2(ux - 1).$$

If $u > 1$, then $u^2(ux - 1) > ux - 1$, which contradicts (1). Similarly, we cannot have $u < 1$. It follows that $u = 1$ and therefore that $y = x$. It then follows from $(*)$ that $x^3 = 1/2$ and therefore that $x = y = (1/2)^{1/3}$. The only solutions of the system are therefore

$$(0, 1), \qquad (1, 0), \qquad \left(\frac{1}{2^{1/3}}, \frac{1}{2^{1/3}} \right).$$

(841) (*MMAG, Vol. 52, 1979, p. 47*). If one of the numbers is 1, the other must also be equal to 1. Assume that (x, y) is a solution with $x \geq 2$ and $y \geq 2$. Then, $x^y = y^{x-y} > 1$ and therefore $x > y$. Dividing both sides of the equation by y^y, we obtain $(x/y)^y = y^{x-2y}$. Since $x/y > 1$, we have $(x/y)^y = y^{x-2y} > 1$. It follows that $x - 2y$ is a positive integer and thus $x/y > 2$, so that $(x/y)^y$ is a positive integer. This implies that x/y is a positive integer. Since the function $f(x) = 2^x - 4^2$ is strictly increasing for $x \geq 5$, it follows that $2^x > 4x$ and therefore for $x/y \geq 5$, we have

$$\frac{x}{y} = y^{(x/y)-2} \geq 2^{(x/y)-2} > \frac{x}{y},$$

a contradiction. On the other hand, when $2 < x/y < 5$ we obtain that x/y must be equal to 3 or 4. Since $x/y = y^{(x/y)-2}$, it follows that by choosing $x/y = 3$, we have $y = 3$ and $x = 9$, and choosing $x/y = 4$, we have $y = 2$ and $x = 8$. Therefore, the only solutions are $(1, 1)$, $(9, 3)$ and $(8, 2)$.

(842) (*MMAG, Vol. 63, 1990, p. 190*). Since $1 + x + x^2 > 0$, $1 + y + y^2 > 0$ and $1 + z + z^2 > 0$, it follows that x, y, z are positive integers. Without any loss in generality, we may assume that $x \geq y \geq z$. Then, $2x(1 + x + x^2) \geq 3(1 + x^4)$, so that $(x - 1)^2(3x^2 + 4x + 3) \leq 0$. Therefore, $x = 1$, which yields the only real solution $x = y = z = 1$.

(843) (*MMAG, Vol. 63, 1990, p. 190*). This follows from the fact that any integer $n > 2$ satisfies the identity $(n^2 + n)! \, (n-1)! = (n^2 + n - 1)! \, (n+1)!$ and the chain of inequalities $n^2 + n > n^2 + n - 1 > n + 1 > n - 1$.

(844) Since $m^3 \equiv 0, 1$ or $8 \pmod 9$, it follows that

$$x^3 + y^3 + z^3 \equiv 0, 1, 2, 3, 6, 7 \text{ or } 8 \pmod 9.$$

We conclude that neither of these two equations is solvable in integers.

(845) The answer is YES. Indeed, this Diophantine equation can be written successively as

$$\begin{aligned} x^4 &= 4y^2 + 4y + 1 - 81, \\ x^4 + 81 &= 4y^2 + 4y + 1, \\ x^4 + 3^4 &= (2y + 1)^2, \end{aligned}$$

this last equation having integer solutions only if $x = 0$, in which case we obtain $2y + 1 = \pm 9$, that is $y = 4$ or -5. We then have only two solutions, namely $(x, y) = (0, 4)$ and $(x, y) = (0, -5)$.

(846) The answer is NO. Indeed, given an arbitrary integer a, we always have $a^4 \equiv 0$ or $1 \pmod 5$. Therefore, the only possible values of $x^4 + y^4 + z^4$ modulo 5 are 0, 1, 2 or 3. Since $363932239 \equiv 4 \pmod 5$, there is no hope for a solution.

(847) The answer is NO. The reason is that $(303, 57) = 3$, while 3 never divides $a^2 + 1$.

(848) The answer is NO. Indeed, this Diophantine equation can be written as
$$x^4 + 2^4 = (2y + 1)^2.$$

But we know that the Diophantine equation $X^4 + Y^4 = Z^2$ has integer solutions only if $X = 0$ or $Y = 0$. Here, $Y = 2$, and therefore $X = 0$. It then follows that $16 = (2y + 1)^2$, which makes no sense.

(849) The answer is YES. It is enough to take $x = 0$, $y = 1$ and $z = 9$, in which case we do have
$$x^4 + (2y + 1)^4 = 0^4 + 3^4 = 9^2.$$

And this is the only solution in nonnegative integers.

(850) Since $x^2 - y^4 = 8$, we have $(x - y^2)(x + y^2) = 8$, which means that the only two possible cases are

$$\begin{cases} x - y^2 = 1, \\ x + y^2 = 8 \end{cases} \qquad \text{and} \qquad \begin{cases} x - y^2 = 2, \\ x + y^2 = 4. \end{cases}$$

The first of these two systems has no solutions, while the second implies that $x = 3$ and $y = 1$. The only positive solution of this Diophantine equation is therefore $(x, y) = (3, 1)$.

(851) Since $x^2 - y^4 = pq$, we have $(x - y^2)(x + y^2) = pq$, which gives rise to the systems of equations

$$\begin{cases} x - y^2 = 1, \\ x + y^2 = pq \end{cases} \qquad \text{and} \qquad \begin{cases} x - y^2 = p, \\ x + y^2 = q. \end{cases}$$

The second system has the solution
$$x = \frac{p + q}{2}, \qquad y^2 = \frac{q - p}{2} = \frac{8}{2} = 4,$$

which provides the following solutions to the Diophantine equation: $x = \frac{p+q}{2}$, $y = \pm 2$.

The first system implies that $2y^2 + 1 = pq = p(p+8)$; that is $2y^2 + 17 = (p+4)^2$ with $p+8$ prime. This last equation may have solutions depending on the value of p. When $p = 3$ with $q = p+8 = 11$, we obtain the solutions $x = \pm 7$ and $y = \pm 2$, and therefore we have found in this case more than one solution.

(852) The answer is NO. Indeed, this Diophantine equation can be written as
$$(x + 1)^2 + (y + 2)^2 = -4z + 3.$$

Assume that this equation has a solution (x, y, z). Since the left-hand side of this equation is congruent to 0, 1 or 2 modulo 4, while the right-hand side is congruent to 3 modulo 4, we have reached a contradiction.

(853) The answer is NO. This follows immediately from the fact that, since the geometric mean is no larger than the arithmetic mean (see Theorem 5),

$$\frac{x^4 + y^4 + z^4 + u^4}{4} \geq (x^4 y^4 z^4 u^4)^{1/4} = |xyzu|.$$

(854) The answer is NO. To prove it, we use the method of infinite descent of Fermat. Indeed, assume that this Diophantine equation has solutions. Let $x = x_0$ be the value corresponding to the smallest positive value of x for which this equation has a solution, say (x_0, y_0, z_0). We then have

(1) $$x_0^3 + 2y_0^3 = 4z_0^3.$$

It is clear from (1) that $2|x_0^3$, which implies that $8|x_0^3$; hence $x_0 = 2X$ for a certain positive integer X. The equation can therefore be rewritten as $8X^3 + 2y_0^3 = 4z_0^3$, that is

(2) $$4X^3 + y_0^3 = 2z_0^3.$$

It follows from (2) that $2|y_0^3$ and therefore that $8|y_0^3$, and equation (2) becomes $4X^3 + 8Y^3 = 2z_0^3$, with $2Z = z_0$, that is

(3) $$2X^3 + 4Y^3 = z_0^3.$$

It follows from (3) that $2|z_0^3$ and therefore that $8|z_0^3$, and equation (3) becomes $2X^3 + 4Y^3 = 8Z^3$, that is

$$X^3 + 2Y^3 = 4Z^3,$$

which is not possible, since we would have thus obtained a solution (X, Y, Z) to the equation $x^3 + 2y^3 = 4z^3$, with $0 < X < x_0$, thereby contradicting the minimal choice of x_0.

(855) If $m = 4k + r$, $0 \leq r \leq 3$, then $m^2 \equiv 0, 1$ or 4 (mod 8). Consequently, $x^2 + y^2 \equiv 0, 1, 2, 4$ or 5 (mod 8) while $8z + 7 \equiv 7$ (mod 8). Therefore, we conclude that this Diophantine equation has no integer solutions.

(856) We may assume that the numbers x and y are not divisible by 7. Consequently, these numbers are of the form $7k \pm 1$, $7k \pm 2$ or $7k \pm 3$. Since $(7k \pm 1)^2 = 7(7k^2 \pm 2k) + 1$, $(7k \pm 2)^2 = 7(7k^2 \pm 4k) + 4$, $(7k \pm 3)^2 = 7(7k^2 \pm 6k + 1) + 2$, it follows that

$$(7k \pm 1)^4 = 7M + 1, \quad (7k \pm 2)^4 = 7N + 2, \quad (7k \pm 3) = 7K + 4$$

and therefore that

$$x^4 + y^4 \equiv 1, 2, 3, 4, 5, 6 \pmod 7,$$

while $7z^2 \equiv 0$ (mod 7). Hence, equation $x^4 + y^4 = 7z^2$ has no integral solution.

For the second equation, the answer is again NO. In this case, we only need to reduce the given equation modulo 5.

(857) We easily see that the left-hand side of the equation is congruent to 0, 2 or 4 modulo 8, while the right-hand side is congruent to 5 or 6 modulo 8. It follows that the equation has no integer solutions.

(858) We have $z^n = z^2 z^{n-2} = z^{n-2}(x^2 + y^2) > x^n + y^n$, a contradiction.

(859) We write the initial relation as

$$X^3 + 3Y^3 = 9Z^3.$$

We proceed by contradiction by first assuming that amongst all the solutions with $Z > 0$, the smallest one (in Z) is x, y, z. From the above equation, we derive that x is a multiple of 3, and we write $x = 3n$, which we immediately substitute in the relation. We then obtain that $9n^3 + y^3 = 3z^3$, which implies that y is also a multiple of 3. We then set $y = 3m$, which gives rise to the relation $9n^3 + 27m^3 = 3z^3$, which in turn implies that $3n^3 + 9m^3 = z^3$. Hence, $z = 3k$ and $3n^3 + 9m^3 = 27k^3$, which is equivalent to $n^3 + 3m^3 = 9k^3$. But this relation is of the same form as the initial equation. But $k = z/3$, which contradicts the minimal choice of z.

(860) The answer is NO. Indeed, since $p|x$, we can set $x = x_0 p$ for some positive integer x_0. Substituting in the original equation, we obtain the equation

$$p^3 x_0^4 + y^4 + p z^4 = p^2 w^4.$$

This equation implies that $p|y$. As above, we then write $y = y_0 p$, which yields the equation

$$p^2 x_0^4 + p^3 y_0^4 + z^4 = p w^4.$$

It follows that $p|z$. Hence, write $z = z_0 p$, giving rise to the equation

$$p x_0^4 + p^2 y_0^4 + p^3 z_0^4 = w^4.$$

This implies that $p|w$. Writing $w = w_0 p$, we obtain

$$x_0^4 + p y_0^4 + p^2 z_0^4 = p^3 w^4.$$

Now, this equation is of the same type as the original equation, but the integers x_0, y_0, z_0, w_0 are respectively strictly smaller than x, y, z, w. Therefore, the method of infinite descent of Fermat then guarantees that the original equation has no integer solution.

(861) We proceed by contradiction by assuming that such a solution x, y, z exists, with positive integers x, y, z. First assume that x is odd and that y and z are even. We have

$$x^2 + y^2 + z^2 \equiv 1 \pmod 4 \quad \text{while} \quad 2xyz \equiv 0 \pmod 4,$$

which makes no sense. Similarly, x and y cannot be odd with z even. We can then finally show that x, y, z are odd. We have thus arrived at the conclusion that the three integers x, y, z must be even, say $x = 2x_1$, $y = 2y_1$, $z = 2z_1$, so that

$$x_1^2 + y_1^2 + z_1^2 = 4x_1 y_1 z_1 \equiv 0 \pmod 4,$$

which again implies that x_1, y_1 and z_1 must in turn be even. Continuing this process, we build triples (x_2, y_2, z_2), (x_3, y_3, z_3), ... delivering each time smaller and smaller even integers, which is impossible. This proves that there exist no integer solutions x, y, z.

(862) Let (x, y) be a solution of

(1) $$x^3 + y^3 = x^2 + y^2.$$

If $x = 0$, we find the two solutions $(x, y) = (0, 0)$ and $(x, y) = (0, 1)$. On the other hand, if $x \neq 0$, set $a = y/x$. Since, from (1), we always have $y \neq -x$, it follows that $a \neq -1$. Substituting $y = ax$ in (1), we find $x = \frac{1+a^2}{1+a^3}$. Similarly, substituting $x = y/a$ in (1), we find $y = \frac{a(1+a^2)}{1+a^3}$. We have thus established that any solution (x, y) of (1) with $x \neq 0$ is of the form

$$(2) \qquad x = \frac{1 + a^2}{1 + a^3}, \qquad y = \frac{a(1 + a^2)}{1 + a^3} \qquad (a \neq -1).$$

Reciprocally, one easily verifies that (2) produces a solution (x, y) of (1).

(863) (*Problem due to Leo Moser*). We will provide a particular solution. Set $x_1 = 2$ and then, for each r, $2 \leq r \leq n$, let

$$x_r = x_1 x_2 \cdots x_{r-1} + 1.$$

Of course the equation is satisfied for $n = 1$. Hence, assume that the set $\{x_1, x_2, \ldots, x_r\}$ satisfies the equation for $n = r$; we will show that the corresponding set with $n = r + 1$ satisfies the equation as well. But for $n = r + 1$, we have

$$\frac{1}{x_1} + \frac{1}{x_2} + \cdots + \frac{1}{x_{r+1}} + \frac{1}{x_1 x_2 \cdots x_{r+1}} = 1 - \frac{1}{x_1 x_2 \cdots x_r}$$
$$+ \frac{1}{x_{r+1}} + \frac{1}{x_1 x_2 \cdots x_{r+1}}$$
$$= 1 + \frac{1}{x_{r+1}} - \frac{x_{r+1} - 1}{x_1 x_2 \cdots x_{r+1}} = 1 + \frac{1}{x_{r+1}} - \frac{x_1 \cdots x_r}{x_1 \cdots x_r x_{r+1}} = 1,$$

as required.

REMARK: The sequence $2, 3, 7, 43, 1807, 3\,263\,443, \ldots$ is also the subject of Problem 479.

(864) We proceed by contradiction by assuming that such a solution x, y, z exists with positive integers x, y, z. First assume that x is odd and that y and z are even. We thus have

$$x^2 + y^2 + z^2 \equiv 1 \pmod 4 \quad \text{while} \quad x^2 y^2 \equiv 0 \pmod 4,$$

which makes no sense. Similarly, one can show x and y cannot be odd while z is even. Finally, one can show that x, y, z cannot be odd. We therefore arrive at the conclusion that all three integers x, y, z must be even, say $x = 2x_1$, $y = 2y_1$, $z = 2z_1$, so that we have

$$x_1^2 + y_1^2 + z_1^2 = 4x_1^2 y_1^2 \equiv 0 \pmod 4,$$

which again implies that x_1, y_1 and z_1 must also be even. Continuing, we construct triples (x_2, y_2, z_2), (x_3, y_3, z_3), $\ldots$ each time made up of even numbers getting smaller and smaller, an endless process, which makes no sense. This argument implies that there is no integer solution x, y, z.

(865) Assuming that x is even or odd, we reach a contradiction.

(866) (*Problem introduced by Johann Walter*). Assume that such odd integers x, y, z exist. Then

$$(x^2 + 2xy + y^2) + (x^2 + 2xz + z^2) = y^2 + 2yz + z^2$$

and therefore

$$x^2 + xy + xz = yz.$$

By adding yz on each side, we obtain

$$(*) \qquad (x + y)(x + z) = 2yz,$$

which is impossible because each of the expressions $x+y$ and $x+z$ is even, so that the left member of $(*)$ is divisible by 4, while its right member is not, y and z being odd.

(867) One easily checks that $x = (s^2 - pr^2)/2$, $y = rs$, $z = (s^2 + pr^2)/2$ is a solution. Conversely, if x, y, z is a primitive solution, then $y^2 = (z^2 - x^2)/p$ and thus $p|(z \pm x)$. Setting $s^2 = z \mp x$ and $r^2 = (z \pm x)/p$, we obtain the result.

(868) Let $n, m \in \mathbb{N}$. Setting $x = n(n^2 - 12m^2)$, $y = m(4m^2 - 3n^2)$, we obtain $z = n^2 + 4m^2$, which yields infinitely many solutions.

(869) The only solutions are $(x, y, n) = (0, 0, n)$ (with arbitrary n) and $(x, y, n) = (2, 2, 1)$. Indeed, first consider the case $n = 1$. The equation $x + y = xy$ becomes $x = (x-1)y$, which means that $x = 0$ (and $y = 0$) or that $x-1|x$, or in other words that $x - 1 = 1$ or -1 (since $x - 1$ and x are two consecutive integers). If $x - 1 = 1$, then $x = 2$ and $y = 2$. If $x - 1 = -1$, then $x = 0$ and $y = 0$. The second case is the one where $n \geq 2$. If x and y are positive, assume that $x > y > 0$; we then have $xy = x^n + y^n > x^n \geq x^2 \geq xy$, a contradiction.

Let us now examine the case where at least one of x, y is negative; clearly both cannot be negative. Assume that $x < 0$ and $y > 0$. If n is even, we are back to the above case. On the other hand, if n is odd, $n \geq 3$, then we can write $x = -a$, with $a > 0$, and say $y = b$. Then, the equation $x^n + y^n = xy$ becomes $a^n - b^n = ab$. Setting $a = b + r$, we have

$$a^n - b^n = (b+r)^n - b^n > nb^{n-1}r + \binom{n}{2} b^{n-2}r^2 \geq 3b^2r + 3br^2$$
$$> b^2 + br = ab,$$

a contradiction.

(870) Since $z \geq \max(x, y)$, we derive from the equation that $n^x|n^y$ and $n^y|n^x$. Consequently, $x = y$ and it follows that $z = x + 1$ and $n = 2$.

(871) Assume that $w \geq \max\{x, y\}$. Then, $n^x|n^w$. Since $n^x|n^z$, we have $n^x|n^y$. By the symmetry of the problem, we also have $n^y|n^x$. Therefore, $x = y$ and the equation we need to solve is reduced to $(*)$ $2n^x + n^w = n^z$. In this case, we derive that $n^w|2n^x$, so that $2n^{x-w}$ is an integer for $x \leq w$. If $n > 2$, then $x = w$ and $(*)$ becomes $3n^x = n^z$, which implies that $n = 3$ and the solution is $x = y = w = z - 1$. For $n = 2$, the solution is $x = y = w - 1 = z - 2$.

(872) (*Contribution of A. Ivić, Belgrade*). Assume that the equation $x^p + y^q = z^r$ has a solution in positive integers x, y, z. Let $A = x^p$, $B = y^q$, $C = z^r$, so that

$$\prod_{p|ABC} p = \prod_{p|x^p y^q z^r} p = \prod_{p|xyz} p \leq xyz.$$

But since $x \leq z^{r/p}$, $y \leq z^{r/q}$, then according to the abc conjecture, for all $\varepsilon > 0$, there exists a positive constant $M = M(\varepsilon)$ such that

$$z^r \leq M \cdot \left(\prod_{p|ABC} p \right)^{1+\varepsilon} = M \cdot (xyz)^{1+\varepsilon} \leq M \cdot (z^r)^{(1+\varepsilon)(\frac{1}{p} + \frac{1}{q} + \frac{1}{r})}.$$

If $z \geq z_0$, we obtain

$$1 \leq (1+\varepsilon)\left(\frac{1}{p} + \frac{1}{q} + \frac{1}{r}\right),$$

which contradicts $(*)$ for ε sufficiently small.

(873) In order to prove the result, we first observe that if $1 < a < b < c$ are three consecutive integers, then $ac + 1 = b^2$.

Now, since ac, 1 and b^2 are relatively prime, it follows from the abc conjecture that, for each $\varepsilon > 0$, there exists a positive constant $M = M(\varepsilon)$ such that

$$(*) \qquad b^2 \leq M \cdot (\gamma(abc))^{1+\varepsilon} \leq M \cdot \left(\sqrt{abc}\right)^{1+\varepsilon} \leq M \cdot b^{3(1+\varepsilon)/2},$$

where $\gamma(n)$ stands for the product of the prime numbers dividing n and where we used the fact that $ac < b^2$. It then follows from $(*)$ that

$$b^{(1-3\varepsilon)/2} \leq M.$$

Choosing ε small enough, we find that b as well as a and c are bounded, which proves the result.

(874) Assume that the number $m = n^3 + 1$ is powerful. Then, according to the abc conjecture, we have that for each $\varepsilon > 0$ there exists a positive constant $M = M(\varepsilon)$ such that

$$m < M \cdot \gamma(mn^3)^{1+\varepsilon},$$

where $\gamma(a)$ is the product of the prime numbers dividing a. Since $(m,n) = 1$, $\gamma(m) \leq \sqrt{m}$ and $n < m^{1/3}$, it follows that

$$m < M \cdot \gamma(nm)^{1+\varepsilon} < M \cdot (m^{1/3}m^{1/2})^{1+\varepsilon} = M \cdot m^{5(1+\varepsilon)/6},$$

so that

$$m^{\frac{1}{6} - \frac{5\varepsilon}{6}} < M.$$

Taking ε sufficiently small, we find that m is bounded, which proves the result.

The numbers $n = 2$ and $n = 23$ are the two smallest numbers (and possibly the only ones) such that the corresponding number $n^3 + 1$ is powerful: $2^3 + 1 = 3^2$ and $23^3 + 1 = 2^3 \cdot 3^2 \cdot 13^2$.

(875) Assume that x, y, z are three 4-powerful numbers relatively prime and verifying $x + y = z$. We apply the abc conjecture to the triple (x, y, z) so that

$$z \leq M \cdot r(xyz)^{1+\varepsilon} \leq M \cdot (xyz)^{(1+\varepsilon)/4} \leq M \cdot z^{3(1+\varepsilon)/4}.$$

It follows that

$$z^{(1-3\varepsilon)/4} \leq M$$

and therefore that z is bounded, and similarly for x and y.

(876) Let $a + b = c$, where c is 4-powerful and where a and b are 3-powerful, $(a,b) = 1$. For each $\varepsilon > 0$, we have $c < M(\varepsilon) \cdot \gamma(abc)^{1+\varepsilon}$. By hypothesis, we have

$$\gamma(a) \leq a^{1/3}, \quad \gamma(b) \leq b^{1/3}, \quad \gamma(c) \leq c^{1/3},$$

which implies that, using the abc conjecture,

$$c < M(\varepsilon)\left(a^{1/3}b^{1/3}c^{1/3}\right)^{1+\varepsilon} < M(\varepsilon)(c^{11/12})^{1+\varepsilon},$$

an inequality which cannot hold if ε is sufficiently small and c large enough. This clearly proves that only a finite number of such a, b, c integers can exist.

(877) Let $y > 0$ be fixed and let $\varepsilon > 0$ be fixed and sufficiently small. Let also $p_1, p_2, \ldots, p_r$ be the list of all prime numbers $\leq y$. If $P(p^2 - 1) \leq y$ for a certain prime number p, then there exist nonnegative integers $\alpha_1, \ldots, \alpha_r$ such that

$$p^2 - 1 = p_1^{\alpha_1} \cdots p_r^{\alpha_r}$$

and therefore

$$p^2 = p_1^{\alpha_1} \cdots p_r^{\alpha_r} + 1.$$

It follows from the *abc* conjecture that for all $\varepsilon > 0$, there exists a positive constant $M = M(\varepsilon) > 0$ such that

$$p^2 < M \cdot (p_1 p_2 \cdots p_r p)^{1+\varepsilon},$$

so that

$$p^{1-\varepsilon} < M \cdot (p_1 p_2 \cdots p_r)^{1+\varepsilon} < M \cdot y^{r(1+\varepsilon)}.$$

Since ε is small and $r = \pi(y)$ is fixed (as well as y), it follows that p is bounded, and the result is proved.

REMARK: For each odd prime number $y \leq 19$, here is the conjectured value of the largest element $p_* = p_*(y)$ of the set of prime numbers A_y:

$y =$	3	5	7	11	13	17	19
$p_* =$	17	31	4801	4801	8191	388961	1419263

Let us mention that although $P(p^2 - 1) > 11$ for each prime number $p > 4801$, the largest prime number p such that $P(p^2 - 1) = 11$ is $p = 881$ (in fact $P(4801^2 - 1) = 7$). For the other prime numbers y listed above, we have $P(p_*^2 - 1) = y$.

(878) Since $p \equiv 1 \pmod 4$, we have $q \equiv 1 \pmod 8$, so that $\left(\frac{2}{q}\right) = 1$ and therefore, by Euler's Criterion,

$$2^{p-1} = 2^{\frac{q-1}{2}} \equiv \left(\frac{2}{q}\right) = 1 \pmod q.$$

On the other hand, since

$$n - 1 = pq - 1 = 2p^2 - p - 1 = (p-1)(2p+1) \quad \text{and} \quad 2^{p-1} \equiv 1 \pmod p,$$

it follows that

$$2^{n-1} = \left(2^{p-1}\right)^{2p+1} \equiv 1 \pmod p,$$
$$2^{n-1} = \left(2^{p-1}\right)^{2p+1} \equiv 1 \pmod q,$$

which implies that $2^{n-1} \equiv 1 \pmod{pq}$, as required.

(879) First we define the function

$$M_0(\varepsilon) = \max_{\delta \geq \varepsilon} M(\delta),$$

which is decreasing for all $\varepsilon > 0$ and is such that $M(\varepsilon) \leq M_0(\varepsilon)$ for each $\varepsilon > 0$.

It follows from the *abc* conjecture that, for $i = 1, 2, 3$,

$$x_i \leq M(\varepsilon/3) \cdot (\gamma(x_1 x_2 x_3))^{1+\varepsilon/3} \leq M_0(\varepsilon/3) \cdot (\gamma(x_1 x_2 x_3))^{1+\varepsilon/3}$$

and therefore that

$$x_1 x_2 x_3 \leq M_0(\varepsilon/3)^3 \cdot (\gamma(x_1 x_2 x_3))^{3+\varepsilon}.$$

Hence, if the conclusion $(*)$ is false, then

$$x_i > M_0(\varepsilon) \cdot \gamma(x_i)^{3+\varepsilon} \qquad (i = 1, 2, 3)$$

and therefore

$$x_1 x_2 x_3 > M_0(\varepsilon)^3 \cdot (\gamma(x_1 x_2 x_3))^{3+\varepsilon},$$

in which case we would have

$$M_0(\varepsilon)^3 \cdot \gamma(x_1 x_2 x_3)^{3+\varepsilon} < x_1 x_2 x_3 \leq M_0(\varepsilon/3)^3 \cdot \gamma(x_1 x_2 x_3)^{3+\varepsilon},$$

and therefore

$$M_0(\varepsilon) < M_0(\varepsilon/3),$$

which is impossible since M_0 is decreasing.

(880) *(Math. Intelligencer* **18** *(1996), p. 58).* It is false. Indeed, choosing $n = 3$, $x = 10$, $y = 16$, $z = 17$, we obtain a contradiction. This counter-example is due to Roger Apéry, the famous mathematician who proved the irrationality of $\zeta(3) = \sum_{n=1}^{\infty} 1/n^3$.

(881) From Wilson's Theorem, we have $(p-1)! \equiv -1 \pmod{p}$ so that

$$\xi := \frac{(p-1)! + 1}{p} \qquad \text{is an integer .}$$

It follows that

$$r^{(p-1)!+1} = \underbrace{r^{(p-1)!} + \cdots + r^{(p-1)!}}_{r}.$$

Hence,

$$\left(r^\xi\right)^p = \left(r^{(p-1)!/\alpha_1}\right)^{\alpha_1} + \left(r^{(p-1)!/\alpha_2}\right)^{\alpha_2} + \cdots + \left(r^{(p-1)!/\alpha_r}\right)^{\alpha_r}.$$

The result follows by setting

$$n = r^\xi \quad \text{and} \quad x_i = r^{(p-1)!/\alpha_i} \text{ for } i = 1, 2, \ldots, r.$$

(882) From Wilson's Theorem, $(p-1)! \equiv -1 \pmod{p}$ and this is why $\xi := \dfrac{(p-1)! + 1}{p}$ is a positive integer. It follows that

$$\left(2^\xi\right)^p = 2^{(p-1)!+1} = 2^{(p-1)!} + 2^{(p-1)!} = \left(2^{(p-1)!/(p-1)}\right)^{p-1}$$
$$+ \left(2^{(p-1)!/(p-1)}\right)^{p-1}.$$

The result then follows by choosing

$$x = 2^{(p-2)!}, \qquad y = 2^{(p-2)!}, \qquad z = 2^\xi.$$

(883) It is all the prime numbers p satisfying one of the congruences $p \equiv 1, 5, 7, 9, 19, 25, 35, 37, 39, 43 \pmod{44}$.

(884) (a) This congruence has the solutions $x \equiv \pm 1 \pmod 3$.

(b) This congruence has no solutions, because $x^2 \equiv 1 \pmod 3$ for each integer x such that $(x,3)=1$, while $x^2 \equiv 0 \pmod 3$ if $3|x$.

(c) This congruence can be modified as follows:
$$
\begin{aligned}
x^2 + 4x + 4 &\equiv -4 \pmod 3, \\
(x+2)^2 &\equiv -1 \pmod 3, \\
y^2 &\equiv -1 \pmod 3,
\end{aligned}
$$
a congruence which is of the form (b) and therefore has no solutions.

(d) This congruence can be modified as follows:
$$
\begin{aligned}
x^2 + 8x + 16 &\equiv -1 \pmod{17}, \\
(x+4)^2 &\equiv -1 \pmod{17}, \\
y^2 &\equiv -1 \pmod{17},
\end{aligned}
$$
a congruence which has solutions, since $17 \equiv 1 \pmod 4$. In fact, a solution is given by $y = \frac{17-1}{2}! = 8! \equiv 13 \pmod{17}$, which implies that $x = y - 4 \equiv 9 \pmod{17}$. The other solution is therefore $y = 17 - 13 = 4$, that is $x = y - 4 \equiv 0 \pmod{17}$.

(885) Multiplying the congruence by 8, we obtain successively
$$
\begin{aligned}
16x^2 + 24x + 8 &\equiv 0 \pmod 7, \\
16x^2 + 24x + 9 &\equiv 9 - 8 \pmod 7, \\
(4x+3)^2 &\equiv 1 \pmod 7, \\
y^2 &\equiv 1 \pmod 7,
\end{aligned}
$$
a congruence which has solutions, namely $y \equiv 1$ and $6 \pmod 7$. To find the corresponding values of x, we must solve separately the congruences
$$
4x + 3 \equiv 1 \pmod 7 \quad \text{and} \quad 4x + 3 \equiv 6 \pmod 7.
$$
The solutions are $x \equiv 3 \pmod 7$ and $x \equiv 6 \pmod 7$.

(886) Let $N = (1!)^2 + (2!)^2 + \cdots + (n!)^2$. Then,
$$
N \equiv (1!)^2 + (2!)^2 + (3!)^2 + (4!)^2 \equiv 617 \equiv 2 \pmod 5.
$$
If there exists a positive integer m such that $m^2 = N$, then $m^2 \equiv 2 \pmod 5$; and since $\left(\dfrac{2}{5}\right) = -1$, then 2 is a nonquadratic residue $\pmod 5$.
Hence, there exist no integers m such that $m^2 = N$.

(887) The congruence $n^2 + 1 \equiv 0 \pmod p$ has solutions only if $p \equiv 1 \pmod 4$. Hence, p must be of the form $p = 4\ell + 1$. On the other hand, since 3 does not divide $n^2 + 1$ whatever the positive integer n, we must have that any prime divisor of n is also of the form $3m \pm 1$.

But a number that is of the form $4\ell + 1$ and also of the form $3m + 1$ is of the form $12k + 1$. On the other hand, a number which is of the form $4\ell + 1$ and of the form $3m - 1$ is of the form $12k + 5$. Hence, the result.

(888) The quadratic residues modulo p are congruent to $1^2, 2^2, \ldots, ((p-1)/2)^2$, so that
$$
1^2 + 2^2 + \cdots + \frac{(p-1)^2}{2^2} = \frac{p(p+1)(p-1)}{24}.
$$

Since $p \neq 2, 3$, it follows that p divides the sum of the quadratic residues modulo p.

(889) By hypothesis, we have $q \equiv -1 \pmod 8$. It follows that
$$\left(\frac{2}{q}\right) = 1.$$
Therefore, by Euler's Criterion, we have that
$$2^p = 2^{\frac{q-1}{2}} \equiv \left(\frac{2}{q}\right) = 1 \pmod q.$$
We then have $q | 2^p - 1 = M_p$. To prove the second part, we observe that
$$p = 4 \cdot 280\,664 + 3 \quad \text{and} \quad q = 2p + 1 = 2\,245\,319$$
and that q is prime. By the first part, we may then conclude that $2\,245\,319$ divides $M_{1\,122\,659}$ and therefore that $M_{1\,122\,659}$ is composite.

(890) Since $9239 \equiv 7 \pmod 8$, we have $\left(\dfrac{2}{9239}\right) = 1$. Using Euler's Criterion, we have that
$$1 = \left(\frac{2}{9239}\right) \equiv 2^{4619} \pmod{9239},$$
and therefore $9239 | 2^{4619} - 1$.

(891) Let $p = 6^n + 1$. Then, $p \equiv 2 \pmod 5$, and using the law of quadratic reciprocity, we have
$$\left(\frac{5}{p}\right) = \left(\frac{p}{5}\right)(-1)^{\frac{5-1}{2} \cdot \frac{6^n}{2}} = \left(\frac{2}{5}\right) = -1.$$

(892) In order to have $1997k - 1 = n^2$, we must have $n^2 \equiv -1 \pmod{1997}$. But $1997 \equiv 1 \pmod 4$, so that $\left(\dfrac{-1}{1997}\right) = 1$, which confirms that such an integer n exists. In fact, for $n = 412$ and $k = 85$, the equation $1997k - 1 = n^2$ is verified.

(893) Assume that there exist only a finite number of such numbers, say $q_1, q_2, \ldots, q_r$, and then consider the number $N = (2q_1 q_2 \cdots q_r)^2 + 3$. It is clear that we have
$$N \equiv 4 + 3 \equiv 1 \pmod 3,$$
so that N is of the form $3k+1$. Hence, if N is prime, the proof is complete because we will have found a prime number of the form $3k+1$ larger than q_r. If N is composite, we can prove that each of its prime divisors is of the form $3k + 1$. Indeed, if $p|N$ and $p \equiv 2 \pmod 3$, then
$$(2q_1 q_2 \cdots q_r)^2 + 3 \equiv 0 \pmod p,$$
which would mean that the congruence $x^2 \equiv -3 \pmod p$ has a solution for a prime number $p \equiv 2 \pmod 3$, in which case we would have
$$1 = \left(\frac{-3}{p}\right) = \left(\frac{-1}{p}\right)\left(\frac{3}{p}\right) = (-1)^{\frac{p-1}{2}}\left(\frac{p}{3}\right)(-1)^{\frac{p-1}{2}} = \left(\frac{p}{3}\right) = \left(\frac{2}{3}\right) = -1,$$
a contradiction. This is why N (being odd and nondivisible by 3) must have a prime divisor $p \equiv 1 \pmod 3$, in which case $p = q_i$ for a certain $i \in [1, r]$, which means that $q_i | 3$, a contradiction.

(894) The answer is NO. Let $K = 1! + 2! + \cdots + k!$. Then, $K \equiv 1! + 2! + 3! + 4! \equiv 3$ (mod 5). Hence, if $K = n^2$, we must have $n^2 \equiv 3$ (mod 5), which is impossible since 3 is a nonquadratic residue modulo 5.

Another solution is as follows.

Since the last digit of $k!$ is 0 for each $k \geq 5$, the last digit of the number on the left-hand side is 3 for each $k \geq 5$. Obviously, if $K = n^2$, then n must be an odd number. Since the last digit of an odd perfect square must be 1, 5 or 9, there are no solutions for $k \geq 5$. Examining the cases for $k \leq 4$, we obtain that there are exactly two solutions, namely $k = n = 1$ and $k = n = 3$.

(895) (*AMM, Vol. 85, 1978, p. 497*). Assume that $A_n = 2^n - 1$ divides $B_n = 3^n - 1$ for some integer $n > 1$. Clearly, $3 \nmid B_n$. If n is even, then $3 | A_n$. Therefore, n must be odd, say $n = 2m - 1$, $m \geq 2$. Since $2^{2m-1} \equiv 0$ (mod 4) and $2^{2m-1} \equiv 2$ (mod 3) for each $m \geq 2$, we have $A_n \equiv -5$ (mod 12). Since each prime number > 3 is congruent to ± 1 or ± 5 (mod 12), there exists at least a prime divisor p of A_n such that $p \equiv \pm 5$ (mod 12). Since $p | 3B_n = 3^{2m} - 3$, it follows that $3^{2m} \equiv 3$ (mod p), that is that 3 is a quadratic residue modulo p. On the other hand, in light of the law of quadratic reciprocity, since $p \equiv \pm 5$ (mod 12), we have

$$\left(\frac{3}{p}\right)\left(\frac{p}{3}\right) = (-1)^{\frac{p-1}{2}\frac{3-1}{2}} = 1,$$

so that, for a certain integer $r \geq 0$,

$$\left(\frac{3}{p}\right) = \left(\frac{p}{3}\right) = \left(\frac{12r \pm 5}{3}\right) = \left(\frac{5}{3}\right) = \left(\frac{2}{3}\right) = (-1)^{\frac{3^2-1}{8}} = -1.$$

(896) The answer is YES. Indeed, we have

$$\left(\frac{p}{q}\right) = \left(\frac{q+4a}{q}\right) = \left(\frac{4a}{q}\right) = \left(\frac{4}{q}\right)\left(\frac{a}{q}\right) = \left(\frac{a}{q}\right).$$

(897) The answer is YES. Indeed, we have

$$\left(\frac{3}{p}\right) = \left(\frac{p}{3}\right)(-1)^{\frac{p-1}{2}\cdot\frac{3-1}{2}} = \left(\frac{p}{3}\right) = \left(\frac{1}{3}\right) = 1.$$

(898) This congruence will have solutions if each of the following two congruences has solutions:

$$x^2 \equiv 52 \pmod 3 \quad \text{and} \quad x^2 \equiv 52 \pmod{53}.$$

The first can be written as $x^2 \equiv 1$ (mod 3) and therefore has solutions. The second one can be written as $x^2 \equiv -1$ (mod 53) and has solutions since $53 \equiv 1$ (mod 4). The stated congruence is therefore solvable.

(899) The answer is YES. Indeed, since $p - 2q = 1$, we have,

$$\left(\frac{q}{p}\right) = \left(\frac{p}{q}\right)(-1)^{\frac{p-1}{2}\frac{q-1}{2}} = \left(\frac{p}{q}\right)(-1)^{2k(4k+1)} = \left(\frac{p}{q}\right)$$

$$= \left(\frac{p-2q}{q}\right) = \left(\frac{1}{q}\right) = 1.$$

(900) Let $q = 2^p - 1$ be a Mersenne prime (where p is an odd prime). Then $\frac{q-1}{2} = 2^{p-1} - 1$ and $q \equiv 1 \pmod 3$, so that

$$\left(\frac{3}{q}\right) = \left(\frac{q}{3}\right)(-1)^{\frac{q-1}{2}\frac{3-1}{2}} = -\left(\frac{q}{3}\right) = -\left(\frac{1}{3}\right) = -1.$$

(901) Since $p \equiv 7 \pmod 8$, it follows that $\left(\frac{2}{p}\right) = 1$, which means that there exists x_0 such that $x_0^2 \equiv 2 \pmod p$. Raising both sides of this congruence to the power $\frac{p-1}{2}$, we obtain

$$(*) \qquad \left(x_0^2\right)^{\frac{p-1}{2}} \equiv 2^{\frac{p-1}{2}} \pmod p.$$

Since, from Fermat's Little Theorem, we have $x_0^{p-1} \equiv 1 \pmod p$, it follows from $(*)$ that

$$1 \equiv x_0^{p-1} \equiv 2^{\frac{p-1}{2}} \pmod p,$$

thus the result.

(902) The answer is NO. Indeed, we first observe that $231 = 3 \cdot 7 \cdot 11$. Thus, for the congruence $x^2 \equiv 2 \pmod{231}$ to have solutions, we must have that each of the congruences $x^2 \equiv 2 \pmod 3$, $x^2 \equiv 2 \pmod 7$ and $x^2 \equiv 2 \pmod{11}$ is solvable. But the first and third of these congruences are not solvable, since $\left(\frac{2}{3}\right) = -1$ and $\left(\frac{2}{11}\right) = -1$.

(903) The answer is NO. Indeed, since $p = 100k + 3 \equiv 3 \pmod 4$, the congruence $x^2 \equiv -1 \pmod p$ has no solutions.

(904) The answer is YES. Indeed, solving the problem boils down to finding whether the congruence $x^2 + 14x + 47 \equiv 0 \pmod{23}$ has solutions. But this congruence can also be written as $x^2 + 14x + 49 \equiv 2 \pmod{23}$, that is

$$(*) \qquad (x + 7)^2 \equiv 2 \pmod{23}.$$

Since $\left(\frac{2}{23}\right) = 1$, it means that the congruence $y^2 \equiv 2 \pmod{23}$ has solutions, and so does $(*)$.

(905) The answer is NO. It is enough to prove that the congruence

$$(*) \qquad x^2 - 3x - 1 \equiv 0 \pmod{541}$$

has no solutions. But this congruence is successively equivalent to

$$\begin{aligned} 4x^2 - 12x - 4 &\equiv 0 \pmod{541}, \\ (2x - 3)^2 &\equiv 13 \pmod{541}. \end{aligned}$$

But the congruence $y^2 \equiv 13 \pmod{541}$ has no solutions. Indeed,

$$\left(\frac{13}{541}\right) = \left(\frac{541}{13}\right)(-1)^{6 \cdot 270} = \left(\frac{8}{13}\right) = \left(\frac{2}{13}\right) = (-1)^{12 \cdot 14/8} = -1.$$

(906) The answer is YES. Indeed, since $p = 24k + 1$, we have that $\frac{p^2-1}{8}$ is even and therefore that

$$\begin{aligned} \left(\frac{6}{p}\right) &= \left(\frac{2}{p}\right)\left(\frac{3}{p}\right) = (-1)^{\frac{p^2-1}{8}}\left(\frac{3}{p}\right) = \left(\frac{3}{p}\right) \\ &= \left(\frac{p}{3}\right)(-1)^{\frac{p-1}{2}} = \left(\frac{p}{3}\right) = \left(\frac{1}{3}\right) = 1. \end{aligned}$$

(907) The answer is NO. Indeed, since $4^n + 1 \equiv 2 \pmod{3}$, we have

$$\left(\frac{3}{p}\right) = \left(\frac{p}{3}\right)(-1)^{4^n/2} = \left(\frac{p}{3}\right) = \left(\frac{4^n+1}{3}\right) = \left(\frac{2}{3}\right) = -1.$$

(908) First observe that if p is odd, then

$$\left(\frac{11}{p}\right) = \left(\frac{p}{11}\right)(-1)^{\frac{11-1}{2}\frac{p-1}{2}} = \left(\frac{p}{11}\right)(-1)^{\frac{p-1}{2}}.$$

This last expression is equal to 1 if any one of the following situations occurs:

  (i) $\left(\frac{p}{11}\right) = 1$ and $p \equiv 1 \pmod{4}$,
  (ii) $\left(\frac{p}{11}\right) = -1$ and $p \equiv 3 \pmod{4}$.

One easily checks that (i) takes place when $p \equiv 1, 3, 4, 5, 9 \pmod{11}$ and $p \equiv 1 \pmod{4}$, that is when $p \equiv 1, 5, 9, 25, 37 \pmod{44}$; while (ii) takes place when $p \equiv 2, 6, 7, 8, 10 \pmod{11}$ and $p \equiv 3 \pmod{4}$, that is when $p \equiv 7, 19, 35, 39, 43 \pmod{44}$. Finally, when $p = 2$, we have $2 \equiv 2 \pmod{44}$ and it follows that $2 \in A$. Therefore,

$$A = \{1, 2, 5, 7, 9, 19, 25, 35, 37, 39, 43\}.$$

(909) Since

$$\left(\frac{5}{p}\right) = \left(\frac{p}{5}\right) = \left\{ \begin{array}{ll} 1 & \text{if } p \equiv \pm 1 \pmod{5}, \\ -1 & \text{if } p \equiv \pm 2 \pmod{5}, \end{array} \right.$$

it follows that $\left(\dfrac{5}{p}\right) = -1$ if and only if $p \equiv \pm 2 \pmod{5}$.

(910) We have

$$\left(\frac{p}{q}\right) = \left(\frac{q+4a}{q}\right) = \left(\frac{4a}{q}\right) = \left(\frac{a}{q}\right)$$

and

$$\left(\frac{q}{p}\right) = \left(\frac{p-4a}{p}\right) = \left(\frac{-4a}{p}\right) = (-1)^{(p-1)/2}\left(\frac{a}{p}\right),$$

and since $(p-1)/2 = (q-1)/2 + 2a$, it follows that $(p-1)/2$ is even if and only if $(q-1)/2$ is even. Consequently, $(-1)^{(p-1)/2} = (-1)^{((p-1)/2)((q-1)/2)}$, in which case

$$\left(\frac{a}{q}\right) = \left(\frac{p}{q}\right) = \left(\frac{q}{p}\right)(-1)^{\frac{(p-1)}{2}\frac{(q-1)}{2}} = \left(\frac{a}{p}\right).$$

(911) They are the prime numbers $p \equiv 1, 3 \pmod{8}$.

(912) We have that

$$\left(\frac{2}{p}\right) = (-1)^{(p^2-1)/8} = 1 \iff p \equiv \pm 1 \pmod{8} \iff x^2 \equiv 2 \pmod{p}$$

has solutions. Since $\left(\dfrac{2}{p}\right) = 2^{(p-1)/2} \equiv 1 \pmod{p}$, we have

$$2^{4n+3} \equiv 1 \pmod{8n+7}.$$

This shows that 263 is a divisor of the Mersenne number $2^{131} - 1$.

(913) For the congruence $x^2 \equiv 1237 \pmod{2717}$ to have solutions, the three expressions $\left(\dfrac{1237}{11}\right)$, $\left(\dfrac{1237}{13}\right)$ and $\left(\dfrac{1237}{19}\right)$ must be equal to 1. But the last two expressions are equal to -1. It follows that the given congruence has no solutions.

(914) We must have $\left(\dfrac{-1}{p}\right) = +1$. Therefore, the required numbers are the prime numbers p satisfying $p \equiv 1 \pmod 4$.

(915) It has no solutions since

$$\left(\frac{131313}{1987}\right) = \left(\frac{19}{1987}\right) = \left(\frac{2}{19}\right) = -1.$$

(916) (Sierpinski, [**39**], Problem #193) Assume that an integer solution $\{x, y\}$ exists. Clearly, $y > 0$. We then consider separately the cases "y even" and "y odd". First of all, if y is even, there exists a positive integer k such that $y = 2k$, in which case $x^2 = 8k^3 + 7$, which is impossible since no perfect square has this form. If y is odd, there exists a positive integer k such that $y = 2k + 1$. We then have

$$x^2 + 1 = y^3 + 2^3 = (y+2)(y^2 - 2y + 4) = (y+2)((y-1)^2 + 3)$$
$$= (2k+3)(4k^2 + 3).$$

It follows that $(*)$ $(2k)^2 + 3 | x^2 + 1$. Since $(2k)^2 + 3$ certainly has a prime divisor of the form $4n + 3$, it follows from $(*)$ that $x^2 + 1$ also has a prime divisor p of the form $p = 4n + 3$. But this is impossible since in this case the congruence $x^2 \equiv -1 \pmod p$ would be solvable, which is not so because $\left(\frac{-1}{p}\right) = -1$.

(917) The number 15 is a quadratic residue modulo p if and only if $\left(\dfrac{3}{p}\right)\left(\dfrac{5}{p}\right) = 1$. Since

$$\left(\frac{3}{p}\right) = \begin{cases} +1 & \text{if } p \equiv \pm 1 \pmod{12}, \\ -1 & \text{if } p \equiv \pm 5 \pmod{12} \end{cases}$$

and

$$\left(\frac{5}{p}\right) = \begin{cases} +1 & \text{if } p \equiv \pm 1 \pmod 5, \\ -1 & \text{if } p \equiv \pm 2 \pmod 5, \end{cases}$$

we conclude that

$$\left(\frac{15}{p}\right) = \begin{cases} +1 & \text{if } p \equiv \pm 1, \pm 7, \pm 11, \pm 17 \pmod{60}, \\ -1 & \text{if } p \equiv \pm 13, \pm 19, \pm 23, \pm 29 \pmod{60}. \end{cases}$$

(918) Since $\left(\dfrac{ab}{p}\right) = \left(\dfrac{a}{p}\right)\left(\dfrac{b}{p}\right)$, we can write

$$\left(\frac{(-1)^{(q-1)/2}q}{p}\right) = \left(\frac{(-1)^{(q-1)/2}}{p}\right)\left(\frac{q}{p}\right).$$

Using this same identity, we have

$$\left(\frac{(-1)^{(q-1)/2}}{p}\right) = \left(\frac{-1}{p}\right)^{(q-1)/2} = (-1)^{\frac{p-1}{2}\frac{q-1}{2}},$$

which gives the result.

(919) It has two solutions, because $\left(\dfrac{34561}{1234577}\right) = +1$.

(920) Let r be a quadratic residue modulo m and let k be such that

$$k^2 \equiv r \pmod{m}.$$

By Euler's Theorem (see Theorem 22),

$$k^{\phi(m)} \equiv 1 \pmod{m}.$$

Combining these two equations and using the fact that $\phi(m)$ is even for $m > 2$, we have

$$r^{\phi(m)/2} \equiv (k^2)^{\phi(m)/2} \equiv k^{\phi(m)} \equiv 1 \pmod{m},$$

as required.

(921) Since $a^n - 1 \equiv 0 \pmod{(a^n - 1)}$, we have $a^n \equiv 1 \pmod{(a^n - 1)}$. It follows that the smallest number r such that $a^r \equiv 1 \pmod{(a^n - 1)}$ is n, which implies that $n \mid \phi(a^n - 1)$.

(922) Since there are as many quadratic residues as quadratic nonresidues, the result follows.

(923) Since $(k, p) = 1$, there exists an integer $1 \le k' \le p - 1$ such that $kk' \equiv 1 \pmod{p}$. Therefore,

$$\left(\frac{k(k+1)}{p}\right) = \left(\frac{k(k+kk')}{p}\right) = \left(\frac{k^2(1+k')}{p}\right) = \left(\frac{1+k'}{p}\right)$$

and

$$\sum_{k=1}^{p-2}\left(\frac{k(k+1)}{p}\right) = \sum_{k'=1}^{p-2}\left(\frac{1+k'}{p}\right) = \sum_{m=1}^{p-1}\left(\frac{m}{p}\right) - \left(\frac{1}{p}\right).$$

In light of Problem 922 and the fact that $\left(\dfrac{1}{p}\right) = 1$, the result follows.

(924) Assume that no two consecutive integers are quadratic residues modulo p; then $\left(\dfrac{k}{p}\right)\left(\dfrac{k+1}{p}\right) = -1$ for each positive integer k. Hence, $\left(\dfrac{k(k+1)}{p}\right) = -1$ for each positive integer k, which contradicts the statement of Problem 923. A similar argument is used to prove that no two consecutive integers are quadratic nonresidues modulo p.

(925) Assume that $5p + 1 = a^2$, in which case $a^2 \equiv 1 \pmod{5}$, a solvable congruence. We then have $5p = (a - 1)(a + 1)$, which is possible only if $a = 6$ or 4. Hence, the possible prime numbers p are $p = 7$ and $p = 3$.

As for the second question, it is easy to see that there exist no such prime numbers p.

(926) (T.M. Apostol [1], page 201). Since $E = \{k \mid k = 0, 1, \ldots, p - 1\}$ is a complete residue system modulo p, $A = \{ak + b \mid k = 0, 1, \ldots, p - 1\}$ is also a complete residue system modulo p if $(a, p) = 1$. Moreover, $ak + b \equiv r \pmod{p}$ implies $f(ak + b) \equiv f(r) \pmod{p}$ and therefore

$$\left(\frac{f(ak+b)}{p}\right) = \left(\frac{f(r)}{p}\right).$$

To prove the second part, we set $f(x) = x$ in the first part, in which case, we obtain

$$\sum_{k=0}^{p-1}\left(\frac{ak+b}{p}\right) = \sum_{k=0}^{p-1}\left(\frac{k}{p}\right) = 0.$$

(927) Set

$$a_p(k) = \begin{cases} 1 & \text{if } \left(\frac{k}{p}\right) = a \text{ and } \left(\frac{k+1}{p}\right) = b, \\ 0 & \text{otherwise,} \end{cases}$$

so that

$$N(a,b) = \sum_{k=1}^{p-2} a_p(k).$$

Since $(a^2 = b^2 = 1)$ and $a_p(k) = 0$ if $\left(\frac{k}{p}\right) \neq a$ or $\left(\frac{k+1}{p}\right) \neq b$, it follows that

$$a_p(k) = \frac{1}{4}\left(1 + a\left(\frac{k}{p}\right)\right)\left(1 + b\left(\frac{k+1}{p}\right)\right).$$

Therefore,

$$
\begin{aligned}
4N(a,b) &= \sum_{k=1}^{p-2}\left(1 + a\left(\frac{k}{p}\right) + b\left(\frac{k+1}{p}\right) + ab\left(\frac{k}{p}\right)\left(\frac{k+1}{p}\right)\right) \\
&= \sum_{k=1}^{p-2}1 + a\sum_{k=1}^{p-2}\left(\frac{k}{p}\right) + b\sum_{k=1}^{p-2}\left(\frac{k+1}{p}\right) + ab\sum_{k=1}^{p-2}\left(\frac{k}{p}\right)\left(\frac{k+1}{p}\right).
\end{aligned}
$$

Since $\sum_{k=1}^{p-1}\left(\frac{k}{p}\right) = 0$, we have

$$\sum_{k=1}^{p-2}\left(\frac{k}{p}\right) = -\left(\frac{p-1}{p}\right) = -\left(\frac{-1}{p}\right) = -(-1)^{(p-1)/2}$$

and

$$\sum_{k=1}^{p-2}\left(\frac{k+1}{p}\right) = -\left(\frac{1}{p}\right) = -1.$$

Since from Problem 923,

$$\sum_{k=1}^{p-2}\left(\frac{k(k+1)}{p}\right) = -1,$$

we obtain that

$$4N(a,b) = p - 2 - b - ab - a\left(\frac{-1}{p}\right),$$

and in particular,

$$N(1,1) = \frac{p - 4 - (-1)^{(p-1)/2}}{4}.$$

(928) It is sufficient to observe that $\left(\frac{j}{p}\right) = \left(\frac{p-j}{p}\right).$

(929) Since $p - k$ runs through all the numbers $1, 2, \ldots, p - 1$ as k runs through these same numbers, it follows that

$$\sum_{k=1}^{p-1} k \left(\frac{k}{p} \right) = \sum_{k=1}^{p-1} (p - k) \left(\frac{p - k}{p} \right) = \sum_{k=1}^{p-1} (p - k) \left(\frac{-k}{p} \right)$$

$$= \sum_{k=1}^{p-1} (p - k) \left(\frac{-1}{p} \right) \left(\frac{k}{p} \right).$$

Therefore,

$$\sum_{k=1}^{p-1} k \left(\frac{k}{p} \right) = p(-1)^{(p-1)/2} \sum_{k=1}^{p-1} \left(\frac{k}{p} \right) - (-1)^{(p-1)/2} \sum_{k=1}^{p-1} k \left(\frac{k}{p} \right).$$

Since $\displaystyle\sum_{k=1}^{p-1} \left(\frac{k}{p} \right) = 0$ for $p \equiv 1 \pmod 4$, the result follows.

(930) Since $p \equiv 1 \pmod 4$, we have

$$\sum_{\substack{k=1 \\ \left(\frac{k}{p} \right)=1}}^{p-1} k = \sum_{\substack{k=1 \\ \left(\frac{p-k}{p} \right)=1}}^{p-1} (p - k) = \sum_{\substack{k=1 \\ \left(\frac{k}{p} \right)=1}}^{p-1} (p - k) = p \sum_{\substack{k=1 \\ \left(\frac{k}{p} \right)=1}}^{p-1} 1 - \sum_{\substack{k=1 \\ \left(\frac{k}{p} \right)=1}}^{p-1} k.$$

Using the fact that there are $(p-1)/2$ quadratic residues, we obtain that

$$\sum_{\substack{k=1 \\ \left(\frac{k}{p} \right)=1}}^{p-1} 1 = \frac{p - 1}{2}$$

and therefore that

$$2 \sum_{\substack{k=1 \\ \left(\frac{k}{p} \right)=1}}^{p-1} k = \frac{p(p - 1)}{2}.$$

(931) We have

$$\sum_{k=1}^{p-1} k^2 \left(\frac{k}{p} \right) = \sum_{k=1}^{p-1} (p - k)^2 \left(\frac{p - k}{p} \right) = -\sum_{k=1}^{p-1} (p - k)^2 \left(\frac{k}{p} \right)$$

$$= -\sum_{k=1}^{p-1} (p^2 - 2pk + k^2) \left(\frac{k}{p} \right) = 2p \sum_{k=1}^{p-1} k \left(\frac{k}{p} \right) - \sum_{k=1}^{p-1} k^2 \left(\frac{k}{p} \right),$$

and the result follows if $p \equiv 3 \pmod 4$.

(932) If there exists a solution, then $x^2 = 5 + 33 y^2$; that is $x^2 \equiv 5 \pmod{33}$ and in particular $x^2 \equiv 5 \pmod 3$. Since 5 is a nonquadratic residue modulo 3, we conclude that there is no integer solution.

(933) The continued fractions are $[1, \overline{2}]$ and $[0, 1, \overline{2}]$.

(934) (a) $x = -26$, $y = 65$; (b) $x = 3$, $y = 2$.

(935) It is easy to see that it is the number $1 + 2\sqrt{2}$.

(936) We have

$$\frac{1}{\alpha} = 0 + \frac{1}{\alpha} = 0 + \cfrac{1}{a_1 + \cfrac{1}{\ddots}} = [0, a_1, a_2, \dots].$$

(937) The convergents of $\sqrt{5}$ are

$$C_2 = \frac{9}{4}, \quad C_3 = \frac{38}{17}, \quad C_4 = \frac{161}{72}, \quad C_5 = \frac{682}{305}.$$

The rational number $682/305$ will therefore serve our purpose.

(938) The result can easily be obtained by induction on n.

(939) This comes from the fact that $ax^2 = bx + c$ and therefore that

$$x = \frac{b}{a} + \frac{c}{a}\frac{1}{x} = \frac{b}{a} + \cfrac{1}{\cfrac{b}{c} + \cfrac{1}{x}}.$$

(940) It is obvious that the two roots are $(5 + \sqrt{57})/4$ and $(5 - \sqrt{57})/4$. Using Problem 939, we have that a root is given by $[\frac{5}{2}, \frac{5}{4}]$. We obtain the following convergents:

$$C_2 = \frac{33}{10} \approx 3.3, \quad C_3 = \frac{205}{66} \approx 3.106, \quad C_4 = \frac{1289}{410} \approx 3.143,$$

$$C_5 = \frac{8085}{2578} \approx 3.136, \quad C_6 = \frac{50737}{16170} \approx 3.1377,$$

and we can say that one of the roots is approximately 3.14. Since the sum of the roots is $5/2$, the other root is approximately -0.64.

(941) We only need to observe that

$$\sqrt{n^2 + 1} - n = \cfrac{1}{n + \sqrt{n^2 + 1}} = \cfrac{1}{2n + \sqrt{n^2 + 1} - n}$$

$$= \cfrac{1}{2n + \cfrac{1}{2n + \sqrt{n^2 + 1} - n}}.$$

(942) Since $n - 1 < \sqrt{n^2 - 1} < n$, it is clear that $[\sqrt{n^2 - 1}] = n - 1$. Therefore,

$$\sqrt{n^2 - 1} = n - 1 + \sqrt{n^2 - 1} - (n - 1) = n - 1 + \cfrac{1}{\cfrac{\sqrt{n^2 - 1} + (n - 1)}{2n - 2}}.$$

Since

$$\frac{\sqrt{n^2 - 1} + (n - 1)}{2n - 2} = 1 + \cfrac{1}{(2n - 2) + \sqrt{n^2 - 1} - (n - 1)},$$

the result follows.

(943) The result follows using the fact that

$$\sqrt{n^2+2} - n = \frac{1}{(\sqrt{n^2+2}+n)/2} = \cfrac{1}{n + \cfrac{1}{\cfrac{1}{\sqrt{n^2+2}+n}}}$$

$$= \cfrac{1}{n + \cfrac{1}{\cfrac{1}{2n + \sqrt{n^2+2} - n}}}.$$

(944) For $n \geq 2$, we have $n-1 < \sqrt{n^2-2} < n$, in which case

(1) $\sqrt{n^2-2} = n-1 + \sqrt{n^2-2} - n + 1 = n-1 + \cfrac{1}{\cfrac{\sqrt{n^2-2}+n-1}{2n-3}}$,

and since $1 < \dfrac{\sqrt{n^2-2}+n-1}{2n-3} < 2$, equation (1) becomes

(2) $\sqrt{n^2-2} = n-1 + \cfrac{1}{1 + \cfrac{\sqrt{n^2-2}-n+2}{2n-3}}$

$$= n-1 + \cfrac{1}{1 + \cfrac{1}{\cfrac{\sqrt{n^2-2}+n-2}{2}}}.$$

Since $n-2 < \dfrac{\sqrt{n^2-2}+n-2}{2} < n-1$, equation (2) can be written as

(3) $\sqrt{n^2-2} = n-1 + \cfrac{1}{1 + \cfrac{1}{n-2 + \cfrac{\sqrt{n^2-2}-n+2}{2}}}$

$$= n-1 + \cfrac{1}{1 + \cfrac{1}{n-2 + \cfrac{1}{\cfrac{\sqrt{n^2-2}+n-2}{2n-3}}}}.$$

Since $1 < \dfrac{\sqrt{n^2-2}+n-2}{2n-3} < 2$, (3) becomes

(4) $\sqrt{n^2-2} = n-1 + \cfrac{1}{1 + \cfrac{1}{n-2 + \cfrac{1}{1 + \cfrac{\sqrt{n^2-2}-n+1}{2n-3}}}}$

$$= n-1 + \cfrac{1}{1 + \cfrac{1}{n-2 + \cfrac{1}{1 + \cfrac{1}{\sqrt{n^2-2}+n-1}}}}.$$

Since $2n - 2 < \sqrt{n^2 - 2} + n - 1 < 2n - 1$, that is

$$\sqrt{n^2 - 2} + n - 1 = 2n - 2 + \sqrt{n^2 - 2} - n + 1,$$

it follows that

$$\sqrt{n^2 - 2} = n - 1 + \cfrac{1}{1 + \cfrac{1}{n - 2 + \cfrac{1}{1 + \cfrac{1}{2n - 2 + (\sqrt{n^2 - 2} - n + 1)}}}},$$

and we obtain equation (1), thus the result.

(945) Since $38 = 6^2 + 2$, it follows by Problem 943 that $\sqrt{38} = [6, \overline{6, 12}]$. Since $47 = 7^2 - 2$, we have by Problem 944 that $\sqrt{47} = [6, \overline{1, 5, 1, 12}]$. Finally, since $120 = 11^2 - 1$, it follows by Problem 942 that $\sqrt{121} = [10, \overline{1, 20}]$.

(946) Since $n < \sqrt{n^2 + n} < n + 1$, we have

$$\sqrt{n^2 + n} = n + \sqrt{n^2 + n} - n = n + \cfrac{1}{\cfrac{\sqrt{n^2 + n} + n}{n}}.$$

Since the integer part of $\dfrac{\sqrt{n^2 + n} + n}{n}$ is 2, it follows that

$$\sqrt{n^2 + n} = n + \cfrac{1}{2 + \cfrac{1}{\sqrt{n^2 + n} + n}} = n + \cfrac{1}{2 + \cfrac{1}{2n + \sqrt{n^2 + n} - n}},$$

and the result follows.

(947) For $n > 1$, we have $n - 1 < \sqrt{n^2 - n} < n$, and therefore

(1) $\quad \sqrt{n^2 - n} = n - 1 + \sqrt{n^2 - n} - n + 1 = n - 1 + \cfrac{1}{\cfrac{\sqrt{n^2 - n} + n - 1}{n - 1}}.$

Since $2 < \dfrac{\sqrt{n^2 - n} + n - 1}{n - 1} < 3$, it follows that (1) can be written as

(2) $\quad\quad\quad\quad \sqrt{n^2 - n} = n - 1 + \cfrac{1}{2 + \cfrac{1}{\sqrt{n^2 - n} + n - 1}}.$

Since $2n - 2 < \sqrt{n^2 - n} + n - 1 < 2n - 1$, it follows that (2) becomes

$$\sqrt{n^2 - n} = n - 1 + \cfrac{1}{2 + \cfrac{1}{2n - 2 + (\sqrt{n^2 - n} - n + 1)}},$$

and using (1), we get the result.

(948) For $n > 1$, $3n < \sqrt{9n^2 + 3} < 3n + 1$, so that

(1) $\quad\quad\quad \sqrt{9n^2 + 3} = 3n + \sqrt{9n^2 + 3} - 3n = 3n + \cfrac{1}{\cfrac{\sqrt{9n^2 + 3} + 3n}{3}}.$

Since $2n < \dfrac{\sqrt{9n^2+3}+3n}{3} < 2n+1$, it follows that (1) takes the form

$$(2)\quad \sqrt{9n^2+3} = 3n + \cfrac{1}{2n + \cfrac{\sqrt{9n^2+3}-3n}{3}} = 3n + \cfrac{1}{2n + \cfrac{1}{\sqrt{9n^2+3}+3n}}.$$

Since $6n < \sqrt{9n^2+3}+3n < 6n+1$, (2) becomes

$$\sqrt{9n^2+3} = 3n + \cfrac{1}{2n + \cfrac{1}{6n + (\sqrt{9n^2+3}-3n)}}.$$

Using (1), we get the result.

(949) The representations of $q+1$ and $q-1$ are

$$\begin{aligned} q+1 &= [2,\overline{1,2}], \\ q-1 &= [0,\overline{1,2}]. \end{aligned}$$

It follows that if we let q_0 be the number $[\overline{1,2}]$, then we have

$$\begin{aligned} q+1 &= [2,q_0] = 2 \cdot q_0, \\ q-1 &= [0,q_0] = \frac{1}{q_0}. \end{aligned}$$

Therefore, $(q+1)(q-1) = 2$, so that the required number is $r = q = \sqrt{3}$.

(950) The infinite continued fraction of π is $[3,7,15,1,292,1,1,\ldots]$. This way, we build the following table:

n	0	1	2	3	4	5	6
a_n		3	7	15	1	292	1
p_n	1	3	22	333	355	103993	104348
q_n	0	1	7	106	113	33102	33215

and p_4/q_4 is the best rational approximation of π amongst all the rational numbers whose denominator does not exceed 1000.

Since $e = [2,1,2,1,1,4,1,1,6,1,1,8,1,\ldots]$, we have the table

n	0	1	2	3	4	5	6	7	8	9	10	11
a_n		2	1	2	1	1	4	1	1	6	1	1
p_n	1	2	3	8	11	19	87	106	193	1264	1457	2721
q_n	0	1	1	3	4	7	32	39	71	465	536	1001

and p_{10}/q_{10} is the best rational approximation of e amongst all the rational numbers whose denominator does not exceed 1000.

Since $\sqrt{5} = [2,\overline{4}]$, we have the table

n	0	1	2	3	4	5	6
a_n		2	4	4	4	4	4
p_n	1	2	9	38	161	682	2889
q_n	0	1	4	17	72	305	1292

and p_5/q_5 is the best rational approximation of $\sqrt{5}$ amongst all the rational numbers whose denominator does not exceed 1000.

(951) The number $9976/6961$ will serve the purpose.

(952) The number $1264/465$ will serve the purpose.

(953) The convergents of π are

$$\frac{3}{1}, \frac{22}{7}, \frac{333}{106}, \frac{355}{113}, \frac{103993}{33102}, \cdots,$$

and we thus find

$$\left|\pi - \frac{355}{113}\right| < \frac{1}{113 \cdot 33102} < \frac{1}{10^6}.$$

(954) A good approximation is 4.3589.

(955) We have

$$0 < |a/b - p_k/q_k| \le |a/b - \alpha| + |\alpha - p_k/q_k| < 2|\alpha - p_k/q_k| < 2/q_k q_{k+1}.$$

Multiplying these inequalities by bq_k, we obtain $0 < |aq_k - bp_k| < 2b/q_{k+1}$. Since $|aq_k - bp_k|$ is a positive integer, we must have $2b/q_{k+1} > 1$, in which case $b > q_{k+1}/2$.

(956) Since $\sqrt{3} = [1, \overline{1, 2}]$, we have the table

n	0	1	2	3	4	5	6	7
a_n		1	1	2	1	2	1	2
p_n	1	1	2	5	7	19	26	71
q_n	0	1	1	3	4	11	15	41

and therefore

$$|\sqrt{3} - a/b| \le |\sqrt{3} - p_6/q_6|.$$

It follows from Problem 955 that $b > q_7/2 = 20.5$, and this is why $b \ge 21$.

(957) We have

$$0 < \left|\frac{a}{b} - \frac{p_k}{q_k}\right| < \left|\alpha - \frac{p_k}{q_k}\right| < \frac{1}{q_k q_{k+1}}.$$

Multiplying this last equation by bq_k, we obtain $0 < |aq_k - bp_k| < b/q_{k+1}$. Since $|aq_k - bp_k|$ is a positive integer, we have $b/q_{k+1} > 1$, that is $b > q_{k+1}$.

(958) This follows immediately from Problem 957.

(959) Observe that if $|\pi - a/b| < |\pi - 333/106|$, then by Problem 955, we have $b > 113/2 = 56.5$ (since the convergent following $333/106$ is $355/113$). If $b \le 56$, we have $|\pi - 333/106| < |\pi - a/b|$.

(960) We shall prove this inequality using induction. We have

$$q_3 = a_2 a_3 + 1 \ge 2 \ge 2^1, \quad q_4 = a_2 a_3 a_4 + a_2 + a_4 \ge 3 \ge 2^{3/2}.$$

Assume that the inequality is true for some n and let us prove it for $n+1$. Using the hypothesis induction, we have

$$q_{n+1} = a_{n+1} q_n + q_{n-1} \ge q_n + q_{n-1} \ge 2^{\frac{n}{2} - \frac{1}{2}} + 2^{\frac{n}{2} - 1}$$

$$= 2^{\frac{n}{2}} \left(\frac{1}{\sqrt{2}} + \frac{1}{2}\right) > 2^{\frac{(n+1)-1}{2}},$$

which proves the result.

(961) Since for $n \ge 1$, we have $p_n q_{n-1} - q_n p_{n-1} = (-1)^n$, it follows that $p_{n-4} q_{n-3} - q_{n-4} p_{n-3} = (-1)^n$, and therefore it is enough to show that for $n \ge 4$, we have

$$p_n q_{n-3} - q_n p_{n-3} = (-1)^n (a_n a_{n-1} + 1).$$

This result can then easily be obtained by induction on n.

(962) Let

$$\alpha = \left[a_1, a_2, \ldots, a_{n-1} + \frac{1}{\alpha'} \right] = \frac{\alpha' p_{n-1} + p_{n-2}}{\alpha' q_{n-1} + q_{n-2}},$$

$$\beta = \left[a_1, a_2, \ldots, a_{n-1} + \frac{1}{\beta'} \right] = \frac{\beta' p_{n-1} + p_{n-2}}{\beta' q_{n-1} + q_{n-2}}.$$

Set $\alpha' = a_n + \alpha''$, $\beta' = b_n + \beta''$, where $0 < \alpha'' < 1$, $0 < \beta'' < 1$. We then obtain that

$$\beta' - \alpha' = b_n - a_n + \beta'' - \alpha'' \geq 1 + \beta'' - \alpha'' > 0.$$

Hence,

$$
\begin{aligned}
\alpha - \beta &= \frac{\alpha' p_{n-1} + p_{n-2}}{\alpha' q_{n-1} + q_{n-2}} - \frac{\beta' p_{n-1} + p_{n-2}}{\beta' q_{n-1} + q_{n-2}} \\
&= \frac{(\alpha' - \beta')(p_{n-1} q_{n-2} - p_{n-2} q_{n-1})}{(\alpha' q_{n-1} + q_{n-2})(\beta' q_{n-1} + q_{n-2})} \\
&= \frac{(\beta' - \alpha')(-1)^n}{(\alpha' q_{n-1} + q_{n-2})(\beta' q_{n-1} + q_{n-2})} = \begin{cases} < 0 & \text{if } n \text{ is odd,} \\ > 0 & \text{if } n \text{ is even.} \end{cases}
\end{aligned}
$$

(963) Let $\theta = [d_1, d_2, d_3, \ldots]$, where $d_{2j-1} = c_{2j-1}$ and $d_{2j} = b_{2j}$. If $\alpha := [a_1, a_2, a_3, \ldots] < \theta$, then using Problem 962, we see that if k is the first position where θ differs from α, then $a_k < d_k$ for k odd and $a_k > d_k$ for k even. If $k = 2n$, then $a_{2n} > d_{2n} = b_{2n}$, while if $k = 2n - 1$, then $a_{2n-1} < d_{2n-1} = c_{2n-1}$, which contradicts our hypothesis. In this case, we must have $\alpha \geq \theta$. The inequality on the right can be obtained in a similar manner.

(964) Using Problem 963 with $c_n = 1$ and $b_n = 2$ for each $n \geq 1$, we obtain

$$\frac{1 + \sqrt{3}}{2} = \overline{[1, 2]} \leq \alpha \leq \overline{[2, 1]} = 1 + \sqrt{3}.$$

(965) Since $20926/86400 = [0, 4, 7, 1, 3, 5, 64]$, we easily find the convergents

$$\frac{1}{4}, \quad \frac{7}{29}, \quad \frac{8}{33}, \quad \frac{31}{128}, \quad \frac{163}{673}, \quad \frac{10463}{43200}.$$

Although $97/400$ is not a convergent of $20926/86400$, it is easy to see, in this case, that $8/33$ provides a better approximation than $97/400$. This means that by adding 8 days every 33 years would provide a better approximation than adding 97 days every 400 years. Finally, observe that the fourth convergent $31/128$ (which could be obtained for example by removing a leap year every 128 years) provides the length of an actual year with a precision of four decimals.

(966) We only need to expand this determinant with respect to the last column and then use induction on k. The value of q_k can be obtained from p_k by crossing out the first column and the first row.

(967) Let $\alpha = \overline{[a_1, a_2, \ldots, a_n]} = [a_1, a_2, \ldots, a_n, \alpha]$. Let p_k/q_k be the k–th convergent of α, so that

$$\alpha = \frac{\alpha p_n + p_{n-1}}{\alpha q_n + q_{n-1}}.$$

We therefore have a quadratic equation in α:

$$q_n \alpha^2 + (q_{n-1} - p_n)\alpha - p_{n-1} = 0.$$

Since α is not rational, the expansion of the continued fraction is infinite and α is a quadratic irrational number.

(968) Let $\alpha = [a_1, a_2, \ldots, a_n, \overline{b_1, b_2, \ldots, b_m}]$ and let $\beta = [\overline{b_1, b_2, \ldots, b_m}]$. In light of Problem 967, β is a quadratic irrational number. If p_k/q_k is the k–th convergent of α, then

$$\alpha = [a_1, a_2, \ldots, a_n, \beta] = \frac{\beta p_n + p_{n-1}}{\beta q_n + q_{n-1}}.$$

Since p_{n-1}, p_n, q_{n-1} and q_n are nonzero rational numbers and β is a quadratic irrational number, we have that α is a quadratic irrational number or simply a rational number. Since α is an infinite continued fraction, it follows that α is a quadratic irrational number.

(969) If p_n/q_n is the n–th convergent of α, then

$$\alpha = \frac{p_n \alpha_n + p_{n-1}}{q_n \alpha_n + q_{n-1}}.$$

Substituting this value of α in $a\alpha^2 + b\alpha + c = 0$ and rearranging the coefficients, we obtain

$$A_n \alpha_n^2 + B_n \alpha_n + C_n = 0,$$

where A_n, B_n and C_n are defined in the statement of the problem. Elementary computations allow one to obtain

$$B_n^2 - 4A_n C_n = (b^2 - 4ac)(p_n q_{n-1} - q_n p_{n-1})^2 = b^2 - 4ac.$$

For the second part, we observe that, since $f(\alpha) = 0$ and since this number α is located between p_{n-1}/q_{n-1} and p_n/q_n, it follows that the values of f at these points must be of opposite signs, the reason being that the other root of this quadratic equation is not located between p_{n-1}/q_{n-1} and p_n/q_n. This shows that $A_n C_n < 0$.

(970) Let α_n be a root of $A_n x^2 + B_n x + C_n = 0$, where the coefficients are given in the statement of Problem 969. Since

$$|\alpha q_n - p_n| < \frac{1}{q_n},$$

we can write

$$p_n = \alpha q_n + \frac{\varepsilon}{q_n}, \qquad |\varepsilon| < 1,$$

(where ε, of course, depends on n). Substituting this value in the expression of A_n, we have

$$
\begin{aligned}
A_n &= a\left(\alpha q_n + \frac{\varepsilon}{q_n}\right)^2 + b\left(\alpha q_n + \frac{\varepsilon}{q_n}\right) q_n + c q_n^2 \\
&= (a\alpha^2 + b\alpha + c)q_n^2 + \left(2a\alpha + b + \frac{a\varepsilon}{q_n^2}\right) \cdot \varepsilon \\
&= \left(2a\alpha + b + \frac{a\varepsilon}{q_n^2}\right) \cdot \varepsilon.
\end{aligned}
$$

Therefore, we have $|A_n| < |2a\alpha + b| + |a|$, which implies that all the A_n's are built from a finite set of integers. Since $C_n = A_{n-1}$, we obtain a result similar for C_n. Moreover, we have

$$B_n^2 - 4A_n C_n = b^2 - 4ac,$$

and then

$$B_n^2 = |4A_nC_n + b^2 - 4ac| < 4\{|2a\alpha + b| + |a|\}^2 + |b^2 - 4ac|,$$

which means that the B_n's are bounded. Since there exist only a finite number of choices for A_n, B_n and C_n, we conclude that there exist only a finite number of distinct polynomials $A_n x^2 + B_n x + C_n = 0$ each having α_n as a root.

(971) Let $\alpha = [a_1, a_2, \ldots, a_n, \alpha_n]$ be a quadratic irrational number. Since each α_n is a root of one of the quadratic equations

$$A_1 x^2 + B_1 x + C_1 = 0,$$
$$A_2 x^2 + B_2 x + C_2 = 0,$$
$$\ldots$$
$$A_N x^2 + B_N x + C_N = 0,$$

one of these polynomials must have at least three of these α_n's as a root (from Problem 970, the triple (A_n, B_n, C_n) therefore takes infinitely many times the same value). Since a quadratic equation can have at most two distinct roots, it follows that two of the α_n's must be equal, say $\alpha_k = \alpha_{k+m}$. From the algorithm outlined in the proof that an irrational number can be written as an infinite continued fraction, we have

$$a_{k+1} = \alpha_{k+m+1}, \ a_{k+2} = \alpha_{k+m+2}, \ldots, a_{k+j} = \alpha_{k+m+j}, \ldots.$$

Hence,

$$\alpha = [a_1, a_2, \ldots, a_k, \overline{a_{k+1}, \ldots, a_{k+m}}].$$

(972) We have $\dfrac{3 + \sqrt{23}}{7} > 1$, $-1 < \dfrac{3 - \sqrt{23}}{7} < 0$ and $\dfrac{3 + \sqrt{23}}{7} = [\overline{1, 8, 1, 3}]$;

$2 + \sqrt{7} > 1$, $-1 < 2 - \sqrt{7} < 0$ and $2 + \sqrt{7} = [\overline{4, 1, 1, 1}]$;

$\dfrac{5 + \sqrt{37}}{3} > 1$, $-1 < \dfrac{5 - \sqrt{37}}{3} < 0$ and $\dfrac{5 + \sqrt{37}}{3} = [\overline{3, 1, 2}]$.

(973) Assume that $\sqrt{D} = [a_1, a_2, a_3, \ldots,]$. Since a_1 is the largest integer smaller than $\sqrt{D}$, it follows that $a_1 + \sqrt{D} > 1$ and $-1 < a_1 - \sqrt{D} < 0$. Then, using the result stated in Problem 972, we have that $a_1 + \sqrt{D}$ is represented by a periodic continued fraction. Hence,

$$\begin{aligned} a_1 + \sqrt{D} &= [\overline{2a_1, a_2, a_3, \ldots, a_n}] \\ \sqrt{D} &= -a_1 + [\overline{2a_1, a_2, a_3, \ldots, a_n}] \\ \sqrt{D} &= -a_1 + [2a_1, \overline{a_2, a_3, \ldots, a_n, 2a_1}] \\ \sqrt{D} &= [a_1, \overline{a_2, a_3, \ldots, a_n, 2a_1}]. \end{aligned}$$

(974) (*CRUX, 1988*, solution by Ed Doolittle). First of all, we observe that the binary representation of $\sqrt{2}$ contains infinitely many 1's

$$\sqrt{2} = 1.01101\ldots$$

(otherwise, from some point on, we would have only 1's, thereby implying that $\sqrt{2}$ would be rational). In base 2, multiplication by 2 moves the dot by one position to the right, so that

$$2\sqrt{2} = 10.1101\ldots, \qquad 2^2\sqrt{2} = 101.101\ldots,$$

and so on. Since $\sqrt{2}$ contains infinitely many 1's, there exist infinitely many integers n such that the binary representation of $2^n\sqrt{2}$ has a "1" to the right of the dot, with eventually another "1" to its right (since there are infinitely many 1's). It follows from this that the fractional part of $2^n\sqrt{2}$ exceeds $0.10000 = \frac{1}{2}$.

Using the notation $\{x\}$ for the fractional part of x, we have thus proved that there exists a set $A \subset \mathbb{N}$ containing infinitely many integers n such that

$$\{2^n\sqrt{2}\} > \frac{1}{2}.$$

Since $1 - \frac{1}{\sqrt{2}} < \frac{1}{2}$, we have that if $n \in A$,

$$\{2^n\sqrt{2}\} > 1 - \frac{1}{\sqrt{2}},$$

in which case

$$\frac{1}{\sqrt{2}} > 1 - \{2^n\sqrt{2}\}$$

and therefore

$$0 < (1 - \{2^n\sqrt{2}\})\sqrt{2} < 1.$$

Since $[m + x] = m$ if m is an integer and $x \in (0, 1)$, we can write

$$\left[2^{n+1} + (1 - \{2^n\sqrt{2}\})\sqrt{2}\right] = 2^{n+1}.$$

Since $2^{n+1} = \sqrt{2}(2^n\sqrt{2})$, we have

$$\left[(2^n\sqrt{2} + 1 - \{2^n\sqrt{2}\})\sqrt{2}\right] = 2^{n+1},$$

which we can write as

$$\left[(2^n\sqrt{2} - \{2^n\sqrt{2}\} + 1)\sqrt{2}\right] = 2^{n+1}.$$

Since we always have $x - \{x\} = [x]$, this last relation can be written as

$$\left[([2^n\sqrt{2}] + 1)\sqrt{2}\right] = 2^{n+1}.$$

Finally, since $[x] + 1 = [x + 1]$, we have

$$\left[[2^n\sqrt{2} + 1]\sqrt{2}\right] = 2^{n+1}.$$

This last relation means that if $k = \left[2^n\sqrt{2} + 1\right]$, then

$$[k\sqrt{2}] = 2^{n+1},$$

a power of 2. Since the distinct values of n give distinct values for k, the infinite set A generates the desired infinite sequence of powers of 2.

(975) It is clear that it is enough to prove that

$$\left|\frac{r+2}{r+1} - \sqrt{2}\right| < |r - \sqrt{2}|.$$

But this follows from the fact that

$$\left|\frac{r+2}{r+1} - \sqrt{2}\right| = \left|\frac{(r+2) - \sqrt{2}(r+1)}{r+1}\right| = \frac{(\sqrt{2}-1)|r - \sqrt{2}|}{r+1}$$

$$< (\sqrt{2}-1)|r - \sqrt{2}| < |r - \sqrt{2}|.$$

REMARK: If $0 < r \in \mathbb{Q}$ is given as an approximation of $\sqrt{3}$, it is easy to prove that the number $\frac{r+3}{r+1}$ represents a better approximation.

(976) (a) We have that $676 = 2^2 \cdot 13^2$. Hence, $\sqrt{676} = 26$ is a rational number.
(b) Let $x_0 = \sqrt{75} + \sqrt{2}$. We have $x_0 = 5\sqrt{3} + \sqrt{2}$. Since $x = x_0$ is not an integer and is a solution of the equation $x^4 - 154x^2 + 77^2 - 600 = 0$, it follows that it must be an irrational number.

(977) We have $a = 2^2 \cdot 3$, $b = 5 \cdot 7^2$, $c = 3 \cdot 11^2$, $d = 3 \cdot 5^3$. It follows that
  (a) $\sqrt{ab} = 14\sqrt{15}$, an irrational number
  (b) $\sqrt{ac} = 66$, a rational number
  (c) $(6ad)^{1/3} = 30$, a rational number
  (d) $\log 12$, an irrational number.

(978) If x is rational, say $x = a/b$ with $a, b \in \mathbb{Z}$, $b \neq 0$, then $m!x = m!a/b$ is an even integer for each integer $m \geq b + 2$, in which case $\cos m!x\pi = 1$ and therefore $(\cos m!x\pi)^n = 1$ for $n \geq 1$, $m \geq b + 2$. This proves that $f(x) = 1$ if x is rational. On the other hand, if x is irrational, then mx is never an integer whatever the value of the integer $m \geq 1$, in which case we always have $|\cos(m!\pi x)| < 1$, so that $|\cos(m!\pi x)|^n$ tends to 0 when $n \to \infty$. It follows that $(\cos(m!\pi x))^n$ tends to 0 when $n \to \infty$, and the result is proved.

(979) Assume that such a solution $x = \frac{a}{b}$, with $(a,b) = 1$, exists. In that case, we obtain

$$a^5 = 10b^5 - ab^4.$$

But since b divides the right-hand side of this equation, it must also divide a^5, which means that $(a,b) > 1$, thereby contradicting the initial hypothesis and establishing the result.

(980) Let $x_0 = a/b$, with $(a,b) = 1$. The equation becomes $\frac{a^2}{b^2} + r\frac{a}{b} + s = 0$, which implies that $a^2 + rab + sb^2 = 0$. It follows from this last equation that $b|a^2$, which is possible only if $b = 1$, because $(a,b) = 1$. Therefore, it follows that $x_0 \in \mathbb{Z}$.

(981) The answer is: for $p = q$ and $m = -n$ or, of course, for $m = n = 0$, but in no other cases. Indeed, if $p \neq q$, then if $m\sqrt{p} + n\sqrt{q}$ were an integer, we would have that $(m\sqrt{p} + n\sqrt{q})^2 = m^2p + n^2q + 2mn\sqrt{pq}$ would also be an integer, in which case $2mn\sqrt{pq}$ would also be an integer, which is not possible, because $\sqrt{pq}$ is irrational.

(982) (This is Theorem 137 in the book by Hardy and Wright [**18**].) If the number α were rational, there would be a period in the decimal expansion of α. In this case, we observe that the digit 1 appears at positions r, $r+s$, $r+2s$, ... There would therefore exist a function $f(n) = sn+r$ such that $f(n)$ is prime for each sufficiently large integer n. But no polynomial with integer coefficient, which is not constant, can be prime for each sufficiently large value of n (see De Koninck and Mercier [**8**], page 31).

(983) Let $x = \sqrt{p} + \sqrt{q}$. Then,

$$\begin{aligned}
x^2 &= p + q + 2\sqrt{pq}, \\
x^4 &= (p+q)^2 + 4pq + 4(p+q)\sqrt{pq}, \\
x^4 &= p^2 + q^2 + 6pq + 4(p+q)\sqrt{pq},
\end{aligned}$$

so that
$$x^4 - 2(p+q)x^2 = p^2 + q^2 + 6pq - 2(p+q)(p+q) = -(p-q)^2.$$

Therefore, the real number x satisfies
$$x^4 - 2(p+q)x^2 + (p-q)^2 = 0.$$

Calling upon Theorem 44, the result follows.

(984) It is clear that $\alpha = 1 + \dfrac{1}{\alpha}$, in which case $\alpha^2 - \alpha - 1 = 0$, meaning that $\alpha = \dfrac{1+\sqrt{5}}{2}$. On the other hand, one easily establishes that $\beta^2 = \beta + 1$, so that $\beta = \alpha$. By substituting these values in the given equation, we easily obtain that $t = 2$.

(985) Let $x = 2^{1/3} + 3^{1/3}$. Then, $x - 2^{1/3} = 3^{1/3}$ and this is why $(x - 2^{1/3})^3 = 3$. Expanding this last equality, we obtain
$$x^3 - 3x^2 \cdot 2^{1/3} + 3x \cdot 2^{2/3} - 2 = 3$$

and therefore successively
$$x^3 - 5 = 3x^2 \cdot 2^{1/3} - 3x \cdot 2^{2/3},$$
$$x^3 - 5 = 3x \cdot 2^{1/3}\left(x - 2^{1/3}\right),$$
$$x^3 - 5 = 3x \cdot 2^{1/3}3^{1/3},$$
$$(x^3 - 5)^3 = 6 \cdot 3^3 \cdot x^3,$$

an expression which can be reduced to
$$x^9 - 15x^6 - 87x^3 - 125 = 0.$$

We have thus proved that x is a root of a polynomial of degree 9 with integer coefficients, and this is why we may conclude that this number x is irrational, since of course it is not an integer.

(986) The answer is YES. Indeed, assuming that there exist a and $b \in \mathbb{N}$ such that $\log_{10} 2 = \dfrac{a}{b}$, we would have $10^{a/b} = 2$, so that $10^a = 2^b$, which would imply that $5|2$, which is nonsense.

(987) Let $\xi = \frac{a}{b}$. Then,
$$\frac{1}{q^2} > \left|\frac{p}{q} - \frac{a}{b}\right| = \frac{|pb - aq|}{bq} \geq \frac{1}{bq},$$

so that $q^2 < bq$ and therefore that $1 \leq q < b$. There are therefore a finite number of possible choices for q and hence a finite number of possible choices for $1 < p < q$. There are therefore only a finite number of choices for p and q.

(988) Let $\theta = \frac{\log 3}{\log 2}$. Assume that θ is algebraic; we will show that a contradiction will follow. Indeed, if such is the case, then, using a result of Gel'fond and Schneider, we have that 2^θ is transcendental. Since it is clear that $2^\theta = 3$, an algebraic number, a contradiction follows.

(989) Of course, it is enough to show that if $m = a/b$, with $(a,b) = 1$, $a, b > 0$, then $m = 1$. But if the number
$$m + \frac{1}{m} = \frac{a^2 + b^2}{ab}$$

is an integer, then $a|a^2+b^2$ and $b|a^2+b^2$, which implies that $a|b^2$ and $b|a^2$. Since $(a,b) = 1$, this means that $a = b = 1$ and therefore that $m = 1$.

(990) Setting $x = \sqrt{2} + \sqrt{7}$, we obtain $x^2 = 9 + 2\sqrt{14}$ and $x^4 = 137 + 36\sqrt{14}$. Hence, the polynomial $x^4 - 18x^2 + 25 = 0$. It then follows from Theorem 44 that x is irrational. It is then of course an algebraic number.

(991) We quickly notice that $x = x_1 = -1$ is a root of $p(x)$. Hence, $p(x) = (x+1)(x^2+x-1)$. Now the zeros of $x^2 + x - 1 = 0$ are $x_2 = \frac{\sqrt{5}-1}{2}$ and $x_3 = \frac{-\sqrt{5}-1}{2}$. Since the polynomial has only integer coefficients and since x_2 and x_3 are not integers, it follows that they must be irrationals. We have therefore found a rational root, namely $x_1 = -1$, and two irrational roots, namely x_2 and x_3.

(992) The answer is NO. Indeed, if such integers existed, that would mean that $ae^3 + be^2 = 16$ and therefore that e is an algebraic number, which is not possible since it is transcendental.

(993) The answer is YES. Indeed, if it did not contain any, we would have that the interval $I = [\frac{7}{2}, \frac{9}{2}]$ contains only algebraic numbers. But this interval contains an uncountable quantity of real numbers, while the set of algebraic numbers contained in I is countable, which makes no sense.

(994) The number $\sqrt{2}^{\sqrt{2}}$ is either rational or irrational. In the first case, by choosing $\alpha = \beta = \sqrt{2}$, we have found two algebraic numbers α, β such that α^β is rational, as required. It remains to consider the second case. In the second case, by choosing $\alpha = \sqrt{2}^{\sqrt{2}}$ and $\beta = \sqrt{2}$, and observing that $\left(\sqrt{2}^{\sqrt{2}}\right)^{\sqrt{2}} = 2$, we have that α^β is rational, as required. Hence, in both cases, we have obtained the required configuration.

(995) (a) Let $a, b, n \in \mathbb{N}$ and consider the function

$$(1) \qquad f(x) = \frac{x^n(a-bx)^n}{n!}.$$

Observe that for $0 < x < a/b$, we have

$$(2) \qquad 0 < f(x) < \frac{a^{2n}}{n! b^n}.$$

Using the Binomial Theorem, we have

$$f(x) = \frac{(ax - bx^2)^n}{n!} = \frac{1}{n!} \sum_{j=0}^{n} \binom{n}{j} (ax)^j (-b)^{n-j} (x^2)^{n-j}$$

$$= \frac{1}{n!} \sum_{j=0}^{n} \binom{n}{j} a^j (-b)^{n-j} x^{2n-j},$$

and setting $2n - j = m$, we obtain

$$f(x) = \frac{1}{n!} \sum_{m=n}^{2n} \binom{n}{2n-m} a^{2n-m} (-b)^{m-n} x^m = \frac{1}{n!} \sum_{m=n}^{2n} c_m x^m,$$

where obviously the coefficients c_m are integers.

We have $f(0) = 0$ and $f^{(k)}(0) = 0$ if $k < n$ or $k > 2n$. Moreover, for $n \leq k \leq 2n$, we find

$$f^{(k)}(0) = \frac{k!}{n!} c_k.$$

Hence, $f^{(k)}(0)$ is an integer for each integer $k \geq 0$, and since $f(x) = f(\frac{a}{b} - x)$, then $f^{(k)}(\frac{a}{b})$ is also an integer for each $k \geq 0$.

The above remarks will now allow us to solve the problem. Observe that if y is a rational number, that is $y = \frac{c}{d}$, and if e^y is a rational number, then $e^{dy} = e^c$ is also a rational number. Moreover, if e^{-d} is rational, then e^d is also rational, and therefore it is enough to show that if m is a positive integer, then e^m cannot be a rational number.

Assume the contrary, that is assume $e^m = \frac{h}{k}$ where $h, k \in \mathbb{N}$, and consider the function

$$F(x) = m^{2n} f(x) - m^{2n-1} f'(x) + \cdots - m f^{(2n-1)}(x) + f^{(2n)}(x),$$

where $f(x)$ is the function defined above with $a = b = 1$. In this case, $F(0)$ and $F(1)$ are integers and we have

$$\frac{d}{dx}\{e^{mx} F(x)\} = e^{mx}\{mF(x) + F'(x)\} = m^{2n+1} e^{mx} f(x).$$

Consequently,

$$k \int_0^1 m^{2n+1} e^{mx} f(x) = k(e^{mx} F(x)) \Big|_0^1 dx$$
$$= ke^m F(1) - kF(0) = hF(1) - kF(0)$$

is an integer. Using (2), with $a = b = 1$, we obtain

$$0 < k \int_0^1 m^{2n+1} e^{mx} f(x)\, dx < k \frac{m^{2n+1}}{n!} \int_0^1 e^{mx}\, dx < \frac{km^{2n} e^m}{n!},$$

and since $km^{2n} e^m / n! < 1$ for n sufficiently large, we obtain a contradiction.

(b) Assume the contrary, that is that $\pi = a/b$, where a and b are positive integers. Consider the function f defined by equation (1) in the solution of (a) and consider the function

$$F(x) = f(x) - f''(x) + f^{(4)}(x) + \cdots + (-1)^n f^{(2n)}(x).$$

Since $f(0)$ and $f(\pi)$ are integers (see the remarks on $f(x)$ stated in the solution of (a)), it follows that $F(0)$ and $F(\pi)$ are also integers.

Since

$$\frac{d}{dx}\{F'(x)\sin x - F(x)\cos x\} = \{F''(x) + F(x)\}\sin x = f(x)\sin x,$$

it follows that

$$\int_0^\pi f(x)\sin x\, dx = \{F'(x)\sin x - F(x)\cos x\} \Big|_0^\pi = F(0) + F(\pi).$$

This integral therefore represents an integer. Using the equation (2) that shows up in the solution of (a) for $0 < x < \pi = \frac{a}{b}$, we have that

$$0 < f(x) \sin x < \frac{\pi^n}{n!} a^n,$$

so that

$$0 < \int_0^\pi f(x) \sin x \, dx < \frac{\pi^n}{n!} a^n \pi.$$

Since it is possible to choose n such that $\frac{\pi^n}{n!} a^n \pi < 1$, we obtain a contradiction.

(996) Set $y = \log 2$. We have of course

$$e^y = e^{\log 2} = e^{a/b} = 2,$$

a rational number. Hence, using Problem 995(a), we have, since e^y is rational, that y is irrational, thus the result. The same argument applies for $\log r$.

(997) We easily show that the minimal polynomial is

$$x^3 - \frac{3}{2}x^2 + \frac{3}{4}x - 1.$$

(998) The polynomial is

$$x^4 - 4x^3 - 4x^2 + 16x - 8.$$

(999) It is an irrational number. Moreover, it is also algebraic and therefore not transcendental. Let $x = 2^{1/2} + 3^{1/3}$. We have $x - 2^{1/2} = 3^{1/3}$, so that $(x - 2^{1/2})^3 = 3$ and also that, successively, $x^3 - 3x^2\sqrt{2} + 6x - 2\sqrt{2} = 3$, $x^3 + 6x - 3 = (3x^2 + 2)\sqrt{2}$, $(x^3 + 6x - 3)^2 = 2(3x^2 + 2)^2$, $x^6 - 6x^4 - 6x^3 + 12x^2 - 36x + 1 = 0$. Therefore, since x is not an integer, it must be irrational and in fact algebraic (of degree 6).

(1000) Let $p(x)$ be the polynomial

$$x^5 + 39x^4 + 83x^3 + 325x^2 - 348x - 1924.$$

Since $p(x)$ has integer coefficients, it follows from Theorem 44 that each root of $p(x)$ is an integer or an irrational number; therefore, each rational root of $p(x)$ must be an integer. We must therefore examine the divisors of $1924 = 2^2 \cdot 37 \cdot 13$. We easily find that $p(2) = p(-2) = p(-37) = 0$, so that

$$p(x) = (x - 2)(x + 2)(x + 37)(x^2 + 2x + 13),$$

and given that the polynomial $x^2 + 2x + 13$ has only complex roots (its discriminant being negative), it follows that the only rational roots of $p(x)$ are $x = 2$, $x = -2$ and $x = -37$.

(1001) Such a rational number does not exist, since if it did, it is easy to see that the number π would then be a root of a polynomial of degree 5 with integer coefficients, thus contradicting the fact that π is a transcendental number.

Bibliography

[1] T.M. Apostol, *Introduction to Analytic Number Theory*, Springer-Verlag, 1976.

[2] W.S. Anglin, *Mathematics: A Concise History and Philosophy*, Undergraduate Texts in Mathematics, Readings in Mathematics, Springer-Verlag, N.Y., 1994.

[3] E.J. Barbeau, M.S. Klamkin & W.O.S. Moser, *Five Hundred Mathematical Challenges*, MAA, 1995.

[4] J. Brillhart, D.H. Lehmer, J.L. Selfridge, B. Tuckerman & S.S. Wagstaff Jr., *Factorisations of $b^n \pm 1$, $b = 2, 3, 5, 6, 7, 10, 11, 12$ up to high powers*, Contemporary Mathematics, AMS, Vol. 22, 2nd edition, 1988.

[5] S. Cavior, *The subgroups of dihedral groups*, Math. Mag. **48** (1975), 107.

[6] J.M. De Koninck & N. Doyon, *On a very thin set of integers*, Annales Rolands Eötrös Nominatae **20** (2001), 157–177.

[7] J.M. De Koninck & J. Grah, *L'enseignement mathématique* **42** (1996), 97–123.

[8] J.M. De Koninck & A. Mercier, *Introduction à la théorie des nombres*, Modulo, 2nd edition, 1997.

[9] H.E. Dudeney, *Amusements in Mathematics*, Nelson, London, 1951.

[10] P. Erdős, *On the irrationality of certain series: problems and results*, New Advances in Transcendence Theory, Cambridge University Press, 1988, 102–109.

[11] P. Erdős & G. Szekeres, *Some Number Theoretic Problems on Binomial Coefficients*, Austr. Math. Soc. Gaz. **5** (1978), 97–99.

[12] R. Finkelstein & H. London, *On D.J. Lewis's equation $x^3 - 117y^3 = 5$*, Can. Math. Bull. **14** (1971), 111.

[13] S.I. Gelfand, *Sequences and Combinatorial Problems*, Gordon & Breach Science Pub., 1968.

[14] P. Giblin, *Primes and Programming*, Cambridge University Press, 1994.

[15] S.W. Golomb, *Equality among number-theoretic functions*, Abstracts Amer. Math. Soc. **14** (1993), 415–416.

[16] R.K. Guy, *Unsolved Problems in Number Theory*, Springer-Verlag, 1994.

[17] F. Halter-Koch, *Letter to the editor*, Can. Math. Bull. **16** (1973), 299.

[18] G.H. Hardy & E.M. Wright, *An Introduction to the Theory of Numbers*, Oxford University Press, 5nd edition, 1979.

[19] E. Hlawka, J. Schoissengeier & R. Taschner, *Geometric and Analytic Number Theory*, Springer-Verlag, 1991.

[20] A. Ivić & Z. Mijajlović, *On Kurepa's problems in number theory*, Publ. Inst. Math (Belgrade) **55** (1995), 19–28.

[21] L. Jones, *A polynomial approach to a Diophantine problem*, Math. Mag. **72** (1999), 52–55.

[22] D.J. Lewis, *Studies in Number Theory* (edited by W.J. LeVeque).

[23] P.J. McCarthy, *Introduction to Arithmetical Functions*, Springer-Verlag, 1986.

[24] D.J. Newman, *Analytic Number Theory*, Graduate Texts in Mathematics, vol. 177, Springer, 1998.

[25] I. Niven, H.S. Zuckerman & H.L. Montgomery, *An Introduction to the Theory of Numbers*, John Wiley, 1991.

[26] J. Pach, *Two places at once: a remembrance of Paul Erdős*, Math. Intelligencer **19**, no. 2 (April 1997), 38–48.

[27] C. Pomerance, *A tale of two sieves*, Notices AMS **43** (1996), 1473–1485.

[28] P. Ribenboim, *The Little Book of Big Primes*, Springer-Verlag, 1991.

[29] P. Ribenboim, *Nombres premiers : mystères et records*, Les Presses Universitaires de France, 1994.

[30] P. Ribenboim, *The New Book of Prime Number Records*, Springer, 1996.

[31] H. Riesel, *Prime Numbers and Computer Methods for Factorization*, Birkhäuser, 2nd edition, 1994.

[32] J. Roberts, *The Lore of the Integers*, MAA, Washington, 1992.

[33] V. St. Udrescu, *On D.J. Lewis's equation $x^3 - 117y^3 = 5$*, Revue Roumaine Math. Pures Appl. **18** (1973), 473.

[34] J.W. Sander, *On the value distribution of arithmetic functions*, J. Number Theory **66** (1997), 51–69.

[35] A. Schinzel & W. Sierpinski, *Sur certaines hypothèses concernant les nombres premiers*, Acta Arith. **4** (1958), 185–208; *Corrigendum*, ibid. **5** (1959), 259.

[36] A. Schlafly & S. Wagon, *Carmichael's conjecture on the Euler function is valid below $10^{10,000,000}$*, Math. Comp. **63** (1994), 415–419.

[37] E.D. Schwab & L. Tóth, *On some elementary number theoretic inequalities involving the Dirichlet convolution*, Universitatea Din Timisoara, Romania, 1990.

[38] H.N. Shapiro, *Introduction to the Theory of Numbers*, John Wiley & Sons, 1983.

[39] W. Sierpinski, *250 Problems in Elementary Number Theory*, Polish Scientific Publishers, Warsaw, 1970.

[40] Eric W. Weisstein, *CRC Concise Encyclopedia of Mathematics*, CRC Press, Boca Raton, FL, 1999.

[41] H.C. Williams, *Édouard Lucas and Primality Testing*, Canadian Mathematical Society Ser. Monogr. Adv. Texts, Wiley, 1998.

Subject Index

Index of Authors